Statutory Valuations
Fourth edition

Andrew Baum, Gary Sams, Jennifer Ellis,
Claire Hampson, Douglas Stevens

2007

EG Books

A division of Reed Business Information

Estates Gazett
1 Procter Stre

©Andrew Baum, Gary Sams, Jennifer Ellis, Claire Hampson, Douglas Stevens, 2007

First published in 1983 by Routledge and Kegan Paul Ltd
Reprinted 1986, 1987
Second edition published by Routledge
Reprinted 1991, 1992
Third edition published by Routledge 1997
Reformatted edition published by EG Books in January 2005

ISBN 978-0-7282-0504-8

Cover design by Ted Masters. Photo of residential block courtesy of Ian Gordon
Typeset in Palatino 10/12 by Amy Boyle
Printed by Cromwell Press, Trowbridge, Wiltshire

Contents

Part 2: National and Local Taxation

Part 3: Compulsory Purchase and Planning Compensation

Preface to the Fourth Edition

Ever since the first edition of this book was published in 1983, there has been a legitimate criticism that a book calling itself *Statutory Valuations* must be incomplete if it does not cover one of the main areas in which the law interferes in the valuation process; rating. This omission has been remedied in this edition by the addition of a rating chapter, contributed by Claire Hampson. Another new addition is a chapter by Patrick Bond covering a range of valuations for taxation purposes including income tax, corporation tax, capital gains tax, inheritance tax, stamp duty land tax and VAT.

The first three editions were published at seven yearly intervals, which makes this edition well overdue after a gap of 10 years. The main reason for this is that we have been awaiting a promised comprehensive revision of the archaic and confused body of legislation and case law, which forms the compensation code for compulsory purchase purposes. Unfortunately after a lengthy process which appeared to be developing into a consensus as to the form which the new legislation would take, the government tired of its task and announced that "there are no quick and easy solutions and moving towards a simpler and more readily accessible set of laws would still require substantial further work". Given its other priorities "implementing the Law Commission's proposals is not a practical proposition for the forseeable future". The compensation chapters of this book have, therefore been updated, rather than completely re-written. The revisions include the changes introduced in the Planning and Compulsory Purchase Act of 2004; in particular the introduction of basic and occupier's loss payments. There is also considerable new case law, some of which actually makes reference to the now redundant reform proposals.

The first chapter relating to landlord and tenant has been extensively revised by Douglas Stevens. It includes changes to the Landlord and Tenant Acts introduced by the Regulatory Reform (Business Tenancies) (England & Wales) Order 2003. These aim to make the Acts more user friendly in relation to notices, contracting out and interim rent and expediting the court procedures for lease renewals, with the intention of reducing the potential for litigation. Chapter 2 on residential property has been updated and includes comments on changes to the assessment of fair rents as a result of the *Spath Holme* case and the Rent Acts (Maximum Fair Rent) Order 1999.

The main legislative change in leasehold reform was the introduction of the 2002 Leasehold and Commonhold Act, which eased the qualification requirements and abolished payment of marriage value where the leases have more than 80 years unexpired. Valuations have been profoundly altered by two Lands Tribunal decisions — *Arbib* and *Sportelli* — which considerably reduced the conventional deferment rates, thus increasing the prices payable for freeholds and extended leases throughout the country. To reflect these and other changes Jennifer Ellis has completely re-written

Chapters 3 and 4 which take on a new format. Whereas in previous editions Chapter 3 covered the legislation relating to houses, and Chapter 4 to flats, the new Chapter 3 describes the various legal rights under which leasehold enfranchisement may arise, while Chapter 4 sets out the valuation processes under those rights.

No doubt further revisions will be required in the future, particularly in the compensation field where the government appears to have abrogated its responsibilities, and delegated to the courts the task of moulding a body of legislation which now dates back over 150 years, into a shape suitable for the 21st century.

Gary Sams and Andrew Baum
Preston and London

Preface to the First Edition

For a number of years I have been aware of the need for a textbook dealing with those areas of property valuation which are customarily studied in the later years of a degree or professional course and which might broadly be described as those areas in which the effect of the law has to be reflected in the valuation approach.

This book is intended to satisfy that need.

That the law affects property values is beyond dispute. Perhaps the most famous example of this is the much-vaunted slaughter of tenanted residential property values said to be the result of the Rent Acts. This book is not, however, concerned with such emotive or philosophical argument. It is concerned with practice, and is designed to aid the student and practitioner in his attempt to produce property valuations which are in accordance with the current state of the law.

It is arguable that a book concerned with statutory valuations should include a consideration of valuation for taxation. I have set this aside as it appears to me that a separate book should be written on the subject. I have also omitted any consideration whatsoever of agricultural property due partly to my own ignorance in the field and partly to the fact that valuers tend to find their time devoted either to the commercial/residential sectors of the market (mode of dress: largely blue or grey suits) or to the agricultural sector (mode of dress: browns or greens, and not always suits). It is for the blues and greys, and aspiring blues and greys, that this book is intended.

The book is, therefore, designed for both students and practitioners. I have attempted to recommend specific approaches to valuation problems and to interpret the legal framework behind these approaches in a clear and yet concise manner. I have been somewhat latitudinarian in my interpretation of this legal framework, which includes both statute and case law, and which is particularly stretched to justify my consideration of rent reviews in Chapter 1.

It is necessary that the reader of this book can afford to buy some other textbooks in addition. A knowledge of valuation techniques is essential, and the book is intended to a certain extent as a partner to *The Income Approach to Property Valuation* (Routledge, 3rd ed, 1989), to which several references are made where some consideration of basic techniques may be necessary. And this is not a law book. For each of the nine chapters there is a fuller description of the law to be found in a specialized law textbook, and the practitioner must be prepared for the awkward problem where such a full description will be required. There is a comprehensive bibliography to be found at the rear of the book which should point the baffled valuer towards further reading. Where such further reading is recommended in notes a full reference is provided in the bibliography.

Throughout the text the valuer is referred to in the male gender only. To female readers, I apologize for this misleading generalisation.

The book is divided into two parts. Part I deals with legislation affecting valuations of commercial and residential property, while part II deals with compensation for compulsory purchase and planning. I hope that my efforts have been worthwhile and that a need has been fulfilled.

Andrew Baum
Reading University

Table of Cases

Table of Statutes

Part 1

Commercial and Residential Property

Commercial Property

1.1 Introduction: the fragmentation of ownership

Behind the facade of every commercial building in the UK may lie a complex pattern of relationships created by the unique nature of property as an economic commodity.

The market for property is not necessarily driven by the desire to own land and buildings. It is instead a market for rights in the product which may have many tiers of ownership. Consequently, a single building might represent the property rights or interests of several different parties.

For a given unit of property, there is a basic dichotomy of rights. These are the right of ownership, which may be fragmented, and the right of occupation which (allowing for the existence of time shares and joint tenancies) is usually vested in a single legal person. In fact, the given unit is commonly delineated by its exclusive occupation by an organisation, family or individual. The following example will serve to illustrate the tiers of ownership and the multitudinous interests which can exist in one property.

Fifty years ago, A acquired a piece of undeveloped land within the town boundary as a speculation. Ten years later, the land had become ripe for development: it was now profitable to construct a factory unit upon it. A, however, was not a builder and consequently let the land to B on a building lease for 99 years. B constructed a small factory on the site. Thirty-seven years ago B sub-let the completed building to C on a 42-year lease. Ten years afterwards A died and his interest in the property passed to his son, D. Fifteen years ago C assigned his lease to an insurance company, E, who sublet the property to the then occupier, F.

Ten years later E obtained a new 25-year lease from B, who immediately sold out to G. This year F assigned his interest to H, who is the current occupier of the factory.

The current vertical division of rights is illustrated in Figure 1.1. A current contractual relationship exists between four parties whose interests are united in the single property unit. H has the right to exclusive occupation of the factory, while D, G, E and H share in its ownership. D has a freehold interest whose value depends upon the income he can expect to receive from G and his successors in title. G, E, and H have leasehold interests, the values of which depend upon the receipt of a profit rent created by the difference between the rents actually or potentially received and the rent paid to the immediately superior landlord.

Such a relationship is typical of the manner in which UK property ownership is fragmented. It is especially typical in the case of commercial property (for the purposes of this chapter this includes all

Figure 1.1 The vertical division of rights in commercial property

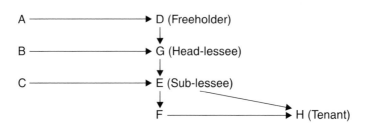

office, shop, industrial and institutional property) where the majority of buildings are 'owned' by non-occupiers, in contrast to residential property, where the rights of ownership and occupation are fused within the same person in over 50% of the stock of buildings in that sector (see Chapters 2 and 3).

This decomposition of the ownership of commercial property produces the probability of the existence of contractual landlord/tenant relationships, as illustrated in Figure 1.1 where such relationships exist between D and G, G and E, and E and H.

1.2 The valuer's role

1.2.1 Introduction

As property advisor, the valuer is often employed as an agent on behalf of one of the parties to negotiate an acquisition or disposal of an interest or a variation thereof. The medium for these negotiations is a contract or a lease, which defines the contractual relationship between the parties. Therefore, the valuer needs to have a full understanding of the legal and financial implications of the provisions of the contract or lease in order to advise the vendor or purchaser or landlord or tenant, as the case may be.

There are many factors contained in a lease which directly affect the valuation of interests in land and buildings. For instance, repairing liabilities, lease length, break clauses, user and assignment restrictions and rent review patterns may all, individually or collectively, impact on capital or rental value and must be taken account of by the valuer.

The valuer's task in such negotiations is to provide a financial interpretation of the terms of the contract.. The investor in property is interested in the return provided by the investment and the valuer is skilled in the quantification of the drivers of that return. His/her advice enables owners and occupiers of property (rights) to make informed decisions concerning the contracts they create and enter into.

We consider below some specific valuation problems which may arise between a landlord and a tenant of a commercial property.

1.2.2 Rental analysis and valuation

Both the landlord and the tenant of commercial property may require advice concerning the rental value of the subject premises in two general sets of circumstances. First, at the beginning of a lease (where new property is to be let for the first time or where a lease runs out and is renewed) the rental value will be assessed primarily by regard to open market rental values based on a standard lease type, and adjusted to take account of the proposed lease terms. Adjustments to the rent and/or the other terms may be made in the negotiation process. Second, at a rent review during the currency of a lease the rental value will be assessed in the light of the extant lease terms, which are not normally changeable.

In either case, evidence derived from recent open market lettings or rent review settlements on comparable properties will be employed and analysed in the estimation of the rental value of the subject property

Example 1.1

10 Station Buildings was recently let on internal repairing terms (IRT) at a rent of £16,000 pa on a 25-year lease with 5-yearly rent reviews. 10 Station Buildings is an office property of 185.8 m².

Calculate the estimated rental value (ERV) of 12 Station Buildings, a similar property of 167.2 m² to be let on full repairing and insuring (FRI) terms on a 20-year lease with 5-yearly reviews.

1. Adjust for different repairing liabilities:

    The comparable property was let at £16,000 pa IRT. Full rental value on FRI terms would be lower pa by:

External repairing costs (say)	£1,000
Insurance costs (say)	£250
Total	£1,250
Therefore ERV (FRI):	£14,750

2. Adjust for different lease length: assume minimal effect (but see p 37).

3. Adjust for different size:

ERV (FRI)	=	£14,750/185.8 m²
	=	£79.39/m²
£79.39 × 167.2 m²	=	£13,275 pa

1.2.3 Capital valuations

The estimation of the capital value of a freehold or leasehold interest in commercial property may be required for loan purposes, for company asset valuation, or upon the purchase or sale of that interest. The estimated rental value of the property will inevitably have to be quantified prior to capitalisation.

Throughout this book it is assumed that the valuation of a leasehold interest should be carried out on a single rate basis (see Baum and Mackmin, 2006, Chapter 7 and Baum and Crosby, 2007, Chapter 7 for a full examination of leasehold valuation methods).

Example 1.2

Referring to Figure 1.1, it appears that G has a leasehold interest, paying a fixed ground rent to D for the remaining 59 years of his lease at a rent of (say) £50 pa. He is currently receiving a rent from E under a lease with five years to run of £1,500 pa. ERV is £15,000 pa All rents are on full repairing and insuring terms. What is this interest worth?

(The valuation approach used is an equivalent yield valuation – see Baum, Mackmin et al, 2006).

Term:

Rent received	£1,500	pa
Rent paid	£50	pa
Profit rent	£1,450	pa
YP (years' purchase) 5 years @ 8%	3.9927	
Term value		£5,789

Reversion:

Rent received	£15,000	pa
Rent paid	£50	pa
Profit rent	£14,950	pa
YP 54 years @ 8%	12.3041	
PV 5 years @ 8%	0.6806	
Reversion value		£125,191
Total value		£130,980

It is arguable that the single rate analysis and valuation of leasehold property should be carried out on a net of tax basis, but in this book it is taken for ease that the investor is interested in the gross return and that gross valuations suffice.

1.2.4 Premiums

In rental analysis and capital valuation the payment of a premium, in addition to or in lieu of rent, is a complicating factor which the valuer must take into account.

In the property world the term premium has two distinct applications. The traditional use of the term is to describe a capital sum paid at the start of a lease by a landlord to a tenant in return for a low rent or some other benefit. The payment of this type of premium can produce advantages for both landlord and tenant. To the landlord, the receipt of an immediate cash sum rather than a flow of income is often more attractive for several reasons, including inflation and liquidity preference. In addition, if the tenant is paying a lower rent the landlord's income stream is likely to be more secure. The tenant, on the other side of the bargain, may prefer to use up capital in return for a reduced rental. Also, rather than holding a lease at a full market rent which might have no disposal value, the tenant will enjoy a profit rent (the difference between the rent he/she actually pays and the full market rent) which might render the leasehold interest valuable in the event that he/she wishes to dispose of it.

It should noted that this situation can apply in reverse where supply exceeds demand, and in such a case the assignor/vendor may be required to pay a premium to a purchaser as an inducement to take over the leasehold interest; such a premium is commonly referred to as a reverse premium. Market conditions will dictate whether a leasehold interest has a positive or negative value, and these conditions will change over a period of time. The premium, or reverse premium, may therefore be a tool by which parties eradicate market imperfections in rental levels as market conditions change. The premium may reflect the fact that the property is under-rented (let at less than the estimated rental value) and the reverse premium may reflect the fact that the property is over-rented. The complication of comparable rental evidence produced by the payment of such a premium at the outset of a lease is a factor often faced by the valuer.

The term premium has a second application. What is often called a premium may be paid upon the assignment of a leasehold interest by the assignee (the purchaser of the lease) to the assignor (the vendor of the lease) because the right to occupation of the property provided by that lease is regarded as valuable, notwithstanding that there may be no profit rent to be enjoyed. In a situation where there is competition to purchase the leasehold interest, a willing purchaser may pay a capital consideration, a premium, for a leasehold interest because (as an example) this will carry with it the right to run a profitable business from the premises. This is often the case with pubs, shops and other properties with locational value in the type of use associated with small business occupiers. The adjustment of comparable rental evidence produced by the payment of such a premium at the outset of a lease would not be appropriate.

Hence, in attempting to estimate rack rental value, the valuer must assess whether the amount of the premium paid represents the difference between contracted rent and the full open market rental value, in which case it may provide useful evidence, or if it does not.

The right of the landlord to increase the rent at review to a rack rental level will end the notional profit rent gained by the tenant who has paid a premium. The valuer must ascertain when the rent will be reviewed and to what level, and must also know whether the rent is to be reviewed to the full open market (rack) rental or to a proportion of it. On occasions the premium will be agreed and the initial rent is to be negotiated; on others the rent will be agreed and the premium will be open to negotiation.

(a) The effect on rent of a premium

When a premium is paid at the outset of a lease, the valuer would estimate the premium required by capitalising the rent lost over the period to the lease end or first rent review, whichever comes first. Each party would be advised, or an independent expert would look at the problem from each party's point of view.

Example 1.3

An initial rent of £10,000 pa is to be charged on a ten-year lease lease with five-yearly reviews to full rental value. Assuming that property has an estimated rental value of £13,500 pa, and a capitalisation rate of 5% would be appropriate, what premium should be paid?

From the landlord's point of view, the lost rent is a fixed amount and the calculation should be carried out using the required return for a fixed income property investment. We adopt the appropriate risk free rate at the date of the valuation, given by the redemption yield on a five-year bond or gilt, plus a premium for risk and illiquidity: see p 9. The risk premium takes particular account of the tenant's covenant strength. It is not appropriate to use a property capitalisation rate derived from comparable evidence.

Assuming a gilt yield of 4.5% and a risk premium of 2.5%, the appropriate discount rate is 7%. The landlord's valuation is as follows:

Rent lost/saved	£3,500 pa
YP 5 yrs @ 7%	4.1002
Premium, say	£14,350

From the tenant's point of view, broadly the same calculation would be applied. As in all leasehold valuations, a single rate valuation should be employed. It is possible to debate whether the discount rate should be the same, because the relevant discount rate for the tenant is the cost of financing the premium (the interest rate charged on a loan). This is likely to be a rate similar to the landlord's required return, which reduces any scope for disagreement over the amount of the premium.

The tenant's discount rate, based on a likely cost of borrowing of 2% over base rate, is 6%.

The tenant's valuation is then as follows:

Rent receivable	£13,500 pa
Rent paid	£10,000 pa
Profit rent	£3,500 pa
YP 5 yrs @ 6%	4.2124
Premium, say	£14,750

A negotiated settlement will not be difficult in this case, as the tenant can afford to pay more than the landlord requires (because his cost of borrowing is less than the landlord's required return). Splitting the difference would produce an agreed premium of £14,550.

Where the premium and rent have both been agreed, the unknown variable becomes the rental value of the property. In analysing this, the premium should be de-capitalised at the average of the landlord's and the tenant's discount rates to find the rent saved or lost which is then added to the agreed rent to find the ERV.

In certain situations the rent might be reviewed to a percentage of full rental value at the review. A similar approach should be adopted with some modification.

(Note that the rent saving or gain is capitalised from year 6 to 15 at a market capitalisation rate of 5% because this is not now a fixed income – it can increase at the year 10 rent review.)

Example 1.4

Take a premium of £15,000 paid in respect of a 15-year lease at £6,500 pa reviewed at five-yearly intervals to 75% of rack (full) rental value. What is the estimated rental value? The required return is 6.5%; the market capitalisation rate is 5%.

Years 0–5:	Rack rental value	$£x$ pa	
	Rent reserved	£6,500 pa	
	Rental gain/loss	$£(x - 6,500)$ pa	
	YP 5 years @ 6.5%	4.1557	
	=		$£4.1557x - £27,012$

Years 6–15:	Rack rental value	$£x$ pa	
	Rent reserved	£0.75x pa	
	Rental gain/loss	£0.25x pa	
	YP 10 years @ 5%	7.7217	
	*PV 5 years @ 5%	0.7835	
		£1.5125 x	

Total capital value of rent gain/loss: $£5.6682x - £27,012$

£15,000	=	$£5.6682x - £27,012$
£5.6682x	=	£42,012
x	=	£7,412 pa
Say	=	£7,500 pa

The example shows that a premium of £15,000 is required by an owner to compensate for the receipt of a rent of £6,500, £1,000 less than rack rental value (£7,500), for five years and then 75% of rack rental value thereafter.

(b) The effect on rent of a reverse premium

Reverse premiums may be paid to encourage the tenant to agree a higher than market rent. Because of upward-only rent reviews, the liability of the tenant to pay an above-market rent may continue past the first review or even later reviews. In such a case, some attempt has to be made to forecast rental values in order to estimate both the extent and length of the tenant liability and a conventional valuation is unlikely to be of benefit. The example below illustrates.

Example 1.5

What premium should be paid to a tenant where the landlord requires a rent of £85,000 pa for a 10-year upward-only lease of a property whose estimated rental value is £80,000 pa?

While the best solution to this type of problem is to use a simulation approach (see Baum and Crosby 2007) a DCF valuation examining the high rent option against the landlord's alternative of letting at rental value (see Baum and Crosby, 2007) is the most appropriate simple route to a solution. A market capitalisation rate of 5% can be compared with the required return of 7% to estimate a market expectation of rental growth using the following formula:

$$k = r - (SF \times p)$$

where k = the market capitalisation rate, r is the required return, p = rental growth over the review period (t) and SF = annual sinking fund to replace £1 over the review period at r.

Then g may then be derived from p, as $(1 + p) = (1 + g)^t$.

To find the implied rental growth rate to justify a 5% initial return when the required return is 7% based upon 5-yearly rent revisions: requires the use of this formula as follows:

k	=	$r - (SF \times p)$
0.05	=	$0.07 - 0.17389p$
p	=	$\dfrac{0.07 - 0.05}{0.17389}$

p	=	0.115015 (11.5% increase in rental value every five years)
1 + p	=	$(1 + g)^t$
1.115015	=	$(1 + g)^5$
g	=	$((1.115015)^{1/5}) - 1$
g	=	0.022 (2.2% pa)

Projected rent at year 5: £80,000 * $(1+.022)^5$ = £89,201

Total value with high rent:
Term:

Rent passing	£85,000	
YP 5 years @ 7%	4.1002	
Capital value		£348,517

Reversion:

Projected ERV		£89,201	
YP perp @ 5.00%		20.0000	
Projected value		£1,784,024	
PV 5 years @ 7.00%		0.7130	
Capital value			£1,271,984
Total value with high rent:			£1,620,500

Total value with normal rent

Term:

Rent passing	£80,000		
YP 5 years @ 7.00%	4.1002		
Capital value			£328,016

Reversion:

Projected ERV		£89,201	
YP perp @ 5.00%		20.0000	
Projected value		£1,784,020	
PV 5 years @ 7.00%		0.7130	
Capital value			£1,271,982

Total value with normal rent:			£1,600,000

The difference in valuations suggests a reverse premium of around £20,500 (also given by £5,000 multiplied by the YP factor for 5 years at 7%, 4.1002, because the reversions are common).

In this case, the penalty rent is cleared at the first review. However, a more punitive going-in rent — say £100,000 — would lead to a different result. In this case the expected rental value at review in five years' time is less than the initial rent, so the penalty is incurred for 10 years. From the valuation below, a reverse premium of nearly £115,000 should be paid in this case (£1,713,453 – £1,600,000).

Term:

Rent passing	£100,000		
YP 10 years @ 7.00%	7.0236		
Capital value			£702,360

Reversion:

Projected ERV (£80,000 * (1.022^10))		£99,449	
YP perp @ 5.00%		20.0000	
Projected value		£1,988,973	
PV 10 years @ 7.00%		0.5083	
Capital value			£1,011,093
Total			£1,713,453

(c) Non-rent effects of premiums

Assignment premiums can cause considerable problems for the valuer. While in certain cases the assignment premium might represent the capitalised value of the rent saving or profit rent or overrentedness, this is often not the case and useful analysis is regularly impossible. It is therefore important that the valuer is able to distinguish between those premiums which are explained *in toto* by a rent saving and those which represent other factors in addition to or instead of a profit rent.

What are those other factors? An assignment premium might represent any or all of the following.

(i) The saving on the full market rent, which is the usual justification for a premium and which, if identified and supported by market evidence, may be useful in analysis.

(ii) An element for fixtures and fittings. Taking a new lease of (for example) shop property will involve the ingoing tenant in considerable expense. The tenant may be required to provide a shop front and will almost certainly be required to provide shelves, other shop fittings, carpets and so on. Taking an assignment of an existing lease on premises already fitted out may save the tenant much of this expense, so part of the premium might represent the second-hand value of these items of fixtures and fittings.

(iii) An amount for tenant's improvements. Improvements to the property may have been carried out by the outgoing tenant, and the value of these may not be reflected in the rent. The ingoing tenant may be prepared to pay a capital sum for the benefit of such improvements (for example an extension to the building) and the amount of that sum will reflect the period which will elapse before the landlord can take them into account in assessing the rent. This, in turn, will depend on the terms of the lease and the provisions of section 34 of the Landlord and Tenant Act 1954 (see p26).

(iv) Key money: this may have many different interpretations. On the assignment of a lease key money will usually constitute at least a part of the premium paid, and will not be justifiable in terms of the investment value of the property. It is, instead, an indication of the division of the market for short leasehold interests into an investment sub-market and an occupation sub-market. Key money is that element of a premium which reflects the trading advantages of the pitch to the assignee. It is an anticipation of a trading profit and is peculiar to the assignee; after all, certain occupiers have more profitable businesses than others, but rack rental value levels remain consistent between different properties. Key money cannot, therefore, be analysed in terms of a rent saving.

Key money is likely to be associated with an undersupply of property available to rent, and is likely to be the result of competition between prospective tenants who, given security of tenure (see later in this chapter), can spread the outlay over many years of trading profits. The key money might even be recoverable upon reassignment. In addition, the goodwill of the assignor's business might be immobile. Even though the assignor might transfer the retail business to larger premises nearby, an element of locational goodwill (consumer loyalty) will remain with the subject property as long as similar goods continue to be sold from the shop. Even potential assignees in a different trade may have to acquire this goodwill by being forced to compete for the lease with other potential assignees in the same trade.

In general, premiums paid on the assignment of leases held at what appear to be market rents are disregarded in assessing rental value. It is accepted that they might represent an indication of the strength of demand for a particular property, but the premium itself is not amortised (converted to an annual rent) and used as evidence of a higher or lower market rental level. It is part of the valuer's professional expertise to be able to make a judgement about what can be a fine point.

Two unreported cases involving rent reviews on large food stores highlight the difficulties involved in considering the reasons why a premium may be paid upon the grant of a new lease and whether that premium reflects payment of rent in advance (the payment of a capital sum to reduce the initial rent). In each case the rent review was decided at arbitration, and the award was the subject of an unsuccessful legal challenge by the aggrieved party. In both cases the arbitrator was presented with comparable evidence of lettings of other food stores where in addition to paying a rent the tenants had also paid a substantial premium to the landlords.

The question was whether the payment of the premium was designed to artificially reduce the initial rent to be paid under the lease or for some other purpose. In the first case the arbitrator decided on the evidence that the rent paid by the food store operator did represent the full open market rent for that property and that the payment of the premium was designed to safeguard their trading position. The premium payment was not therefore taken into account in assessing rental value. In the second case involving a different food store there was evidence that there had been more than one food store operator which had made a rental offer and in addition had offered a substantial premium payment for the property in question. The arbitrator concluded that if several parties were willing to pay a premium in addition to a rent then this indicated that in the absence of that premium the rent of that property would have been fixed at a higher level. Accordingly in those circumstances it was appropriate to analyse the premium and make an upward adjustment to the rental level of that piece of evidence.

1.2.5 Marriage value

The fragmentation of ownership means that a freeholder and a leaseholder of a property may each have a valuable interest in that property. The total value of those interests may often be less than if those interests were merged. The merging of the two interests may give rise to what is called marriage value.

Where an investor or occupier with an interest in certain property is in negotiation for the purchase/sale of the interest to another individual with an interest in the same property, an element of marriage value might be released upon the combination of those interests. It will often be the case that the total market value of the freehold and leasehold interests in a property is less than the market value of the freehold interest with vacant possession. A key reason for this may be different discount rates applied by the respective parties, as shown in Example 1.3 above, or the conventional technique of valuing leasehold interests by dual-rate techniques which can produce an undervaluation which is corrected upon the merger of the interests into a longer term property holding (Baum and Mackmin 2006; Baum and Crosby, 2007).

In order to calculate the marriage value, it is necessary to value each separate existing interest and then to calculate the value of the merged interests. The difference is the marriage value. Typically, if both parties are willing to merge their interests the marriage value is assumed to be split 50/50 – but this may depend upon their respective negotiating strengths.

Consequently, when advising in such negotiations the valuer should have regard not simply to the market value of each interest but also to the gain to be made by the purchaser of the merged interest and how the marriage value is to be split.

1.2.6 Extensions and renewals of leases

Where a tenant holds a lease of property which he feels to be unsatisfactory for some reason, he may approach the landlord, offer to surrender the lease, and request a new lease on more favourable terms. This is often referred to as re-gearing. The requested variation might, for example, be in respect of the lease length, removal of a landlord's break clause, or changes to a repairing liability. A landlord might similarly request a surrender in order to remove a tenant's break clause or to exchange an unfavourably irregular rent review pattern for a lease in the modern style with reviews every five years.

In either case, the valuer may be asked to advise the parties what rent (or rent and premium) should be payable under the new lease to ensure that they are in a similar financial position as previously.

Example 1.6

A tenant whose lease has nine years to run without review at an annual rent of £110,000 is prepared to surrender this lease and to accept a new 20-year lease with five-yearly reviews to ERV in its place. The full rental value of the subject property is £150,000. What initial rent should be charged and paid under the new lease? The required return is 7%; the market capitalisation rate is 5%. An additional 1% is added to the required return for a leasehold.

Again, this problem should be examined from both landlord's and tenant's point of view.

Landlord's present interest:
Rent received	£110,000 pa	
YP 9 years @ 7%	6.5152	
		£716,676
Reversion to:	£150,000 pa	
YP rev. perp. 9 years @ 5%	12.8922	
		£1,933,827
Total		£2,650,503

Landlord's proposed interest:
Rent received	£ x	
YP 5 years @ 7%	4.1002	
	£4.1002x	
Reversion to:	£150,000 pa	
YP rev. perp. 5 years @ 5%	15.6705	
	£2,350,578	
Total		£4.1002x + £2,350,578

£4.1002x + £2,350,578	=	£2,650,503
£4.1002x	=	£299,925
x	=	£73,149 pa

Tenant's present interest:
Term
Rent received	£150,000 pa	
Rent paid	£110,000 pa	
Profit rent	£40,000 pa	
YP 5 years @ 8%	6.2469	
		£249,876

Reversion

Rent received	£167,242 pa	
(£150,000 * 1.022^5: see Example 1.5)).		
Rent paid	£110,000 pa	
Profit rent	£57,242 pa	
YP 4 years @ 8%	3.3122	
PV 5 years @ 8%	0.6806	
	2.2542	
		£129,037
Total		£378,913

Tenant's proposed interest:

Profit rent	£x pa	
YP 5 years @ 8%	3.9927	
x	=	3.9927
£3.9927x	=	£378,913
x	=	£94,901 pa

This is the tenant's required profit rent.

Rental bid = £150,000 − £94,901 = £55,091 pa

Negotiated settlement suggested at, say (£73,149 + £55,091)/2 = £64,120 pa, say £64,000 pa

The foregoing valuation problems are typical of the exercises which are regularly carried out by the valuer advising a client in such a transaction. In practice, however, the completion of such an exercise may not secure agreement between the parties. Such deals are not always mutually beneficial, so the valuer will often be required to interpret the respective negotiating strengths of the parties involved and to endeavour to achieve a financial gain for his client. In addition, external factors may drive the deal: for example, a landlord may be prepared to pay a capital sum to the tenant or accept a rent less than the open market rent if the new lease improves the value of adjoining properties within the ownership. Such an attitude might arise especially when pension funds and other investing institutions insist upon particular lease terms.

The valuer's interpretation of the existing lease terms is vital in the preparation of these exercises.

1.3 The lease

The valuation and management of leasehold property is made significantly more complicated by the fact that almost every lease of commercial property contains different clauses with variations in wording. The interpretation of these leases often gives rise to disputes which are sometimes only resolved by litigation.

Colam (as long ago as 1981) made a plea for standardisation of the lease terms which are found in commercial leases. He identified the following 13 pieces of information which need to be included in a lease.

(a) The name of the landlord.

(b) The name of the tenant.

(c) A description of the demised (let) premises.

(d) The term, that is, the length of the lease.

(e) The dates at which the rent is to be reviewed (commonly at five-year intervals) with a provision for the manner in which any dispute is to be settled.

(f) The repairing liability — whether it be wholly upon the landlord, wholly upon the tenant or divided between them.

(g) The decorating liability, which may be split between internal and external decorations.

(h) The insuring liability.

(i) Restrictions upon the user (use) of the building. This may accord with the planning authority's powers of restriction (for example, where the use is restricted to offices) or may be considerably more restrictive (for example, where the use is restricted to the branch office of a firm of engineers).

(j) The provision regarding alienation, that is sub-letting or assignment of the premises, which normally requires the landlord's consent with an implied condition that such consent is not to be unreasonably withheld (see 1.7).

(k) The provision for a service charge, which in the case of a multi-occupied office or shopping centre may cover landlord's services such as cleaning common parts. At the time there was an increasing momentum towards "clear" leases, in which the service charge is designed to cover all possible expenses incurred by the landlord.

(l) The landlord's covenant promising quiet enjoyment of the premises by the tenant.

(m) Rights (such as access by other users of the property) reserved by the landlord.

Special cases may necessitate special clauses, but the above list constitutes the essence of a typical lease. Variations in the comprehensiveness of the tenant's liabilities mean that the terms are often summarised into one of four categories, as follows:

(a) full repairing and insuring (FRI), where the tenant is responsible for all repairs and the cost of insurance

(b) internal repairing terms (IRT), where the tenant is responsible for internal repairs, but the landlord retains responsibility for external and structural repairs and insurance

(c) an inclusive lease, typical in short-term agreements, where the tenant has no responsibility for repairs or insurance

(d) a clear lease, where a full service charge ensures that the income received by the landlord is net of all expenses.

Despite such efforts to to achieve standardisation there is still no standard form of lease which is universally adopted in the commercial property sector. Even in a shopping centre where there may be 50 similar shop units it is more likely than not that there will be variations in the leases for the units, as each lease will contain variations negotiated by the different solicitors and surveyors acting for the individual tenants. Such variations will be a function of the respective negotiating strengths of the parties when they negotiated the lease. The result is that the valuer must carefully examine every separate lease and cannot rely on any standardisation.

Following the recession in the early 1990s, the government became concerned that commercial leases were inflexible, and that upward only rent reviews coupled with long lease terms and restrictions on alienation were anti-competitive and may leave tenants financially exposed. They threatened to introduce legislation to outlaw upwards-only rent reviews such that at rent review the market rent would become payable even if were lower than the rent currently passing. The British Property Federation reacted by introducing a voluntary Code of Commercial Lease Practice in 1995.

The University of Reading was commissioned by the government in 1999 to review the impact of the Code and to ascertain whether the property industry's attempt at self-regulation was succeeding. They found that market forces had acted to reduce lease lengths and to encourage the introduction of more break clauses, but that upward-only rent reviews were still prevalent. Still unsatisfied that there was sufficient flexibility in commercial lease terms, the then Department of Environment, Transport and the Regions (later the Office of the Deputy Prime Minister) gave the property industry one further chance to introduce more flexibility before legislation was introduced.

In response to the unwelcome prospect of legislation, many of the UK`s largest landlords promoted a revised form of the Code of Commercial Lease Practice (2002) under which every tenant would be offered appropriately-priced leases with a variety of lease terms including varying length and upwards-only or upwards and downwards rent reviews. Once again the University of Reading was commissioned to monitor the implementation of the code. The Crosby Report (ODPM, 2005) concluded that lease lengths had further shortened as a consequence of market forces and many tenants did not push hard to take up the option of upwards and downwards rent reviews when offered to them. Legislation was not introduced, and as a result the Code of Commercial Lease Practice continues to be operated on a voluntary basis. The focus of concern has moved slightly away from lease lengths and upwards/downwards rent reviews to issues of alienation (and in particular the conditions a tenant must satisfy in order to assign or underlet his/her leasehold interest), but the possibility of legislation against upward only rent reviews has not gone away.

1.4 The Landlord and Tenant Acts

1.4.1 Introduction

Statutory interference in the contractual relationship which exists between landlord and tenant affects the three major sectors of the land market — commercial, residential and agricultural — in a broadly similar manner. There are two major targets of statutory control: security of tenure and rent control.

In the commercial sector, the governing statutes are the Landlord and Tenant Acts of 1927 and 1954 and the Law of Property Act 1969. It is the rent control provisions together with regulations concerning the payment of compensation that are of major importance to the valuer, but the valuer cannot negotiate on behalf of the landlord or the tenant without a detailed knowledge of the tenant's statutory rights to security.

The governing statute is Part II of the 1954 Act, which is amended by certain of the 1969 provisions. Parts of the 1927 Act continue to operate with specific reference to compensation for improvements only.

Part II of the 1954 Act has been regarded as one of the more successful of the statutory interferences in the land market. It was introduced post-war to achieve a fair balance between landlord and tenant and to enable the business activity of the country to continue smoothly while allowing property investors to receive a competitive rate of return. Under the provisions of the Act, (subject to certain exceptions) a tenant is entitled to a new lease upon the expiry of the old lease subject to agreement of terms between the landlord and the tenant, or, in the event of non-agreement, fixed by a court.

Despite the success of this legislation, it has been argued for many years that interference in the granting of commercial leases is no longer warranted. As a result the government commissioned Lord Woolf to advise on modernisation of the legislation following consultation with the property and legal professions.

The Regulatory Reform (Business Tenancies) (England & Wales) Order 2003 introduced, with effect from 1st June 2004, significant changes to some of the provisions and operation of the Act, although much of it was left intact. (For convenience the Order is referred to as the RRO.)

Part II of the 1954 Act applies to any tenancy where the property comprised in the tenancy is or includes premises which are occupied by the tenant and are so occupied for the purposes of a business carried on by him or for those and other purposes (section 23(1)): that is business premises or premises in a mixed use including business purposes.

The expression "business" includes a trade, profession or employment and includes any activity carried on by a body of persons, whether corporate or unincorporate (section 23(2)), but excludes certain agricultural and mining tenancies and residential tenancies. An individual and the company he or she controls should be treated as equivalent when assessing qualifications for statutory procedures under the Act.

Where premises are occupied for a mixed use, typically residential and business purposes combined, a landlord is likely to argue that the 1954 Act applies to the letting as the alternative — the Housing Acts and others — can be considerably more onerous upon the landlord in the security offered to the tenant (see Chapter 2). Consequently, disputes have arisen over whether a tenancy is a business or residential tenancy for the purposes of these Acts.

It has been held that the sub-letting of rooms and flats within a residential building could not be said to constitute a business tenancy (*Bagettes Ltd* v *GP Estates* (1956)). In *Royal Life Saving Society* v *Page* (1978) Lord Denning held in the Court of Appeal that it is the importance and degree to which the occupation of the premises is for business purposes which constitutes the acid test. The question of where to draw the line was considered in *Trans-Britannia Properties* v *Darby Properties Ltd* (1986) which concerned lock-up garages, most of which were sub-let, though one was used by the tenants as a store room. The Court of Appeal decided that, although the tenants carried out some maintenance and cleaning, this was not sufficient to bring them within the protection of the Act. See also *Clear Channel UK* v *Manchester CC* (2005).

The main aim of the 1954 Act is to give the tenant a right to renew the lease upon its contractual expiry. This is referred to as the tenant's security of tenure. It is not possible for a landlord and tenant to agree to a lease which removes the tenant's security of tenure, unless both parties agree to contract out of the relevant protective provisions in the Act (sections 24–28). Whereas prior to the RRO it was necessary for the parties to make a joint application to the court, the RRO has simplified this process. The landlord can now serve an effective contracting out notice on the tenant as long as this is clearly marked as a health warning on such notices and the tenant signs a declaration that the notice has been received and its consequences accepted.

Some see upward-only rent reviews and other features of the standard institutional lease as the price paid for 1954 Act protection. Hence shorter contracted-out leases have become more commonplace, as occupiers are often unwilling to commit to long leases and active manager/landlords are increasingly likely to wish to re-develop or refurbish property in the short to medium term. Such short leases and licences are usually intended to fall outside of the reaches of the 1954 Act.

The provisions ensuring that the tenant has security of tenure are contained in sections 24–28 of the 1954 Act and are considered below.

1.4.2 Security of tenure

Under section 24 a tenancy to which Part II of the Act applies continues to run (and at the end of the lease the tenant will continue to hold over) unless it is terminated in accordance with the Act as amended by the RRO. Where the tenant does not wish to renew, this may be effected on the tenant's part by a section 27 notice to quit served three months before the contractual expiry date or simply by vacating on that date with no notification to the landlord. After the contractual expiry date the tenant is required to give three months' notice (section 27).The landlord can bring the tenancy to an end by service of a notice of termination under section 25.

The landlord's power to terminate the tenancy is not the same as its power to end the tenant's occupation, as the service of a valid section 25 notice merely sets in motion the apparatus designed to enable the tenant to be granted a new lease. The notice must be given not more than 12 nor less than six months before the contracted date of termination.

The landlord's section 25 notice must state whether the landlord is opposed to the grant of a new tenancy. If not opposed, the notice must set out the proposals as to the property to be comprised in the tenancy, the rent to be payable and the other terms of the tenancy. If he is opposed to the grant of a new tenancy, the landlord must specify one or more of the grounds under section 30 as the grounds for the opposition. The seven section 30 grounds (see below) are the only bases upon which the landlord might resist the application for a new tenancy made by the tenant.

RRO removed many of the strict time limits within which the landlord and particularly the tenant were required to serve a counter-notice and make an application to the court for a new lease. Now both the landlord and the tenant may apply to the court at any time before the date on which the tenancy will end or they can agree to extend that period.

Service of the section 25 notice by the landlord with the requisite response by the tenant is the most likely overture to the granting of a new lease. In a rising market the landlord is likely to be eager to serve a notice in order that a new tenancy may be commenced as soon as possible at a full market rent. In a falling market where the property may be over-rented the landlord may well wish to allow the tenant to hold over the existing lease terms by delaying service of a section 25 notice.

Alternatively, prior to the landlord's serving of a section 25 notice the tenant might request a new tenancy under section 26 by issuing a notice which states the date of commencement of the new tenancy, the property to be comprised in that tenancy and the proposals as to rent and other terms. The tenant must inform the landlord in the request that if he intends to oppose an application for a new tenancy he must within two months state his grounds of opposition under section 30. If the landlord fails to respond within the two months period then the landlord will lose the right to object to a new tenancy under any of the section 30 grounds.

Section 30 enables the landlord to resist an application for a new tenancy under any one of the following seven grounds.

(a) The tenant has been in serious breach of a repairing obligation.
(b) The tenant has persistently delayed in the payment of rent, see, for example, *Horowitz v Ferrand* (1956).
(c) The tenant has been in serious breach of any other covenant in the lease or may be criticised for its use or management of the holding. In the exercise of this ground and of ground (a) the court has considerable discretion to exercise and will normally err on the side of the tenant. An exception to this is to be found in *Turner and Bell v Searles (Stanford-le-Hope) Ltd* (1977) where the court refused to grant a new tenancy because the tenant's use was in contravention of planning regulations, employing the second limb of this ground.

(d) The landlord is willing and able to provide suitable alternative accommodation on reasonable terms. It is rare that a landlord will be able to rely upon this, due to the considerable room for argument it allows a tenant.

(e) The building is sub-let in parts and the total rent passing is less than if the whole were in single occupation.

(f) The landlord intends to carry out substantial works of demolition or reconstruction and cannot reasonably do so without obtaining possession. The tenant may defeat the landlord's notice under this ground by employing section 31(A) (introduced into the 1954 Act by the 1969 legislation). Where the tenant is prepared to give the landlord access and/or take a new tenancy of an economically separable part of the holding in order that the reconstruction can still be carried out, the court will not give an order for possession. *Price* v *Esso Petroleum Co Ltd* (1980) established that if, under the tenancy agreement, the landlord has a right to enter to carry out the works and did not therefore require possession, this ground could not in any event be relied upon. In *Cerex Jewels Ltd* v *Peachey Property Corporation plc* (1986) the judge at first instance decided that, as the works would require closure of the tenant's business for six weeks, the landlord would require possession and so the tenant could not rely on this provision. This was reversed by the Court of Appeal, which found that only two weeks' closure was necessary and that this would not constitute interference to a substantial extent. Generally speaking, it is incumbent upon the landlord to prove:

    (i) that the property is ripe for development and that it is ready to carry out that redevelopment showing architect's plans, planning permissions, and the appropriate financial resources (see *Turner* v *Wandsworth London Borough Council* (1994), *National Group Car Parks* v *Paternoster Consortium Ltd* (1989), *Marazzi* v *Global Grange Ltd* (2003), *Ivorygrove Ltd* v *Global Grange Ltd* (2003) and *Yoga for Health Foundation* v *Guest* (2002)

    (ii) that possession of the property must be obtained to carry out that redevelopment and

    (iii) that the work could not be carried out even if the tenant offered to take a new lease of a smaller economically viable unit within the building.

(g) The landlord or a party with a controlling interest has owned the superior interest for a period of five years which ends with the termination of the current tenancy and intends to occupy the premises for its own (business or residential) purposes. In order that the landlord can rely upon this ground it must be able to evidence intention to occupy to the court's satisfaction (see, for example, *Lightcliffe and District Cricket and Lawn Tennis Club* v *Walton* (1978) and *Europark (Midlands) Ltd* v *Town Centre Securities plc* (1985) and *Zarvos* v *Pradhan and Hakim* (2003)).

If the landlord is able to satisfy the court that any one (or more) of these grounds is fulfilled then section 31 provides that the court shall not make an order for a grant of a new tenancy. The unfortunate tenant might then be entitled to compensation under two heads, for disturbance and for improvements.

1.4.3 *Compensation for disturbance*

Under section 37, compensation for disturbance is payable by the landlord to the tenant where the landlord has obtained possession under grounds (e), (f) and (g) of section 30. The attitude of Parliament was presumably that a tenant who brings dispossession upon himself (grounds (a), (b) and (c)) should not be compensated for it, that where alternative accommodation is provided the tenant's loss is considerably reduced, and the reasonable terms necessitated by ground (d) may reflect the tenant's liability for removal expenses.

Currently the maximum compensation amount payable is two times the rateable value of the property subject to the following conditions:

(a) that, during the whole of the 14 years immediately preceding the termination of the current tenancy, premises being or comprised in the holding have been occupied for the purposes of a business carried on by the occupier or for these and other purposes
(b) that, if during those 14 years there was a change in the occupier of the premises, the person who was the occupier immediately after the change was the successor to the business carried on by the person who was the occupier immediately before the change.

In any other qualifying case (that is, where occupation has been for less than 14 years) the compensation is limited to one times the rateable value. In *Department of the Environment* v *Royal Insurance plc* (1986) it was held that 13 years and 364 days of occupation were insufficient to qualify for the higher multiplier as the de minimis rules did not apply. (For a case dealing with the correct multiplier where the regulations have changed, see *International Military Services* v *Capital and Counties plc* (1982).)

1.4.4 Compensation for improvements

Compensation for improvements is governed by the Landlord and Tenant Act of 1927, which refers in general to the same group of tenancies that falls within the ambit of the 1954 Act.

Section 1(1) of the 1927 Act provides that a qualifying tenant shall be paid compensation on quitting his holding at the termination of the tenancy in respect of any improvement made by the tenant or its predecessors in title during the current tenancy, as long as that improvement adds to the letting value of the holding (for a dispute over predecessors see *Pelosi* v *Newcastle Arms Brewery Ltd* (1981)). The termination of the tenancy may result from the effluxion of time or a notice to quit served by either party during the lease.

Under section 2 no compensation shall be paid in respect of any improvement made in pursuance of a statutory obligation or as the result of a contractual obligation made with the landlord. In other words, no compensation is payable where the tenant carries out an improvement as a condition of the lease, as it is implicit in the contractual relationship that the tenant will strike a reasonable bargain with the landlord in return, usually by paying a low rent or receiving a reverse premium.

Under section 3 it is only improvements carried out with the landlord's consent which earn compensation (see *Deerfield Travel Services Ltd* v *Leathersellers' Company* (1982)). There are precise rules laid down in section 3 for the procedure to be followed in obtaining such consent. If the landlord offers to do the work the tenant must accept and will then have to pay an increased rent, either agreed between the parties or determined by the court.

The 1927 Act contains no explanation or definition of what constitutes an improvement, although it is apparent that the construction of a new building falls within the Act. It is arguable that the valuer is best qualified to decide upon what is, in effect, a factual matter; if the result of the work is to increase the rental value of the holding, then it is an improvement. An alternative view of this was taken in *Lambert* v *Woolworth (FW) Co Ltd (No 2)* (1938) where it was held that such a test should be applied from the tenant's point of view so that an improvement may be regarded as work which improves the property in some way with respect to the tenant's occupation of the premises.

The amount of compensation to be paid is limited to the lesser of the following two amounts (section 1(1)):

(a) the net addition to the value of the holding as a whole (ie the freehold in possession) which is the direct result of the improvement

(b) the reasonable cost of carrying out the improvement at the termination of the tenancy, subject to a deduction for the cost of putting the improvement into reasonable repair, except so far as such an expense is covered by the tenant's repairing obligations in the lease (and may therefore be recoverable by the landlord as dilapidations).

Example 1.7

T is the new tenant of secondary shop property which is leased from L under a new lease at a rent of £100,000 pa on full repairing and insuring terms. Having obtained L's consent, T wishes to extend the shop (current dimensions 10m × 20m) by adding 10m depth of storage space to the rear at a cost of £150,000. This improvement is not carried out as a condition of the lease. Calculate T's compensation for improvements if T serves a notice to quit upon completion of the improvement.

(a) Net addition to value of the freehold in possession:

Let £x = full rental value per square metre for the front Zone (A) space to a depth of 7m[7]
Let £x/2 = full rental value per square metre for the mid Zone (B) space to a further depth of 7m
Let £x/4 = full rental value per square metre for the rear Zone (C) space (the remainder)

Then £(7 × 10x) + £(7 × 10 x/2) + £(6 × 10 x/4)	=	£100,000 pa
£120 x	=	£100,000 pa
x	=	£833.33 per m²
Addition to rental value	=	100 m² × £833.33/4
say	=	£20,830 pa
YP perp @ 9%	11.111	
	£226,444	
Addition to capital value, say	£230,000	

(b) Current cost of improvements:	£150,000	
Take lesser: compensation =	£150,000	

For a consideration of zoning techniques, see *Jane's (Gowns) Ltd* v *Harlow Development Corporation* (1980); *UDS Tailoring Ltd* v *BL Holdings Ltd* (1982); and *Turone* v *Howard de Walden Estates Ltd* (1982).

It should be noted that compensation for improvements is unlikely to be paid where the landlord has obtained possession to demolish or reconstruct the premises (section 30(1)(f)) as the addition to value of the freehold in possession is likely to be nil in such a case. In general, therefore, the tenant's likely right to compensation may be summarised as shown in Table 1.1 in relation to the ground employed to obtain possession.

Table 1.1 Tenant's likely rights to compensation

Ground for possession	Disturbance compensation	Improvements compensation
(a)	No	Yes
(b)	No	Yes
(c)	No	Yes
(d)	No	Yes
(e)	Yes	Yes
(f)	Yes	No
(g)	Yes	Yes

Note: see Landlord and Tenant Act 1954, section 30

1.5 The renewal of a lease

1.5.1 Introduction

In the majority of cases the landlord will neither wish nor be able to obtain possession of business premises and will agree the terms of a new lease with his tenant without recourse to the court. However, just as the purchase of land by a local authority by agreement is often made to the background of compulsory powers (see Chapter 7), the existence of the Landlord and Tenant Acts will have a major effect upon the bargain struck between even the most fraternal of parties.

Although the parties can normally be relied upon to reach a mutually satisfactory arrangement, the provisions of the 1954 Act for lease renewal err on the side of the tenant in controlling the amount of rent to be paid under the new lease and certain other terms. Any new lease is likely to incorporate these statutory provisions unless a variation is mutually beneficial.

A lease of business premises may include an option for the tenant to renew the lease at the end of current tenancy; this will almost always be in addition to the tenant's statutory rights. The professional adviser should be aware of the advantages and disadvantages of employing the right to renewal given by the Act in preference to the option. For example, while the landlord has no right to oppose the option by employing the section 30 grounds, and the option will usually duplicate the terms of the old lease in the new rather than leave them to be negotiated as in the statutory procedure, a more favourable rent may well be the result of using the Act in preference to the option.

For potential problems in respect of options to renew see *Trane (UK) Ltd* v *Provident Mutual Life Assurance* (1994) regarding timing, and *Bairstow Eves (Securities) Ltd* v *Ripley* (1992) regarding conditions precedent (where a breach of contract may preclude exercise of the option).

1.5.2 Property to be comprised in the new tenancy

Interpreting in tandem the provisions of sections 23(3) and 32, the property to be comprised in the new tenancy will be that part of the property included in the current tenancy and used by the tenant for the purposes of a business. If the whole of the property currently let is not included within this definition then the parties may agree the extent of the property included in the tenancy in writing; in default of such an agreement, the court shall designate that property by reference to the circumstances existing at the date of the order of the new tenancy under section 29.

1.5.3 *Duration of the new tenancy*

As outlined above and given statutory weight in section 33, the new tenancy shall be as agreed between the landlord and tenant and may consequently be for any number of years, regardless of the length of the terminating tenancy. In default of agreement section 33 empowers the court to grant a new tenancy up to a maximum term of 15 years.

The court has discretion to grant any term which it feels is appropriate but will usually have regard to the previous lease term. In cases where the landlord has redevelopment proposals but has narrowly failed to oppose the granting of a new lease under section 30(1)(f) a shorter term may nonetheless be granted by the court to take account of this prospect.

As an alternative, the court may order a break clause or an option to determine in the new lease. This was established in *McCrombie* v *Grand Junction Co Ltd* (1962). In *Amika Motors Ltd* v *Colebrook Holdings Ltd* (1981) the landlord of a car showroom had proposals to redevelop but did not attempt to gain possession at the end of the terminating 14-year lease under section 30(1)(f). Instead, the landlord proposed that a short-term lease of five years with an option to break after three years be granted and the court concurred in this because of the underlying policy of the Act not to restrict landlords in redeveloping their property. In *Adams* v *Green* (1978) a break clause was inserted by the court in a new lease for 14 years at any time given two years' notice of rebuilding by the landlord (see also *National Car Parks* v *Paternoster Consortium Ltd* (1989)).

During the 1970s and 1980s institutional leases of 20 or 25 years became commonplace in all commercial property sectors. In the recessionary climate of the 1990s and indeed in the stronger economic conditions prevailing afterwards many occupiers seeking greater flexibility and lesser liability prefer to take leases of 15 years or 10 years only. A number have been successful in persuading county courts to grant a lease for a term of five years only, notwithstanding that the previous lease term may have been longer. A compromise often reached is for a 10 year term but with a tenant break clause at the fifth year.

Disputes between landlords and tenants over five year terms and in particular whether a higher rent should be payable for a shorter term are becoming more frequent. Following the introduction of the second Commercial Lease Code in 2002 (see section 1.3), the need arose to price lease flexibility. A commercial software package — OPRent — was developed with the support of leading landlords. OPRent (*www.oprent.com*) is a software system that uses simulation to compare the cash flows available to property owners under a variety of different lease structures. This produces an option pricing facility which for the first time allows the various options available to property owners to be priced. OPRent compares the value of a flexible lease with that of a standard lease over the period of the standard lease and helps users to assess appropriate rental values for variable lease lengths, review periods, review options, break clauses and so on. As a result of such innovations, it is possible to price these effects, although it remains to be seen how far courts will be persuaded by such outputs.

Where the tenant holds over after the expiry of the old lease and before the grant of the new, this may also be a factor taken into account by the court in fixing the lease term.

1.5.4 *Terms of the new lease: general*

The terms of a tenancy granted by order of the court under this Part of this Act (other than terms as to the duration thereof and as to the rent payable thereunder) shall be such as may be agreed between the landlord and the tenant or as, in default of such agreement, may be determined by the court; and in determining those terms the court shall have regard to the terms of the current tenancy and to all relevant circumstances (Landlord and Tenant Act 1954, section 35).

Section 35 gives the court a circumscribed discretion to alter the terms of the new lease from those forming the basis of the old. While the court might give weight to "all relevant circumstances" and decide that a change is justified, the underlying assumption is that the terms of the old lease will be repeated in the new.

In *Kirkwood* v *Johnson* (1979) the court found good reason to make a variation. The expiring lease contained an option for the tenant to purchase the landlord's interest within the currency of the five-year lease. This option had obviously not been exercised by the tenant and the court declined to repeat the option in the new five-year lease it granted. The reasons given were that the object of the statutory provisions was to protect the tenant's business rather than its property asset; unlike (for example) a repairing obligation. "Time is of the essence" in an option to purchase: if the correct notice is not served in time then the right is lost (see *Westway Homes* v *Moores* (1991)).

Many leases have covenants imposed on the tenants which restrict the use that the tenant can make of the premises. The court may therefore have to consider whether such a user clause should be incorporated into the new lease or whether the restriction should be varied in any way. Often the tenant will require a wider user clause due to a subtle — or otherwise — change in his business or simply to increase the value of his alienable lease. In *Aldwych Club* v *Copthall Property Co* (1962) the court refused to vary the user clause where the tenant simply required a more saleable asset as, again, the purpose of the Act was to protect the tenant's business rather than to provide a valuable property asset.

The landlord, too, may wish to extend the permitted use of the premises upon renewal simply to increase the value of the asset. A higher rent would normally be payable for a property which can be put to a higher range of uses. In *Charles Clements Ltd* v *Rank City Wall Ltd* (1978) such a variation urged by the landlord was not granted by the court, again because the intention of the Act was to protect the business of the tenant, which in this case would be hindered by the charging of a higher rent.

Ideally a landlord would prefer to impose on to the tenant all the costs of managing a property and grant what might be termed a clear lease (see p15). In the case of *O'May* v *City of London Real Property Co Ltd* (1979) the landlord sought to vary the previous lease terms in this way. It was shown in evidence that the introduction of a clear lease would considerably increase the value of the landlord's interest, even allowing for a reduction in rent, presumably due to a reduction in the risk of the income being reduced by rising costs and hence the yield which would be accepted by a purchasing institution. The landlord urged the court to alter the lease terms on renewal so that a clear lease would be created. The judge formulated four tests which should be considered before deciding whether an alteration was to be permitted.

(a) Has the party seeking a variation shown a reason for doing so?
(b) If there is a variation, will the other party be compensated by a consequential adjustment of the open market rent?
(c) Will the proposed change materially impair the tenant's security in carrying on the business or profession?

(d) If a *prima facie* case is shown regarding the above, is the proposal, in the court's opinion, fair and reasonable between the parties?

The court of first instance allowed the variation of the lease in the manner proposed after considering these four tests. However, the Court of Appeal reversed the decision under grounds (b) and (d) above. Ground (b) was not fulfilled because in the court's opinion:

> The effect of the proposed modification of the terms of the existing tenancy in the present case is to make the lessee (tenant) an insurer of the lessor (landlord) against risks connected with the structure of the building which under the terms of the existing lease have to be borne by the lessor. They are risks which are inherent in the ownership of the building. The reduction in the rental which is said to compensate the lessee for assuming these risks is far from being an indemnity against them.

Commenting with respect to ground (d) the court said:

> The proposed alterations in the terms of the lease . . . alter the mutual relations of the lessor and the lessee so drastically and so adversely to the lessee, that the court could not in the exercise of its discretion properly impose them on the lessee under the Act in the absence of some more cogent reason than can be discovered in this case.

This is an interesting indictment of the normally accepted tenet that an adjustment of rent can compensate accurately for an adjustment of obligation (see Baum 1982) and has elicited protests that contrary to the spirit of the 1954 Act the result of the Court of Appeal decision is an unnecessary interference in the relationship between a landlord and tenant without which the landlord would be under no compulsion to grant a lease under any other terms than those which are normally acceptable in the marketplace.

An appeal to the House of Lords by the landlords was unsuccessful in 1982, thereby confirming the comments made by the Court of Appeal.

1.5.5 The interim rent

There may be a considerable delay between the contractual end of the lease and the granting by the court of a new tenancy or agreement between the parties out of court during which time the rent may have risen (or fallen).

Section 3 of the 1969 Act inserted section 24(A) into the 1954 provisions to enable the landlord to collect an interim rent during this period. Having served a notice under section 25 to terminate the tenancy, or having received a tenant's request for a new tenancy under section 26, the landlord could apply to the court to determine a rent which should be paid under the tenancy while it continues by virtue of section 24 and before the new terms are agreed. The RRO now also permits the tenant to make an application to the court for an interim rent and also removes the cushioning effect whereby the effect of a large rent increase on the tenant was tempered by a reduction (often of 10% or more). As amended, section 24 still protects a landlord against a loss of rent where rents have risen, but also protects the tenant's position if rents have fallen.

Either party must make the application not later than six months after the tenancy was due to be terminated. The interim rent is payable from the earliest date which could have been specified in the section 25 or section 26 notice. The amount of the interim rent payable shall be the same rent as that to be payable under the new tenancy unless:

(a) the landlord or the tenant shows to the satisfaction of the court that the interim rent differs substantially from the rent which would be payable under the new tenancy or

(b) the terms of the new tenancy differ to such an extent from the previous tenancy that the interim rent determined by the court would have been substantially different.

In these cases the court will still have regard to the rent payable under the old lease.

1.5.6 *The rent payable under the new lease*

The assessment of the rent to be paid under the new lease should be the last matter to be agreed by the parties after all other heads of terms have been settled. The rental value of the premises will directly reflect those lease terms.

The valuer, acting either for the landlord or the tenant, will gather relevant comparable rental evidence and then apply that evidence, with any appropriate adjustments, to the subject property. In the majority of cases the valuers acting for the landlord and the tenant will be able to reach a negotiated settlement of the rent to be payable under the new lease. Where they cannot do so the rent payable under the new lease will be decided by the court. If the matter is decided in court, then the valuer may be called to give expert evidence. The valuer's role now becomes that of expert witness. The duties and responsibilities of an expert witness were clearly laid out in the case of *The Ikarian Reefer* (1993). The evidence given shall be an independent, objective, unbiased opinion based on matters within the expert's expertise. It shall not omit material facts.

The assumptions which are to made in the assessment of the rent to be payable under the terms of a new lease are set out in section 34 of the Act.

Section 34 of the 1954 Act, as amended by section 1 and section 2 of the 1969 Act, reads as follows:

(1) The rent payable under a tenancy granted by order of the court under this Part of this Act shall be such as may be agreed between the landlord and the tenant or as, in default of such agreement, may be determined by the court to be that at which, having regard to the terms of the tenancy (other than those relating to rent), the holding might reasonably be expected to be let in the open market by a willing lessor, there being disregarded:

  (a) any effect on rent of the fact that the tenant has or his predecessors in title have been in occupation of the holding

  (b) any goodwill attached to the holding by reason of the carrying on thereat of the business of the tenant (whether by him or by a predecessor of his in that business)

  (c) any effect on rent of any improvement to which this paragraph applies

  (d) (licensed premises).

(2) Paragraph (c) of the foregoing subsection applies to any improvement carried out by a person who at the time it was carried out was the tenant, but only if it was carried out otherwise than in pursuance of an obligation to his immediate landlord, and either it was carried out during the current tenancy or the following conditions are satisfied, that is to say:

  (a) that it was completed not more than 21 years before the application of the new tenancy was made; and

  (b) that the holding or any part of it affected by the improvement has at all times since the completion of the improvement been comprised in tenancies of the description specified in section 23(l) of this Act; and

(c)   that at the termination of each of those tenancies the tenant did not quit.

(3)   Where the rent is determined by the court the court may, if it thinks fit, further determine that the terms of the tenancy shall include such provision for varying the rent as may be specified in the determination.

This section has the following specific effects upon the valuer's task.

(a)   The rent is to be determined after the agreement of the terms of the tenancy, as suggested above.

(b)   The new rent should be assessed in disregard of the old rental level. There should therefore be no automatic cushioning of increases or decreases in rent where rental values change considerably.

(c)   While it is to be assumed that there is a willing lessor there is no statutory assumption of a willing tenant. However, it is to be assumed from the actions of the tenant in reaching this stage that he is willing.

(d)   The valuer should disregard the fact that there may be considerable forces of inertia which might influence a tenant who wishes to remain in these premises, such as investment in fixtures.

(e)   The building up of goodwill should be regarded as a benefit earned and to be enjoyed by the tenant of commercial premises and should not be transferred to the landlord as rent.

(f)   Improvements carried out by the tenant or his predecessors during the current tenancy or the last 21 years should be disregarded in as far as rental value is affected by them, if the improvements were not carried out as a condition of the lease. It has been argued (Leach 1981) that the only logical interpretation of this requirement is that the valuer must ignore the improvement itself rather than simply its effect on rent, in order to value the premises as if the improvement has not been carried out. The valuer should envisage the premises as they were before the improvements were made and value them on that basis. The authors, in common with Bernstein and Reynolds (1991) and others, argue that there is no difference in practice between disregarding the effect on rent of an improvement and disregarding the improvement itself.

(g)   The court has a power to include rent reviews in the new (maximum 15 year) lease. These will normally be in accordance with current market practice (for example, five yearly).

For examples of rental and capital valuations of commercial property incorporating this interpretation, please go to the end of this chapter.

1.6 Rent reviews

1.6.1 Introduction

The rent review clause is an essential feature of a commercial lease designed to ensure that market rents are secured and landlords' investments yield the maximum possible return. A corresponding urge on behalf of the tenants to keep operating expenses to a minimum has ensured that the interpretation of a rent review clause is now a regular springboard for dispute.

Rent reviews, particularly on commercial property, are virtually an industry in themselves. Bernstein and Reynolds' *Handbook of Rent Review* (1991) is regularly updated and some 90% of the cases put to the Royal Institution of Chartered Surveyors requiring an arbitrator's award or an independent expert's determination comprise rent review disputes and commercial property. Both the legal profession and the surveying profession have long since identified the necessity of employing specialists in the field of rent review.

The starting point in any rent review is the wording of the rent review clause and the other lease provisions, and it is the task of the valuer to interpret those provisions so as to give effect to the intentions of the parties. Unlike the position at the renewal of a lease, where the assumptions set out in section 34 remain as a constant, each and every rent review clause is likely to have different wording conferring different assumptions.

The rent review process is started by the service of a rent review notice. The timing and validity of such rent review notices has itself been a fertile source of litigation. In *United Scientific Holdings Ltd v Burnley Borough Council* (1978) it was decided that time is not of the essence in rent reviews unless the lease specifically states or implies that this is so. Time traps can also arise where a party is deemed to have accepted a rent (or lost the opportunity to negotiate or apply for an arbitrator or expert) if they have failed to take a certain action specified in the lease (see *Starmark Enterprises Ltd v CPL Distribution Ltd* (2001), *Iceland Frozen Foods v Dangoor* (2002), *Barclays Bank v Saville Estates* (2003), *First Property Growth Partnership LP v Sun Alliance Property Services Ltd* (2002), and *Lancecrest v Aswaju* (2005).)

In circumstances where one party is not permitted under the terms of the lease to commence the rent review or to have an arbitrator or expert appointed, and can argue that this is financially damaging to them, they can serve a notice making time of the essence (see *Factory Holdings Group Ltd v Leboff International* (1986) and *Panavia Air Cargo Ltd v Southend on Sea BC* (1988)).

1.6.2 *Interpretation of the review clause*

(a) *The rental value*

In its most basic form the rent review clause will require the assessment of the new rent based on a number of assumptions which create a hypothetical scenario wherein the subject property is assumed to be vacant and to let on the open market on a new lease at the rent review date. There are innumerable variations to the wording used and the specified assumptions. This leads not only to lengthy and complicated valuation arguments between the parties to the review but also to litigation over the legal interpretation of those words.

Not all leases provide for the payment of a full open market rent. Rent may be geared, whereby a percentage of full rental value is payable because a premium reflecting this advantage has been paid by the tenant at the start of the lease or where, perhaps in the case of a shop, the tenant pays a base rent (being a percentage of full open market value — often 80%) plus a percentage of turnover. Frequently, however, the lease will still provide for the assessment of the open market rent in order that the geared rent or base rent can be calculated.

In the majority of cases, the review will be upward only (but see *Jane's (Gowns) Ltd v Harlow Development Corporation* (1980), *Forbuoys plc v Newport Borough Council* (1994) and *Royal Bank of Scotland plc v Jennings* (1994)) to the rack rental value, the open market rental value, or, in the RICS model clause, "the yearly rent . . . having regard to the open market rental values current at the relevant Review Date". There is little difficulty in translating such phraseology into the valuer's instructions.

However, the phrase "reasonable rent" has been the subject of dispute. In *Ponsford v HMS Aerosols Ltd* (1978) this was held to represent the rent at which the demised premises might reasonably be expected to let rather than the rent which it would be reasonable for the tenant to pay. A distinction was drawn between reasonable rent and open market rent as it was felt that exceptionally high rental bids might be excluded by the former, but included by the latter.

(b) *The willing tenant*

The RICS model includes an assumption that the "demised premises are available to let by a willing landlord to a willing tenant". This, it may be argued, is unnecessarily harsh upon the unwilling tenant who at review may be forced to pay a rental in excess of the value of the premises to him or anyone else.

This is likely to happen in the case of a functionally obsolescent building, such as an office building specifically designed 20 years ago for computers that have been technologically superseded. The tenant may be unable to alienate the premises at anything other than a loss and the model clause might thus be regarded as overly harsh.

In *FR Evans (Leeds) Ltd* v *English Electric Co Ltd* (1978) unique factory buildings and land in Liverpool providing nearly one million square feet of floor space were the subject of review. There was only one potential lessee of the property — the actual tenant. The English Electric Company Ltd were apparently most unwilling tenants of obsolete property, but the court held that their attitude was irrelevant and that it was the attitude of the hypothetical willing lessee which was of importance. A rent of £515,000 was fixed by the arbitrator after the Court of Appeal had confirmed this decision, while had the attitude of the actual tenants to be taken into account a rental as low as £290,000 might have been fixed.

In *Dennis Robinson Ltd* v *Kiossos Establishment* (1987) the Court of Appeal, notwithstanding the absence of the words "willing tenant", held that reference to the determination of an open market rent necessarily presumed the existence of a willing tenant in order to give meaning to the rent review provisions.

The valuer should not assume that the existing willing tenant would pay more in the open market than any other tenant or that the willing landlord would accept any lesser rent or hold out for any higher rent than could be obtained in the open market taking account of the alternative properties available to the willing tenant.

(c) *Fitness of the premises*

Many rent review clauses, the RICS model included, assume that the demised premises are fit for immediate occupation and use, that the tenant has not carried out any works which have diminished the rental value of the premises, and that the tenant's covenants particularly with regard to repair have been duly performed. These assumptions are designed to effect the overriding principle that "no man may profit by his own wrong".

There are three important principles here, and each has been the subject of litigation. In *Harmsworth Pension Fund Trustees Ltd* v *Charringtons Industrial Holdings Ltd* (1985), where the tenant had failed to comply with the repairing covenants in the lease and argued that the property in its existing state of repair would not command the full open market rent, the High Court had no difficulty in rejecting that argument and held that the premises must be assumed to be in good repair (see also *Ladbroke Hotels Ltd* v *Sandhu* (1995)). Where there is a positive covenant on behalf of the landlords to repair, it must also be assumed for the purposes of rent review that those repairs had been carried out.

The valuer should be aware that a proviso whereby the tenant is assumed not to have carried out any works which have diminished the rental value of the premises may well be contradicted by other terms of the lease such as an obligation to comply with Acts of Parliament (whereby, for example, a tenant may be required under the Fire Precautions Act 1971 to install a fire corridor thus reducing the floor space and the rental value of the property).

The assumption that the premises are fit for immediate occupation and use is the most controversial assumption of all, and has been the subject of much recent litigation. It is commonplace in the letting of industrial and warehouse property and invariably the case in the letting of office and shop property that an ingoing tenant will be granted a rent-free period towards the fitting out of accommodation which is in shell condition or indeed where a previous occupier's fitting out works are now obsolete.

In the case of *London & Leeds Estates Ltd* v *Paribas Ltd* (1993) the rent review contained an assumption that the demised premises which were let as a shell

> are fit for immediate occupation and use, in a state of good repair and condition and that all fitting out and other tenant's works required by such willing tenant have already been completed.

There was also a disregard of any reduction in rent which the absence of a rent-free period or other financial inducement might cause.

The Court of Appeal (1993) reversed the original Chancery Division decision, holding that the wording of this clause was merely to preclude the actual tenant from claiming that the hypothetical tenant will be entitled to a discount from the best open market rent on account of the actual state of repair and condition of the premises (shell condition). This was on the ground that the hypothetical tenant would require further or different works from those carried out by the actual tenant and the cost of these works would be borne by the hypothetical tenant with a corresponding reduction in the rent he was willing to pay. This gave the tenant the opportunity to require the arbitrator determining the rent review on the now-fitted premises to disregard the value of such works under the improvements disregard.

In *Ocean Accident & Guarantee Corporation* v *Next* (1996), the assumption to be made was that the demised premises had been fully fitted out and equipped so as to be ready for immediate use and occupation by such willing tenant for such use. Nothwithstanding this more prescriptive wording, it was held that the premises should be considered not as fitted but as ready to be fitted.

Further litigation on this issue is anticipated. At the present time the position is that the premises are neither to be treated as in shell condition, nor are they to be treated as fully fitted out, but they are to be treated as in a condition whereby they are ready to receive an ingoing tenant's fixtures and fittings. Accordingly the landlord does not achieve an enhanced rent for a fitted unit while the tenant does not achieve a rent less than market rent for fitting out a shell unit. Any wording additional to the basic provisos set out above should be carefully considered as it may, by its literal construction, require an entirely different approach.

(d) Tenant's improvements

The valuer should note that fewer and fewer leases of commercial property contain provisos relating to the disregard of tenants' improvements at rent review which are faithful to those set out in section 34 of the 1954 Act. Rent review clauses will generally require the valuer to assume that improvements to the property carried by the tenant, other than those which the lease may have obligated the tenant to carry out, are to be disregarded in the assessment of the rent. However, innumerable variations to this general provision are to be found in commercial leases. The interpretation of the wide and varied improvements clauses in commercial leases has been the source of much litigation.

First, unless it is clearly stated to be so, there is no implied assumption that works of improvement which a tenant has carried out at its own expense with the landlord's consent will be disregarded at rent review. This will be dependent upon the wording of the improvements disregard clause. In *Ponsford* v *HMS Aerosols Ltd* (1978), for example, the House of Lords felt constrained by the terms of the clause to hold that the rent to be assessed at review should be the full rental value of the premises including the improvements.

Second, even where the wording clearly directs that in assessing the value of the premises the value of the tenants' improvements must be disregarded, this can be over-ridden by a subsequent licence for alterations which specifically provides that the authorised alterations were not to be disregarded at any rent review. The court held that the lease and the licence had to be read together for the purposes of the rent review and therefore the improvements were to be taken into account on review (see *Ivorygate Ltd v Capital City Leisure Ltd* (1993)). This may, however, be distinguished from *Historic Houses Hotels Ltd v Cadogan Estates* (1993) where the Court of Appeal, in similar circumstances, held that the wording of the licence was ambiguous and that the clear intention of the parties was that the premises were to be valued as originally demised, disregarding any tenant improvements properly carried out thereafter.

(e) Tenant's occupation and goodwill

The provisions of section 34 of the 1954 Act relating to tenant's occupation and goodwill are more often than not faithfully repeated in most commercial leases. However, see *Prudential Assurance Co Ltd v Grand Metropolitan Estates Ltd* (1993) in which the rent review provisions of the lease of a public house contained a disregard of the actual tenant's occupation but no specific disregard of goodwill. In this case the Court held that if one is to disregard the effect on rent that the tenant has been in occupation then it follows that goodwill generated by that occupation must also be disregarded.

(f) User clauses

Most rent review clauses make no specific reference to the use to which a premises may be put and state only that the hypothetical lease which is to be assumed in assessing rent for the subject premises is to be "on the same terms and conditions" as the subject lease. A restriction of the use to which a tenant may put the property (stated in the user clause, review clause or both) will customarily force the landlord to accept a lower rent at review; the reduction from full rental value depending upon the nature of the user restriction.

There are three principal types of user clause, which can conveniently be described as absolutely restrictive, qualified and open.

In an absolutely restrictive user clause the use of the premises is restricted to one particular use only, as in the case of *Charles Clements Ltd v Rank City Wall Ltd* (1978) where the use was restricted to the trade of cutlers. A deduction, in this particular case of 16%, was made from full market rental value to reflect this restriction. In the case of *UDS Tailoring v BL Holdings Ltd* (1982) a deduction of 10% was made to reflect the user clause restricting the use to the sale of menswear.

A qualified user is where a property may be used for a specified purpose only subject to landlord's consent for an alternative use. In the case of *Plinth Property Investments Ltd v Mott, Hay and Anderson* (1979), where the use was restricted to consulting engineers, the court held that the landlords, in deciding whether to give consent for another use, could act commercially in reaching that decision, that is they were entitled to say no. The difference in rental value between the value of the premises used as offices and the value restricted to the use as offices for the business of consulting engineers was assessed at 31%.

An open user may take the form of a clause specifying a particular use but with landlord's consent for any other use or any other use within the same use class being subject to landlord's consent which is not to be unreasonably withheld. This is clearly not restrictive and a full open market rent can be obtained at review.

Variations upon the above qualifications are often found, for instance, where the user clause may be open but the landlords can take account of the principles of good estate management or tenant mix when considering whether consent for another use should be granted. This scenario frequently occurs in shopping centres where the landlords are keen to control the mix of retail uses. The courts have been unwilling to allow the good estate management proviso to thwart a tenant's change of use unless the landlord can demonstrate that they have a clearly-worded good estate management/tenant mix policy in place and they they have consistently adopted that policy.

In yet another form the user clause itself may be restrictive but the rent review provisions on which basis the rent is to be assessed may provide that any restriction on the user is ignored thereby ensuring that the landlord maintains strict control over the use of the premises but is still able to achieve a full market rent at review.

The case of *Law Land Co Ltd* v *Consumers Association Ltd* (1980) is interesting in that while the user clause restricted the use to that of the business of the Consumers Association only (of which there was clearly only one such business) the rent review provisions required the assessment of an open market rent. The court (finding that there could not be an open market situation with only one possible occupier) held that the two clauses were at odds and that for the purposes of the rent review the restrictive user should be ignored.

The valuer must therefore look closely to determine whether there is, first, a restriction on user, and second, whether the rent review provisions require that restriction to be taken into account.

The use of user clauses in leases is subject to the provisions of section 19(3) of the Landlord and Tenant Act 1927 which stipulates that in granting an alteration of the user the landlord cannot require that a "fine" or increase of rent shall be payable in respect of such consent (see *Barclays Bank plc* v *Daejan Investments (Grove Hall) Ltd* (1994)).

(g) Planning and illegal user

While there is no direct authority on this point, it appears from the decisions in *Compton Group Ltd* v *Estates Gazette Ltd* (1977) and *Molton Builders Ltd* v *Westminster City Council* (1975) that the valuer should disregard any excess rental value created by the illegal use of the premises by the tenant. A common example of this is the use of part-residential premises wholly as offices without planning consent. In such a case an illegal user should be ignored in the assessment of rent at review unless there appears to be a prospect of permission being forthcoming.

(h) The prospect of continuation

A lease of office premises with rent reviews provided for every five and a half years was the subject of dispute in the Court of Appeal in *Secretary of State for the Environment* v *Pivot Properties Ltd* (1980). The rent review clause defined the market rental value as the "best rent at which the demised premises might reasonably be expected to be let in the open market as a whole for a term not exceeding five years and one half of another year".

The question before the court was whether the effect on rental value of the prospect of continuation of the tenant under the 1954 Act was to be taken into account. It was accepted that if no such account was to be taken, the annual rent was to be fixed at £2.1m; if the prospect of continuation was to be taken into account a rent of £2.925m would be fixed. The court looked to the intentions of the parties and the likely attitude of any bidder for the premises were they to be put on the market. Its reaction to this was clear. Any hypothetical tenant would take the prospect of continuation into account and the higher rent would be paid.

(j) Ignoring future rent reviews

In its most common form the rent review clause may be summarised as requiring the assumption of a hypothetical situation whereby the premises are to let in the open market with vacant possession on the same terms and conditions as the subject lease. Notwithstanding those terms the rent review clause may specify that different terms are to apply to the hypothetical lease and where, as a consequence, there have been substantial valuation implications the literal construction of these additional terms has been the subject of litigation.

In *National Westminster Bank plc* v *Arthur Young McClelland Moores Co* (1985) a proviso to consider a lease on the same terms "save as to rent" was, surprisingly, held to mean that the mechanism for varying the rent, that is the rent review clause itself, was to be ignored thus requiring the rent for the property to be assessed as though there were no further rent reviews. As a consequence an enhanced rent for a period longer than the five-year review pattern in that lease was to be assessed. In a series of cases which followed, this principle was overturned (see, for example, *British Home Stores plc* v *Ranbrook Properties Ltd* (1988)).

Many rent review clauses in commercial leases and the RICS model rent review clause avoid this problem by requiring the assessment of the rent of the premises on the terms of the subject lease to include the provisions for the review of that rent. However, if the wording in the alternative is "excluding the provisions for review or save for this proviso" as in *Pugh* v *Smiths Industries Ltd* (1982) and *Safeway Foodstore Ltd* v *Banderway Ltd* (1983) the literal construction is clear; namely that future rent reviews are to be ignored and the rent to be assessed is for the whole (or, as the case may be, unexpired) term without review. The cases of *Lear* v *Blizzard* (1985) and *Prudential Assurance Co Ltd* v *Salisburys Handbags Ltd* (1992) go some way towards mitigating the impact of the artificial or hypothetical rent review period which was to be assumed in the *Pugh* and *Safeway* cases.

(k) Abnormal review patterns

The five-year rent review is still commonplace in the UK property market. Where the period between reviews is only three or four years, and evidence is drawn from comparables having five-year review patterns, it is typically the case that an adjustment will be made to reflect the fact that the property held on a three-year or four-year pattern will, except in unusual circumstances, be less valuable to the tenant than one held on a five-year pattern.

Since the mid-1970s, landlords seeking to keep pace with inflation have rarely granted leases with review patterns longer than five years but the valuer must nevertheless be able to value property held on historic leases with rent review patterns longer than five years or those which arise from artificial or hypothetical rent review provisions. In general a lease with an actual or assumed rent review pattern longer than five years will be regarded as being worth more rent and accordingly an allowance will be made.

The allowance which is to be made for a long review pattern is variously referred to as uplift, enhancement or overage. While software such as OPRent will produce a scientifically supported estimate, rule of thumb figures are commonly adopted. Typically, an addition of 0.5%–1% would be made for each year that the review pattern exceeds five years, so that, for example 4.5% or 9% overage would be applied to a 14-year review pattern. The overage factor to be applied will take account of market conditions and comparable evidence derived from properties having similar long review patterns. Another material consideration is the quality of the property`s location; a property in a prime location is likely to attract more overage than one in a more secondary location. In some instances 2% pa, and in other instances zero, has been applied. For a more defensible and sound method of estimating overage allowances, see Baum and Crosby (2007), *www.oprent.com* and 1.6.3 below at p37 (the constant rent theory).

(l) Headline rents

In certain circumstances, typically to maintain investment value, landlords may wish to grant leases at rents which exceed full rental value. To induce the tenant to pay an artificially high level of rent the landlord may offer by way of compensation an extended rent-free period, a capital sum as an inducement to sign the lease and/or a capital sum to assist the ingoing tenant with fitting-out costs. This may be beneficial to the tenant who might not otherwise be able to fund the cost of fitting out new premises or who may find that it is easier to establish a business from the premises where the occupation is rent free for a period of time. Alternatively, the bargaining position of the tenant may be such that he can demand such inducements to sign a lease, and in a weak market a landlord will be willing to pay it in order to effect a letting. The rental level set against this background is now commonly referred to as a headline rent, that is the rent which is contractually payable but which would not have been achieved in the open market without an extended rent-free period, capital sum or other inducement. The valuer's task is to calculate what the rent would be without the inducement.

The range of rent-free periods and capital inducements in all sectors of the commercial leasing market is wide, and the motives for deals being struck on this artificial basis are often difficult to determine in each separate case. As yet there is no standard valuation practice for converting a headline rent into an actual rent. There is widespread support in current market conditions for treating the first three months of any rent-free period granted to an ingoing tenant on a new lease as representing standard market practice, so that no adjustment to the rent reserved would be warranted. There is no universal agreement as to how a rent-free period exceeding three months should be treated.

One school of thought amortises this additional sum of rent on a straight-line basis until the next rent review and the annual equivalent derived is deducted from the reserved headline rent to produce the actual rent. Alternative schools of thought amortise the sum of rent over the entire term of the lease granted on a straight-line basis. Yet another school amortises the sum of the additional rent loss over the period to the next rent review at which an increase in rent is expected. Finally, and best, a DCF method may be employed to estimate the expected cash-flow benefit and disbenefit to each side. Simulation-based software, such as OPRent, provides appropriate solutions, but this development is relatively new to the market. Simulation solutions are described in Baum and Crosby (2007). Example 1.8 shows alternative conventional methods of deriving an actual rent from a headline rent.

Example 1.8

A headline rent of £375 per m^2 was agreed in 2005 on the letting under a 15-year lease of a City of London office building. The letting is subject to a 2.25 year rent-free period when in a normal market a quarter's rent-free period would be typical. What is the underlying rental value?

Method 1: amortise over the first review period
(Ignore three months of the rent free period; two years rent free is the inducement.)
Total rent bill for three years: £375 × 3 = £1,125.

This should be spread over the whole of the period until the first review (5 years) on the assumption that after the first review the rent will revert to open market rental value and there will be no advantage to the landlord of agreeing a high rent for the initial period.

£1,125/5 = £225
Underlying rental value (actual rent): £225 per m^2

The weakness in this approach is the possibility that the headline rent will continue to affect the rent paid after the first review. Some valuers have argued that the rent should therefore be amortised over the whole lease.

Method 2: amortise over the whole lease
Total rent bill over the lease: £375 × 13 = £4,875
£4,875/15 = £325
Underlying rental value: £325 per m^2

Method 1 clearly suits the tenant side; method 2 suits the landlord side. The difference is significant and neither method should normally be acceptable.

It is suggested that both sides should attempt to ask at which review the benefit of the higher rent will cease to have effect. For example, let us assume that the parties agree that rental rises will be such that the benefit will disappear at the second review in 10 years' time.

Method 3: amortise until first upward review
Assume the first upward review to be likely in year 10.
Total rent bill to first upward review: £375 × 8 = £3,000
£3,000/10 = £300
Underlying rental value: £300 per m^2

However, the parties may well not agree; and in any case all three methods shown ignore the fundamentals of discounting cash flows.

Method 4: equate the present values of the two alternative expected cash flows over the period of the lease by using a DCF technique
Let us assume that rental values are expected to grow at a rate of 5% pa and that a reasonable discount rate is 12%. The table shows the landlord's expected cash flow under the agreed scenario and on the alternative assumption that the rental value is paid immediately without a rent-free period. To find the underlying rental value, the two present values should be made to equate.

Using trial and error, it can be shown that the rental value which produces the same discounted value under either scenario is £240 per m^2. The present values are £1,938.8 and £1,945.4 for rent free and no rent-free period respectively. Note the effect of the upward-only rent review.

Year	With rent free	Without rent free
1	0	x
2	0	x
3	375	x
4	375	x
5	375	x
6	375 or $x \times (1.05)^5$*	$x \times (1.05)^5$
7	375 or $x \times (1.05)^5$	$x \times (1.05)^5$
8	375 or $x \times (1.05)^5$	$x \times (1.05)^5$
9	375 or $x \times (1.05)^5$	$x \times (1.05)^5$
10	375 or $x \times (1.05)^5$	$x \times (1.05)^5$
11	375 or $x \times (1.05)^{10}$	$x \times (1.05)^{10}$
12	375 or $x \times (1.05)^{10}$	$x \times (1.05)^{10}$
13	375 or $x \times (1.05)^{10}$	$x \times (1.05)^{10}$
14	375 or $x \times (1.05)^{10}$	$x \times (1.05)^{10}$
15	375 or $x \times (1.05)^{10}$	$x \times (1.05)^{10}$

*Note: whichever is the higher

Year	With rent free	Without rent free
1	£0.00	£240.00
2	£0.00	£240.00
3	£375.00	£240.00
4	£375.00	£240.00
5	£375.00	£240.00
5	£375.00	£306.30
7	£375.00	£306.30
8	£375.00	£306.30
9	£375.00	£306.30
10	£375.00	£306.30
12	£390.90	£390.90
13	£390.90	£390.90
14	£390.90	£390.90
15	£390.90	£390.90
Present values at 12%	£1,938.80	£1,945.40

The above examples deal with an inducement in the form of a lengthy rent free period. The payment of a capital sum which is an inducement to pay a headline rent will also be treated on the same alternative bases as above. It is in this regard similar to the reverse premium scenario discussed earlier, but in this instance the landlord is seeking to obtain a rent higher than the market rent by paying an inducement.

In an attempt to preclude a tenant at rent review from arguing that evidence of headline rents on other premises should be disregarded, leases have been drafted with the intention of safeguarding a headline rent at review. This has inevitably led to litigation in cases where there is a substantial differential between actual rents and headline rents.

In November 1994, four cases were brought before the Court of Appeal in conjoined appeals for a declaration as to whether a headline rent should be assessed as evidence of rental value at review without deductions to reflect concessionary or rent free periods other than for fitting out. The four cases are listed in the footnote below.*

Due either to ambiguity in the wording of the rent review provisions or a reliance upon the fitting out element of the rent-free period or concession, the landlords, Co-operative, Scottish Amicable and Prudential failed and only Broadgate succeeded in a declaration for a headline rent. In all four cases leave to appeal to the House of Lords was refused.

The conclusion which may be drawn, notwithstanding its artificiality, is that it is possible to achieve a headline rent at review but only if the construction of the rent review clause is absolutely unambiguous in this regard. The valuer is forewarned that there is considerable difficulty in determining whether any rent review clause which departs from the Broadgate form (set out below) will require a headline rent to be assessed. The arbitrator made alternative awards for the office premises known as 1 and 2 Broadgate at £5,281,365 pa for 1 Broadgate and £3,931.795 pa for 2 Broadgate as an actual open market rent and £7,330,316 pa for no 1 and £5,236,046 pa for no 2 if a headline rent was to apply. The wording of the Broadgate rent review clause (extract) was as follows:

> Open Market Rent means the best yearly rent which would reasonably be expected to become payable in respect of the premises after the expiry of a rent free period of such length as would be negotiated in the open market between a willing landlord and a willing tenant upon a letting of the Premises as a whole by a willing lessor to a willing lessee in the open market at the Relevant Review Date for a term of 10 years or a term equal to the residue then unexpired of the Term (whichever is the longer) with vacant possession without fine or premium.

1.6.3 *The constant rent theory*

In 1.6.2(k) at p33 above (abnormal review patterns) we considered the market approach to the valuation of review periods longer than five years. The constant rent theory is designed to calculate more accurately the difference between standard review patterns and longer review patterns.

With the benefit of a long review pattern the tenant is in a fortunate position in an inflationary economy as his rent is to be fixed for a greater period than is usual, and his ability to enjoy a profit rent, and consequently the value of his interest, is greater as a result. The landlord acknowledges this benefit and wishes to be compensated for the absence of a standard rent pattern by the fixing of a higher rent in excess of the full rental values which are paid for similar properties on normal review patterns. The tenant paying a rent higher than open market rent will then suffer a rental loss in the early years of the review period which is compensated by a profit rent in later years.

Arguments against this approach have been vigorously presented. It may be argued that a tenant has paid a high price on the assignment of such an interest to reflect the review pattern advantage which is now taken away from him. It may also be argued that it was not the intention of the parties to fix higher rents at review than those paid for normally reviewable properties and that the landlord

---

*   *Co-operative Wholesale* v *National Westminster Bank plc* (1994); *Scottish Amicable Life Assurance Society* v *Middleton Potts and Co* (1994); *Broadgate Square plc* v *Lehman Brothers Ltd* (1994); *Prudential Nominees Ltd* v *Greenham Trading Ltd* (1994).

should suffer the consequences of his (or his predecessor's) lack of foresight in not providing for regular reviews. It can also be suggested that inflation may not continue. The current landlord might have paid a low price for an apparently disadvantageous investment which will be dramatically improved by an application of the constant rent theory. And the tenant may not be able to trade viably when paying a front-loaded rent higher than open market rent.

While these general arguments might be persuasive, each situation must be considered having regard to its specific circumstances. If a long review pattern exists it must be considered. This is clear from two court decisions. In *National Westminster Bank Ltd* v *BSC Footwear Ltd* (1981) a lease for 21 years without review contained an option for renewal on the same terms "at the then prevailing market rent". The question to be considered by the Court of Appeal was whether the arbitrator had the power to determine a rent which was subject to periodic review. It was held that he did not, as the lease referred only to the prevailing market rent, and not the prevailing market practice. Consequently the arbitrator was required to decide a single rent which reflected the advantages of the letting, payable throughout the term of the 21-year lease.

In *Bracknell Development Corporation* v *Greenlees Lennards Ltd* (1981) a similar problem arose concerning an option to renew a 21-year lease "at a full and fair market rent", and the court's decision followed the *National Westminster* case.

Where rental values are expected to rise over the period of a review tenants may pay more rent for a longer review period. Valuers may be asked to solve two problems: first, to suggest how much more rent would be appropriate for an abnormally long review; or, alternatively, what rental value on a normal review is implied by evidence of rents agreed for a long review.

In 1979, Jack Rose devised a formula to produce a factor, K, which should be applied to an open market rent in order to calculate what that rent should be with a non-standard rent review pattern.

Example 1.9

Calculate the rental value on an annual tenancy of a shop whose full rental value under a 20-year lease with five-yearly reviews would be £10,000 pa

$$ K = \frac{(1 + r)^n - (1 + g)^n}{(1 + r)^n - 1} \times \frac{(1 + r)^1 - 1}{(1 + r)^t - (1 + g)^t} $$

where:

r = required return
n = number of years between reviews in subject lease
g = annual rental growth expected
t = number of years between reviews in normal cases

The required return is the overall yield (required return or internal rate of return) required by the investor and must as a minimum be based upon the return to be derived from low-risk, no-growth investments such as government bonds. The annual rental growth expected must be a subjective estimate but its estimation may employ an analysis of market transactions (see Baum and Crosby 2007). In this example r is taken as 15% and g is estimated at 8%.

$$ K = \frac{(1 + r)^n - (1 + g)^n}{(1 + r)^n - 1} \times \frac{(1 + r)^1 - 1}{(1 + r)^t - (1 + g)^t} $$

$$= \frac{(1.15)^1 - (1.08)^1}{(1.15)^1 - 1} \qquad \times \qquad \frac{(1.15)^5 - 1}{(1.15)^5 - (1.08)^5}$$

$$= \frac{0.07}{0.15} \qquad \times \qquad \frac{1.0114}{0.5421}$$

$$= 0.4667 \qquad \times \qquad 1.8657$$

$$= 0.8707$$

$$0.087 \times £10,000 \qquad = \qquad £8,707$$

Rental value on an annual tenancy should be (say) £8,700 pa

Example 1.10

A shop is let on a 21-year lease with seven year rent reviews and the first review is now due. The full rental value of similar property let with the normal five-yearly review pattern is £40,000 pa. Rents are expected to rise at 10% pa and a 15% required return is required by the investor in this type of property.
   What is the new rent to be fixed?

$$K \qquad = \qquad \frac{(1 + r)^n - (1 + g)^n}{(1 + r)^n - 1} \qquad \times \qquad \frac{(1 + r)^1 - 1}{(1 + r)^t - (1 + g)^t}$$

$$= \qquad \frac{(1.15)^1 - (1.10)^1}{(1.15)^7 - 1} \qquad \times \qquad \frac{(1.15)^5 - 1}{(1.15)^5 - (1.10)^5}$$

$$= \qquad 0.4285 \qquad \times \qquad 2.5230$$
$$= \qquad 1.081$$

$$\text{Rent} \qquad = \qquad £40,000 \times 1.081 \qquad = \qquad £43,240 \text{ pa}$$

A similar problem may arise, this time with the parties' connivance, at the granting of a new lease. The valuer's problem is less acute, as a failure to agree a rent on behalf of the parties can result in an amendment of the lease terms.

Example 1.11

What rent should be charged for a shop let on lease for 21 years without review? The current full rental value for similar shops let with five-yearly reviews is £100,000 pa. Rental growth of 5% is expected and investors require a 13% return for such investments.
   K in this case is 1.267: the rent should therefore be £126,700 pa.

It is recognised that the most contentious element in this approach is the estimation of the expected annual rental growth. This figure need not be estimated; it can be derived from market evidence (see

Baum and Crosby, 2007). The key input is the market capitalisation rate, often (confusingly) termed the yield.

As a simplification, it can be argued that there are two factors which influence capitalisation rates. The risk of the subject investment forces capitalisation rates upwards; the prospects of rental or capital growth force them downwards.

A flat interest rate yielded by investments showing little or no risk or growth potential may be selected as a datum point. Undated government stock approaches this ideal; yields in 2006 fluctuated around 4.5%. Consequently, any property investment must produce a return in excess of this – say 7% for a relatively prime property – in order to tempt investors. A low initial yield may be accepted only if rental growth prospects will increase that yield at rent reviews so that an overall return of 7% or above is produced.

The greater the growth potential of an investment, the lower the initial yield that will be accepted. Prime London offices, for example, were purchased at prices indicating initial yields as low as 4.0% in 2006, reflecting high rental growth expectations.

The relationship between initial and required returns created by growth expectations is expressed by the formula (see also Example 1.5):

$$k = r - (SF \times p)$$

Where:

k = initial yield or capitalisation rate
r = required return
SF = annual sinking fund over the rent review period at r
p = percentage rental growth over the rent review period

Given k and r for any property investment, the level of p required to make the investment competitive (and by inference the level expected by investors) can be found. Consequently the valuer's problem can be overcome.

Example 1.12

Where prime shop property sells on a 3.5% basis and undated stock yields 14%, what annual rental growth is expected by prime shop investors? Assuming a 2% risk premium, the target rate becomes 16%.

k	=	$r - (SF \times p)$
0.035	=	$0.16 - (0.14541 \times p)$
0.14541p	=	0.125
p	=	0.8596 (a 85.96% rental increase over 5 years)

If £1 will grow to £1.8596 in 5 years and x is the annual rental growth, then:

$(1 + x)^5$	=	1.8596
$1 + x$	=	$\sqrt[5]{1,7586}$
x	=	0.13210
	=	13.21%

Using these assumptions, annual rental growth of 13.21% is required to make an investment in prime shops worthwhile in these circumstances. In constant rent problems, this value can be used in the process of rental adjustment to account for abnormally long review periods.

In practice, the constant rent theory is not commonly applied. As in many areas of valuation, simple rules of thumb appear to be preferred to rational discounting-based process. In Example 1.10, the rational value of a rent fixed on seven-year reviews when the estimated rental value based on five-year reviews was £40,000 turned out, using Rose's constant rent formula, to be 8% higher. The standard rule of thumb, on the other hand, would raise rents by 1 or 2% to £40,400 or £40,800.

No doubt the inertia of common practice in partnership with the vested interests of surveyors representing tenants will continue to be powerful: but common sense, in the form of a simple equation, should prevail (in this case to the advantage of the landlord). Alternatively, superior simulation-based approaches may be used (see OPRent and Baum and Crosby, 2007).

1.6.4 Admissibility of evidence

When negotiating a lease renewal, valuers can take account of rental evidence occurring up to the date on which they finally agree the terms for that lease renewal, or up to the date of the court hearing, as the case may be. However, the position at rent review is different because the valuation date of the rent review is normally specified in the lease. The negotiated settlement of that rent, or the decision by an arbitrator or independent expert, can, however, take place long after the specified rent review date. Is it, therefore, correct to suppose that evidence which is revealed after that date may be employed as evidence and taken into account?

In *Ponsford v HMS Aerosols Ltd* (1977) it was held (obiter) that the valuer could have regard to all transactions up to the relevant date of review but all developments thereafter were to be disregarded. In 1984, the case of *Segama NV v Penny Le Roy Ltd* reversed the principle established in the Ponsford case. The question put to the courts was whether the arbitrator had been right to receive evidence of post-review date comparables. It was held that the arbitrator had been right to do so and the judge said:

> If rent of comparable premises had been agreed on the day after the relevant date, I cannot see that such an agreement would be of no relevance whatsoever to what the market rent was at the relevant date itself. If the lapse of time before the agreement for comparable premises becomes greater then, as the Arbitrator said, the evidence will become progressively unreliable as evidence of rental values at the relevant date. The same is no doubt true of rents agreed some time before the relevant date; but nobody suggested to me that those should be excluded. So, too, political and economic events may have caused a change in the market rents, either before or after the relevant date. All those factors must be considered by the Arbitrator in assessing the weight to be attached to a rent agreed for similar premises, whether before or after the relevant date. It may happen that no rents of comparable premises that were agreed on the relevant date, or for months beforehand, can be found, but a great number very shortly thereafter. It does not seem right to me that the Arbitrators should be bound to disregard them.

The principle of *Segama NV* is widely accepted, and the RICS guidance notes direct arbitrators that such evidence is in general terms admissible and to be accorded appropriate weight in the light of the arbitrator's experience and knowledge of the relevant market. The consideration as to weight is whether the movement in rental values, up or down, suggested by the post-review date evidence could reasonably have been predicted or contemplated by the parties. If there had been a rising, or falling, trend in values then this would be regarded as more predictable. However if the period of post-datedness is too long the late evidence would normally be regarded as less relevant.

There is a question also of what types of evidence are admissible. In the hierarchy of types of evidence, open market lettings are generally accorded most weight, followed by rent review settlements. Determinations by independent experts are generally placed above arbitration awards in this hierarchy.

The admissibility of evidence arising from an arbitrator's award was considered in *Land Securities plc* v *Westminster City Council* (1992). In this case the judge decided that without the consent of both parties the award of an arbitrator in relation to a comparable property is not admissible in evidence. However, because so many rent reviews are settled at arbitration and the awards provide a source of evidence, it has become common practice to have regard to this category of evidence. Valuers will generally have regard to both arbitrator's awards and independent expert determinations treating them as admissible, but having due regard to how much weight to attach to them. (See also *The Ikarian Reefer* (1993) 2 Lloyds Rep. 68.)

1.7 Alienation

1.7.1 Introduction

Alienation is the term used to define the process by which the owner of a leasehold interest in land or property can dispose of that interest. The interest can be alienated, or disposed of, either by an assignment (sale) or by the creation of another interest (sub-letting or under-letting). The party disposing of its lease by assignment is referred to as an assignor (and the purchaser an assignee) while the party disposing by way of the granting of a sub-lease is referred to as a head-lessee or intermediate lessee.

Figure 1.1 (p4) illustrates both means of parting with possession of property in return for a financial inducement. The assignment can (where the interest is valuable) produce a capital sum, and a sub-letting can produce a rental income or profit rent. The disposer's choice of means of alienation might consequently be influenced by his personal tax position, bearing in mind the constraints imposed by the lease.

On alienation, the valuer might be required to advise upon the sale or purchase price of the interest or the rent to be asked or paid upon sub-letting. The right of a tenant to alienate his interest may be in three alternative forms:

(a) absolutely restrictive whereby the lease prohibits assignment and/or subletting throughout the term or for a specified period;

(b) a qualified restriction upon alienation whereby landlord consent to assign or under-let is required. Section 9(1)(A) of the Landlord and Tenant Act 1927 provides that such consent shall be deemed not to be unreasonably withheld;

(c) open alienation provisions whereby the landlord's consent to assign or under-let is not to be unreasonably withheld.

In case (a) the tenant holding the leasehold interest has no right to dispose of that interest. The valuer would reflect this restriction on alienation in the capital valuation of that interest.

By virtue of section 19(1)(A) a qualified alienation provision (b) affords as much flexibility to a disposing tenant as an open alienation provision (c). The valuer will be required to calculate the capital or premium value of the lease on assignment or, in the case of sub-letting, calculate the rent to be reserved in the sub-lease. In the event that the property is over-rented or there is no demand the valuer's task may be to calculate the reverse premium which it might be necessary to pay to the assignee or sub-lessee in order to persuade him or her to take on the over-rented or otherwise undesirable property.

In cases (b) and (c) above the tenant will be required to make an application to the landlord for permission to assign or sub-let the lease. Landlords might not wish to give consent for a number of reasons and this has led to litigation as to whether a landlord has acted unreasonably or reasonably in withholding consent.

The Landlord and Tenant Act 1988 sought to address this issue and switched the onus from the tenant (who previously had to prove that the landlord was unreasonable) to the landlord, such that it is now for the landlord to prove (on the balance of probabilities) that his withholding of consent is reasonable. The 1988 Act also strengthened the tenant`s ability to recover damages if the landlord can be demonstrated not to have acted reasonably. It also applies more pressure to the landlord to reply to the tenant's application fully and in a timely fashion (unspecified — but normally regarded as within one month). For cases on alienation see *Go West Ltd* v *Spigarolo* (2003), *Clinton Cards (Essex) Ltd* v *Sun Alliance and London Assurance Co Ltd* (2002), *Allied Dunbar Assurance plc* v *Homebase Ltd* (2001), *NCR* v *Riverland Portfolio No 1 (No 2)* (2005), and *Crestfort* v *Tesco Stores* (2005).

The valuer must determine whether the alienation provisions are in any way restrictive and adjust the rental or capital value in accordance with market practice.

1.7.2 Assignment

The assignment of a leasehold interest in commercial property can cause considerable problems for the valuer. Two markets for such interests exist — a market for investment and a market for occupation.

The value of a short or medium-term leasehold investment to an investor is approached in a variety of ways. A discounted cash flow (DCF) approach (see Baum and Crosby 2007) is the best means of producing a realistic estimate of value.

The market for occupation is fragmented due to the payment in certain cases of key money (see p11) which can defy logical analysis and valuation.

The value of a leasehold interest in either case will be to some extent affected by statutory interference. The valuer should concern himself with the following:

(a) the assignee's rights to a new tenancy
(b) any improvements carried out by the assignor or his predecessors which may fall to be disregarded upon review or renewal
(c) any transferable rights to compensation.

The right to disturbance compensation will depend upon the assignee carrying on the assignor's business. If he does this, then the period of occupation of the assignor will be taken into account when deciding whether compensation is to be one or two times the rateable value of the holding. The transferability of improvement compensation is unrestricted, as all qualifying improvements carried out during the current tenancy by the tenant's predecessors in title are normally compensated should the tenant lose possession under any one of six of the seven grounds in section 30 of the 1954 Act (see p18).

However, the right to compensation will not form part of the value of the interest unless the parties are aware that the landlord is likely to wish, and to be able, to gain possession at the end of the lease. Only in such a case should the discounted value of the compensation payable be included in the valuation.

1.7.3 Sub-letting

Sub-letting will normally be permitted at no less than the rent paid by the head-lessee. This practice was strongly challenged by The Crosby Report (ODPM, 2005).

Property might also be sub-let at a profit rent after a lease is granted at a rent which disregards the value of improvements carried out by the head or intermediate landlord. However the provisions of section 34 1954 Act (relating to disregard of improvements) will apply once a period of 21 years has elapsed since those improvements were carried out. In such a case, the intermediate landlord can obtain a profit rent from the sub-tenant reflecting the difference in rental terms between the improved and unimproved property, but the profit rent will cease once the intermediate landlord's interest expires or falls to be valued as improved.

Example 1.13

Twenty years ago A let an office suite to B at a rent of £50,000 pa on a 25-year lease with 5-yearly reviews. Ten years ago B carried out improvements to the property with the landlord's consent which have added £30,000 pa to the rental value of the premises (currently £180,000 pa). B is considering sub-letting the premises to C and assigning the interest to E, with A's consent. What rent will C pay? What premium should B ask from E upon assignment?

The sub-tenant C will have the right to obtain a new (maximum term 15 years) lease in 5 year's time at a rent ignoring the value of those improvements carried out within 21 years before the new tenancy application is made or during the current tenancy. If a new 15 year term is granted with two 5-yearly reviews the rent payable by C will ignore the value of improvements at renewal and at the first review, but not at the second review because at this point in time (in 15 years) the improvements will have been carried out outside the last 21 years and therefore the rental value of those improvements can be had regard to. A will then be able to obtain a rent including the value of those improvements.

E will enjoy the profit rent for 5 years only because on the expiry of the lease E will not be a tenant in occupation and will have no right to renew the lease. A once again becomes the landlord. The value of E's interest is therefore based on the receipt of £30,000 in profit rent for 5 years.

Rent received	£180,000 pa	
Rent paid	£150,000 pa	
Profit rent		£30,000 pa
YP 5 yrs @ 11%		4.3295
Premium (say)		£130,000

As the assignment is not to an occupier, there will be no element of key money included in the assignment premium. The valuer should, however, check the valuation by reference to the attitude of the likely purchaser who might be interested in the net of tax return and/or might set up a sinking fund to replace capital.

1.7.4 Surrender

In *Allnatt London Properties Ltd* v *Newton* (1981) the cause of dispute was a requirement in the lease that, should the tenant wish to alienate, he must first offer to surrender his interest to the landlord without consideration. Such a surrender-back clause was held to be void following section 38(1) of the 1954 Act whereunder an agreement to surrender is void (see also *Bocardo SA* v *SM Hotels Ltd* (1979)). However clauses wherein a landlord offers, within a specified period, to pay a consideration for the tenant's interest equal to a bona fide offer received by that tenant from a prospective assignee are common-place, and are now commonly referred to as pre-emption clauses. Until the introduction of the RRO any agreement between a landlord and a tenant to operate the pre-emption had to be approved by the court. Following RRO a court order is not required; the landlord serves a notice and the tenant signs a declaration.

Pre-emptions have become increasingly popular. They give an element of control to the landlord who can choose whether to exercise a pre-emption and buy in a lease rather than permit an assignment to a party he does not wish to have as tenant. The tenant benefits in that a surrender removes the exposure to privity of contract (see below).

1.7.5 Privity of contract

Figure 1.1 (p4) illustrates a vertical division of rights in a property. The contractual relationship which exists between D and G, E and H is one of privity of contract.

In general a party signing a lease is liable to fulfil the terms of the lease during the entire term if the successors in title fail, that is they commit breaches of covenant and/or pass into receivership or liquidation, notwithstanding that the leasehold interest may have been assigned or sub-let.

The Landlord and Tenant (Covenants) Act 1995, which came into force on 1 January 1996, is primarily intended to abolish original tenant liability which occurs under privity of contract. The Act's main thrust is directed at leases granted on or after 1 January 1996, but it also alters, favourably, the position of existing tenants and guarantors.

Section 5 provides that a tenant who assigns "the whole of premises demised to him" is "released from the tenant's covenants and ceases to be entitled to the benefit of the landlord's covenants". Section 16, however, tempers the situation by entitling a landlord to require an assignee to guarantee the performance of the covenants by the assignee tenant. These are referred to as authorised guarantee agreements (AGAs) but will apply to the assignor only until such time as the succeeding assignee is released from his obligations under the lease either by expiry of the term or a further assignment. Disputes have arisen following the introduction of this Act (see *Avonridge Property Co v Mashru* (2005).

Existing tenants are, nonetheless, also affected by the 1995 Act. A landlord will not be able to recover arrears of rent from a former tenant or surety unless within six months of those monies becoming due the landlord has notified the original tenant of the amount due.

1.8 Examples

Example 1.14

The occupying lessee of a shop in a prime position holds an internal repairing lease granted 10 years ago for 14 years at a rent of £240,000 pa Seven years ago the tenant obtained the necessary consents from the landlord and carried out improvements at a cost of £600,000. It is estimated that the current rental value of the shop on full repairing and insuring terms is £500,000 pa of which £50,000 is due to the improvements. The rateable value of the shop is £500,000.

(a) estimate the rent which should be charged if a new 20-year internal repairing lease is granted when the present lease expires,

(b) value the current interests of the landlord and the tenant assuming (i) the landlord will get possession for its own occupation at the end of the present lease and (ii) the tenant will continue in occupation.

It is assumed that there are to be typical reviews in the new lease and that a typical freehold yield for such an investment is 3.5%. The required return is 15%.

(a) At the grant of a new lease the rent to be paid will be the full rental value of the property disregarding the tenant's occupation, goodwill, and improvements carried out during the last lease or the last 21 years (1954 Act, section 34 as amended by section 1 of the 1969 Act). Improvements to be disregarded must have been carried out with the landlord's consent other than as a condition of the lease. In this case, £50,000 is to be left out of the rent on that basis and a rent of £450,000 should be paid on full repairing and insuring terms. This would be greater on internal repairing terms by the cost of external repairs and insurance saved by the tenant.

Rent	£450,000
Add cost of external repairs, say	£50,000
Add cost of insurance, say	£10,000
Rent	£510,000

(b) (i) If the landlord obtains possession at the end of the lease for his own occupation, compensation for both disturbance and improvements will be payable to the tenant. This will affect the value of its interest in the market as any purchaser of this interest will be liable to pay both compensation amounts upon obtaining possession.

Compensation for disturbance: the tenant has been in occupation for 14 years at the end of the lease and must be paid 2 × rateable value:

2 × £500,000 = £1,000,000

Compensation for improvements: under section 1(1) of the 1927 Act this is limited to the lesser of the net addition to the value of the freehold in possession contributed by the improvements and the reasonable cost of carrying out the improvements at the termination of the tenancy.

(1) Net addition to value:
Value with improvements

Rent	£500,000 pa FRI	
YP perp @ 3.5%	28.57	
Capital value		£14,285,000

Value without improvements:

Rent	£450,000	
YP perp @ 3.5%	28.57	
		£12,856,500

Addition to value		£1,428,500

(2) Cost of carrying out improvements: advice should be sought from a quantity surveyor or similar. However, seven years ago the improvements cost £600,000. Assuming increasing costs at 10% pa over the previous seven years of the lease, the cost today would be:

£600,000 × 1.10^7 =	£1,169,230
Take lower	£1,169,230
Add disturbance compensation	£1,050,000
Total compensation payable	£2,219,230

Valuation: landlord's interest

Rent received	£240,000 pa			
Less				
Ext. repairs	£50,000			
Insurance	£10,000			
Management	£20,000			
	£80,000 pa			
Net income	£160,000 pa			
YP 4 years @ 3.5%	3.6731			
		£587,700		
Reversion to ERV (FRI):	£500,000 pa			
YP rev. perp. 4 years @ 3.5%	24.898			
		£12,449,000		
			£13,036,700	
Less compensation	£2,219,230			
PV 4 years @ 10%	0.683			

(10% reflects the possibility of moderate growth in the compensation value).

	£1,515,730
Say	£11,520,000

Valuation: tenant's interest
The transferability of compensation should be considered. If a purchaser of the tenant's interest is likely to carry on the same business, then the current tenant's period of occupation will be taken into account in deciding upon the quantum of disturbance compensation. It is assumed that this is likely to be the case. The right to improvement compensation is fully transferable.

Rent received	£500,000 pa FRI		
Convert to IRT:			
add ext repairs	£50,000		
add insurance	£10,000		
	£560,000 pa IRT		
Rent paid	£240,000 pa IRT		
Profit rent	£320,000 pa		
YP 4 years @ 15%	2.855		
		£913,600	
Add right to compensation		£2,219,230	
PV 4 years @ 10%		0.683	
		£1,515,730	
		£2,429,330	
Say		£2,430,000	

(ii) If the tenant continues in occupation at the end of the current lease, no question of compensation arises and no purchaser would pay a sum in anticipation of it.

The rental value of the improvements is assumed to be disregarded at each review in accordance with the RICS model. These reviews will fall 16, 21, and 26 years after the improvements were carried out. Assuming that the second review falls marginally less than 21 years after the improvements were completed, the rent will not be reviewed to its full level until the third review.

Valuation: landlord's interest

Rent received	£160,000 net (as before)	
YP 4 years @ 3.5%	3.6731	
		£587,700

Reversion to rent at lease renewal: full rental value disregarding the value of improvements:

Rent received	£450,000 pa FRI	
YP 15 years @ 3.5%	11.5175	
PV 4 years @ 3.5%	0.8714	
		£4,516,322

At the third review in the new lease full rental value including the value of the improvements will be paid.

Rent received	£500,000 pa FRI		
YP rev. perp. 19 years @ 3.5%	14.8616		
		£7,430,800	
Total, say			£12,500,000

Valuation: tenant's interest

Rent received	£560,000 pa IRT (as before)		
Rent paid	£240,000 pa		
Profit. rent	£320,000 pa		
YP 4 years @ 15%	2.855		
		£913,600	

At the new lease, a profit rent equal to the addition to rental value made by the improvements will be received.

Profit rent	£50,000 pa		
YP 15 years @ 6%	7.693		
PV 4 yrs @ 6%	0.7921		
		£304,679	
Total, say			£1,220,000

(6% reflects growth potential and is based on the freehold capitalisation rate plus a leasehold risk premium).

Example 1.15

X is interested in the purchase of Y's leasehold interest in an office property now comprising 250m^2 constructed on two floors. There is a user restriction in the lease to "employment agency purposes only"; X wishes to use the property for this purpose.

The lease was granted 6 years ago for a term of 14 years with a single rent review at seven years at an initial rent of £100,000 pa on full repairing and insuring terms.

Two years ago Y demolished a rear extension of 50m^2 in order to construct a modern extension of 75m^2 incorporating a boiler room and provided central heating to the building.

The leasehold interest in similar adjacent property of 200 m^2 was recently sold to a charity at a price of £250,000. The rent payable under this lease was fixed at £150,000 pa on full repairing and insuring terms for the remaining five years of the lease, which was subject to no onerous restrictions. The sub-lessee's rent is subject to immediate review and this property is fully modernised.

Freehold interests in similar property currently sell on a 6% basis. Undated government stock currently yields around 4.5%. Advise X on his bid.

The charity acquiring this interest acted as an investor rather than as an occupier, and the premium paid can consequently be usefully analysed. For interests such as this with negligible growth potential (depending upon the ability of the landlord to review the sub-tenant's rent) pension funds are assumed to require an 8% return. For a leasehold a higher required return (say 10%) is appropriate.

Rent received (ERV)		£ x pa FRI
Rent paid		£150,000 FRI
Profit rent		£ x – £150,000
YP 5 years @ 8%		3.9927

$$£3.9927x – £462,070$$

£3.9927 x – £462,070	=	£250,000
£3.9927 x	=	£712,070
£ x	=	£178,343 FRI

Full rental value/m² = £178,343/200 = £891

Full rental value of subject property (as improved but subject to user restriction) lower than £891: say £800/m²:

250 m² @ £800/m² : £200,000

Full rental value of subject property (ignoring improvements and subject to user restriction) at say £750/m²:

225 m² @ £750/m² : £168,750

X's interest is of value because a profit rent is currently received and because at the rent review in one year's time the value of improvements will be disregarded (assuming a suitable review clause, the receipt of the landlords consent, etc.). The rent will continue to disregard the value of improvements at the renewal in eight years' time, 10 years after completion of the improvement. If five-yearly reviews are incorporated in the new lease, it will not be until the third review that the value of improvements is taken into the rent, at which time the improvements will have been carried out 25 years previously.

At the review in one year's time the rent is to be fixed for a seven-year period and the landlord invoking the constant rent theory might insist upon uplift.

ERV on five-yearly reviews ignoring improvements: £168,750 pa

r	=	9% (8% plus a 1% risk differential)
k	=	5%
n	=	7
t	=	5
g	=	?
p	=	$(1 + g)^5 – 1$
SF	=	ASF to replace £1 in 5 years @ 9%
k	=	$r – (SF \times p)$
0.05	=	0.09 – 0.1671p
0.1671p	=	0.04
p	=	0.2394 (23.94%)
p	=	$(1 + g)^5 – 1$
0.2394	=	$(1 + g)^5 – 1$
$\sqrt[5]{1.2394}$	=	$(1 + g)^5$

$$1.0439 \quad = \quad 1 + g$$
$$g \quad = \quad 0.0439 \ (4.39\% \ pa)$$

This is the level of annual rental growth expected by investors in this type of property.

$$K = \frac{(1 + r)^n - (1 + g)^n}{(1 + r)^n - 1} \quad \times \quad \frac{(1 + r)^t - 1}{(1 + r)^t - (1 + g)^t}$$

$$= \quad \frac{(1.09)^7 - (1.0439)^7}{(1.09)^7 - 1} \quad \times \quad \frac{(1.09)^5 - 1}{(1.09)^5 - (1.0439)^5}$$

$$= \quad 0.5763 \times 1.8015$$

$$= \quad 1.0381$$

Full rental value with seven-yearly reviews (ignoring improvements):

1.0381 × £168,750 pa	=	£175,187
Say		£175,000 pa

Valuation of leasehold interest:

Rent received	£262,500	
Rent paid	£100,000	
Profit rent	£162,500	
YP 1 year @ 10%	0.9091	
		£147,727

At the review the rent payable is increased to £175,000 pa

Rent received	£262,500 pa FRI	
Rent paid	£175,000 pa FRI	
Profit rent	£87,500 pa	
YP 22 years @ 10%	8.7715	
PV 1 year @ 10%	0.9091	
		£697,740
Total		£845,467
Say		£845,000

Added to this figure might be an allowance for goodwill and fixtures and fittings and, depending on the supply of such properties in the market, key money might have to be paid.

Residential Property

2.1 Introduction: residential tenure forms

Those in the UK fortunate enough to have homes to live in are almost certainly representatives of one of three categories: owner-occupiers, council tenants, or tenants of private landlords. It is arguable that tenants of housing associations (be it part-owner/part rented or fully rented occupiers) fall into none of these categories, but they are now treated in much the same way as private tenants. Eccentrics such as those in permanent occupation of a hotel room defy categorisation, while other forms of tenure are particularly unusual.

The 20th century has seen a remarkable shift in the relative importance of the three major categories of tenure. Between 1914 and 2001 owner-occupation increased as the mode of tenure from 10% of all households to 68.4%, the public rented sector grew from 1% to 27.9%; while the private rented sector, housing 80% in 1914, provided accommodation for only 8.7% of 2001 households.

A curious synthesis of forces has created this change in balance. Continuing government persuasion towards owner-occupation is evidenced by income tax relief on mortgages, and freedom from capital gains tax on the sole or main residence. The growth of council housing is explained by the increasing powers and duties of local authorities to provide residential accommodation, beginning with the demand for homes fit for heroes at the termination of the First World War and continuing through the 1950s and 1960s passion for slum clearance.

The decline of the privately rented sector during the last century is surely explained, at least in part, as a result of a decline in demand precipitated by the substitution effects of government policies encouraging owner-occupation, and the security and low cost offered by the public sector. There is, after all, no income tax relief on rents paid, and subsidised council housing will inevitably be cheaper to rent than privately owned dwellings, making the relative net cost of private accommodation unattractively high.

However, the decline of the rented sector was customarily explained as a loss of supply. Private landlords, it was said, were no longer happy to allow tenants to occupy their property due to the increasing body of legislation restricting their ability to charge reasonably high rents and gain vacant possession. It is partly because of this that in recent years further legislation has attempted to reverse this trend by relaxing security of tenure and rent control. This has, in tandem with decreasing interest rates and slow-down of gilt/equity returns, resulted, in part, in the growth in the buy-to-let sector and the creation of some institutional investment vehicles.

It is the effect of this legislation upon residential property in the privately rented sector that forms the basis of this chapter, although passing references will be made to other tenure forms. In no other area of statutory valuation can it be argued that the valuation approach prompted by the relevant legislation has had such a dramatic effect upon the property market and on people's lives.

2.2 The legislative background

The current legislation affecting rented residential property in the private sector is superimposed upon an existing structure of legal principles, collectively known as landlord and tenant law which is common to all types of property. This does not encourage simplicity, and the task of the professional valuer is not eased by the abundance of legislation in the field.

The current statutory control of the residential sector of privately rented property (henceforth referred to for simplicity as residential property) can be traced back to the Increase of Rent and Mortgage Interest (Restrictions) Acts of 1915 and 1920 which applied rent control to a wide range of residential property, subject to certain rateable value limits. These Acts were the result of an excess of demand over supply and the subsequent escalation of rents, coupled with the government of the day's avowed intent to improve the lot of the working man on his return from the battlefield.

Whether as a result of this government intervention or other factors, a virtually unbroken vicious circle of shortage and further rent control has been the aftermath of the First World War. Major measures designed to effect the aims of the restriction of rents and the landlord's ability to evict his tenant became consolidated in the 1977 Rent Act. This Act now covers only a small number of tenancies; those commencing before 1989, and government policy has in recent years been aimed at moving away from rent restriction and security of tenure and towards a situation where market forces can pre-dominate. This policy, coupled with the expectation of strong capital growth for investors, appears to have been successful in halting the decline of the privately rented sector. This policy is principally embodied in the Housing Acts of 1980, 1985, 1988, and 1996, but there has been other important legislation during this period including the Housing and Planning Act 1986, the Landlord and Tenant Act 1987 and the Housing Act 2004.

For the remainder of this chapter reference will be made to the 1977, 1980, 1985, 1988, 1996 and 2004 Acts only unless the full statutory title of the relevant legislation is given.

2.3 Assured tenancies

The concept of assured tenancies was first introduced by the 1980 Act which allowed approved bodies (mainly building societies, pension funds, and insurance companies) to charge open market rent for property built especially for that purpose after the passing of the Act. The scheme was extended by the 1986 Housing and Planning Act to include certain vacant properties which had been substantially improved. It was not, however, conspicuously successful in its declared aim of reviving the privately rented housing sector (despite further governmental tax induced vehicles such as the Business Expansion Schemes) , possibly because of its restricted application.

The 1988 Act has swept away these restrictions to such an extent that the current assured tenancy is best considered as an entirely new form of tenancy, and one which will comprise almost all new lettings of residential property. It should not be confused with assured tenancies under the 1980 Act which, in any event, were automatically converted to the new form on the enactment of the 1988 Act.

The new assured tenancies still give a high degree of security of tenure, though control over the level of rents is greatly reduced. However, the 1996 Act has gone still further by ensuring that new lettings will now normally be assured shorthold tenancies, which provide a fixed term of at least six months. These provide greatly reduced security of tenure, generally restricted to the contractual term with a two month statutory notice period.

2.3.1 The law: general application

All new tenancies under which a dwelling house is let as a separate dwelling will be assured tenancies under the 1988 Housing Act providing that:

(i)   the tenant is an individual and
(ii) the tenant occupies the house as his only or main residence and
(iii) the tenancy is not specifically excluded under the Act.

Existing assured tenancies under the 1980 Act automatically became assured tenancies under the 1988 Act.

The principal exclusions as set out in schedule 1 of the 1988 Act are as follows:

(1) tenancies entered into before the commencement of the Act
(2) a tenancy which is entered into on or after 1 April 1990 (otherwise than, where the house had a rateable value (RV) on 31 March 1990, in pursuance of a contract made before 1 April 1990) and under which the rent payable exceeds £25,000 pa
(3) a tenancy which was entered into before 1 April 1990 and under which the house had an RV on 31 March 1990 which exceeded £1,500 in London and £750 elsewhere
(4) tenancies as a low rent, ie tenancy entered into after 1 April 1990 under which the rent is payable at a rate of £1,000 or less pa in London, or £250 or less elsewhere, or a tenancy entered into before 1 April 1990 under which the rent is less than two-thirds of the RV on 31 March 1990
(5) business tenancies
(6) licensed premises
(7) tenancies of agricultural land
(8) tenancies of agricultural holdings
(9) lettings to students
(10) holiday lettings
(11) lettings by resident landlords
(12) crown tenancies
(13) tenancies of a local authority, development corporation, etc
(14) regulated tenancies, housing association tenancies, secure tenancies, and tenancies in existence before the commencement of the Act.

As a result of section 96 of the Housing Act 1996 which inserts a new section19A into the 1988 Act, all assured tenancies entered into after the 1996 Act comes into force will be assured shorthold tenancies unless they fall within the following specific exceptions which are set out in a new schedule 2A to the 1988 Act:

(1) where a notice is served by the landlord before it commences stating that it will not be an assured shorthold tenancy.

(2) where a notice has been served by the landlord after the tenancy has been entered into stating that it will no longer be an assured shorthold tenancy

(3) where the tenancy contains a provision to the effect that it is not an assured shorthold tenancy

(4) where the tenancy arises as a result of the succession of a regulated tenancy

(5) where the assured tenancy arises on its ceasing to be a secure tenancy

(6) where the assured tenancy arises on the ending of a long tenancy by virtue of schedule 10 to the Local Government and Housing Act 1989

(7) where the tenancy is granted to someone who had an assured tenancy immediately before the granting of the new assured tenancy, by the landlord under the old tenancy, and where the tenant has not, prior to the granting of the tenancy, served notice on the landlord stating that it will be a shorthold tenancy

(8) where the tenancy arises on the coming to an end of assured tenancy and becomes a periodic tenancy

(9) certain agricultural tenancies (schedule 2A, 1988 Act).

Exemption (7) is particularly interesting and appears to be designed to allow landlords to negotiate terms with tenants for changing assured tenancies to assured shortholds.

The overall effect is that whereas previously the majority of private sector tenancies were regulated tenancies under the 1977 Act, these are gradually disappearing to be replaced by assured and assured shorthold tenancies.

The exemptions are broadly similar under both Acts, but there are some significant differences, perhaps the most important being that Housing Association lettings are not excluded from the 1988 Act, and are effectively moved from the public to the private sector. Another variation is that the provision of board will not exclude a tenancy under the 1988 Act. Lettings by a resident landlord are excluded and there is no protection of any sort for a tenant who is living in the same property as their landlord, not even a requirement to obtain a court order for possession.

2.3.2 Security of tenure

By their very nature assured shorthold tenancies have no real security of tenure beyond the initial term, other than a two-month notice period and the procedural requirements for obtaining vacant possession. While assured tenancies under the 1988 Act have a lesser degree of security of tenure than that afforded by the 1977 Act, there is still a large degree of protection available to tenants.

A landlord cannot obtain possession of a dwelling let under an assured tenancy without a court order. Schedule 2 of the Act sets out eight mandatory grounds under which the court must order possession, and nine discretionary grounds under which the court will order possession only if it considers it reasonable to do so.

(1) Mandatory grounds:
  (i) the landlord has given notice before the tenancy commenced that possession might be required under this ground or the court is satisfied that it is reasonable to dispense with notice and
    (a) the landlord has at some time before the tenancy began occupied the dwelling at his home or

      (b) the landlord requires the property at his home

(ii)   a mortgagee requires recovery of possession

(iii)   a holiday home is let out of season for a period not exceeding eight months and the landlord has given advance notice that possession may be required under this ground.

(iv)   a property which has previously been let to students is let for a period not exceeding 12 months and advance notice has been given

(v)   the property is required for a minister of religion and advance notice has been given.

(vi)   the landlord intends to carry out demolition, reconstruction, or substantial works to the house and the work cannot be undertaken without the tenant giving up possession

(vii)   the tenancy is a periodic tenancy or a statutory periodic tenancy and the landlord brings proceedings for possession within 12 months of the death of the tenant and the devolution of the tenancy under his will

(viii) at the date of commencement of proceedings and at the date of the hearing there are arrears of rent amounting to two months or eight weeks if the rent is paid fortnightly.

(2) Discretionary grounds:

(ix)   suitable alternative accommodation is available to the tenant

(x)   rent is in arrears at the date of commencement of proceedings

(xi)   payment of rent has been persistently delayed

(xii)   the tenant has broken his obligation under the tenancy

(xiii) the dwelling has deteriorated due to the neglect or default of the tenant

(xiv) the tenant has been guilty of nuisance or annoyance to other occupiers

(xv)   the condition of furniture has deteriorated due to ill-treatment by the tenant

(xvi) the dwelling was let in consequence of the tenant's employment by the landlord and that employment has ceased

(xvii) the landlord induced the tenant to grant possession by a false statement made knowingly or recklessly.

Possibly the most significant reduction in the tenant's security of tenure as compared with the 1977 Act is ground (vi) above which is similar to the ground which has previously been applicable only to business tenancies under the 1954 Act (see Chapter 1). Ground (vii) is also new and ground (i) has been considerably widened so that, for example, a landlord who has previously occupied the property, no matter how long ago or for how short a period, has a right to possession even if he has no intention of taking up occupation again. While he is required to give notice before the commencement of the tenancy that possession might be recovered on this ground, even this notice can be dispensed with if the court feels it just to do so.

2.3.3 Succession

Under section 17 of the 1988 Act an assured periodic tenancy will be transferred to the spouse on the death of the tenant, providing that tenant was not himself a successor under the section. Therefore, there can be only one succession and this is limited to a spouse which includes a person living with the tenant as his, or her, wife or husband. This is a considerably reduced right of succession from that under the 1977 Act and has improved the marketability of privately rented residential property, which will not as a result be tied up for generations.

2.3.4 Forms of assured tenancy

(a) Assured periodic tenancy

This was originally the basic form of assured tenancy, and there is considerable freedom for the landlord and tenant to agree terms including the rent review pattern and the payment of a premium.

(b) Assured fixed-term tenancy

This is similar to an assured periodic tenancy except that the length of the term is fixed by agreement between the parties. During the term the landlord may only obtain possession on ground (ii) or (viii) of the mandatory grounds or on any of the discretionary grounds other than (ix) and (xvi) and then only if the terms of the tenancy make provision for possession on those grounds. When the term ends the landlord does not regain possession but a statutory periodic tenancy arises, unless the termination date is after the 1996 Act came into force, in which case an assured shorthold tenancy arises.

A fixed-term tenancy gives the tenant added security of tenure for the period of the term and is likely to be useful where the landlord requires that the tenant pays a premium or carries out works of repair or improvement as a condition of the lease.

(c) Assured statutory periodic tenancy

A statutory periodic tenancy arises upon the termination of a fixed-term tenancy through effluxion of time, but only where this occurs before the 1996 Act came into force. The terms are as under the fixed-term tenancy except that the grounds of possession will be as set out in the Act for any periodic tenancy.

(d) Assured shorthold tenancy

An assured shorthold tenancy has quickly become the most common form of assured tenancy due to the ease with which the landlord can regain possession. It is an assured tenancy where the property is let for a fixed term of not less than six months and where the landlord has, in relation to a tenancy which commenced before the 1996 Act, served a notice before the beginning of the tenancy stating that it will be a shorthold tenancy. However, section 96 of the 1996 Act inserts a new section 19A into the 1988 Act which states that any tenancy entered into after the date on which the 1996 Act comes into force, will be an assured shorthold tenancy unless otherwise specified. It differs from a fixed-term assured tenancy in that the landlord has an absolute right to obtain possession at the end of the term. However, at the end of a shorthold tenancy a periodic shorthold tenancy will arise unless, before the last day of the term, the landlord serves two months' notice in writing that possession will be required or he offers the tenant a new shorthold tenancy. Where a periodic shorthold tenancy has arisen the landlord will still be able to gain possession by serving not less than two months' notice.

As a result of section 20A of the 1988 Act (inserted as a result of section 97 of the 1996 Act) a tenant under an assured shorthold tenancy is entitled to a written statement of any term of his tenancy which has not been put in writing and which relates to:

(a) the date of commencement of the tenancy
(b) the rent payable and the dates on which it is due

(c)   any rent review terms and

(d)   the length of the tenancy where it is a fixed term.

While an assured shorthold tenancy does not have to specify any length of term, an order for possession is required to obtain vacant possession and this will not be given to take effect less than six months from the commencement date.

2.3.5 Rent under assured tenancies

(a) Introduction

Assured tenancies are not subject to rent control in the generally accepted sense of the term, and market forces are free to dictate the level of rents. There are, however, constraints in the Act designed to prevent sitting tenants being exploited by the fixing of rents in excess of market value.

(b) Premiums

Unlike regulated tenancies, assured tenancies are not subject to any restriction on the charging of premiums.

It is particularly likely that premiums will be charged at the commencement of a fixed-term or shorthold tenancy as the tenant is likely to require a guaranteed term in return for his capital payment. In theory any premium paid will be reflected in a reduced rental for the period until the first review. However, it may be that in areas or times of particular shortage, tenants may be prepared to pay a key money premium in order to obtain possession, in addition to the payment of an open market rent: see p 00.

When valuing property which is subject to an assured tenancy it is important to establish whether a premium was paid at the commencement of the term and, if so, whether the rent paid is below market levels.

(c) Fixed-term and periodic assured tenancies

At the commencement of the tenancy the landlord and tenant are free to negotiate the terms of the letting, including the commencing rent, the rent review pattern, and where appropriate, the length of the term.

Where periodic tenancies, including statutory periodic tenancies, contain no review provision the landlord may serve a notice on the tenant proposing a new rent under section 13 of the Act. The new rent must take effect at the beginning of a new tenancy period and section 13(3) states that the following minimum period of notice must be given:

(a)   in the case of a yearly tenancy, six months

(b)   in the case of a tenancy where the period is less than a month one month and

(c)   in any other case, a period equal to the period of the tenancy.

If the tenant is unhappy with the proposed rent, he has the right to refer the notice to the rent assessment committee (RAC) which will determine what it considers to be an open market rent for an assured tenancy taking into account the terms of the actual tenancy (section 14(1)). The committee will disregard any special value to the sitting tenant, any reduction in value attributable to the tenant's failure to comply with the terms of the tenancy, and the effect of any improvements carried out by the tenant other than as a condition of the tenancy (section 14(2)). This latter provision has obvious similarities with the protection afforded to business tenants under the Landlord and Tenant Act (1954) (Part I) see Chapter 1.

There is nothing to prevent landlord and tenant agreeing a new rent at any amount before, during, or following the deliberations of the RAC. In the absence of any agreement, however, the landlord cannot serve another notice of rent increase which would take effect less than 12 months from the date on which the last increase took effect (section 13(2)).

(d) Assured shorthold tenancies

A tenant under a shorthold tenancy may apply to the RAC at any time for the determination of a market rent, providing that the rent currently payable has not previously been determined by the committee, and where the tenancy commenced prior to the 1996 Act, providing the application is made within the first six months of the tenancy (section 22, 1988 Act). However, the role of the RAC in these circumstances is somewhat different to that under an assured periodic tenancy.

The committee will only intervene where it considers that there are sufficient similar houses in the locality let on assured tenancies, and that the rent payable is "significantly higher than the rent which the landlord might reasonably be expected to obtain". If these provisions are satisfied it will determine a rent on the same basis as for assured periodic tenancies.

2.3.6 Valuation

As there are open market rents, regular reviews, and security of tenure not much greater than that available to business tenants it could be expected that a pure investment valuation, similar to those set out in Chapter 1, would be appropriate. However, the desire for home ownership, coupled with mortgage interest tax relief creates a situation where the vacant possession value of houses will almost invariably exceed a figure which would result from an investment valuation, particularly where house prices are high such as London and the South-East. It may be, therefore, that a valuation calculated as a percentage of vacant possession value will be preferred, as has generally been the case under earlier forms of residential tenancy. That percentage will vary considerably depending on the demand for both rented and owner-occupied property in any given locality.

The deduction from vacant possession value will reflect the prospects of obtaining vacant possession value in the particular circumstances, and an appropriate deferment factor based upon the anticipated length of the tenancy and the time necessary to obtain a court order.

In the case of shorthold tenancies, the time period will be relatively certain and a simple deferment of vacant possession value would be appropriate. On the other hand, many assured periodic tenancies will be almost as secure as a regulated tenancy and should be valued in much the same way. At this extreme a pure investment valuation plus an uplift to reflect the hope of obtaining vacant possession may be the market practice.

Important points to be borne in mind by the valuer when considering residential property let on assured tenancies are as follows:

(i)   if a premium has been paid the rent passing could be below market levels and careful comparison with similar lettings in the locality will be necessary

(ii)   there is considerable scope for varying the terms of the letting so, for example, it would be dangerous to assume that the previously usual practice of the landlord bearing responsibility for external and structural repairs and insurance will apply. Similarly, an unusual rent review pattern could influence the rent paid and the valuer should take care that the rent of the subject property, or of any comparables, has not been distorted by unusual conditions of this nature

(iii) the income is relatively secure in money terms due to the undersupply of residential property, but with rents at market levels the security of income is not as great as for regulated lettings. The Act appears to be succeeding in its aim of expanding the privately rented sector, and even the situation of undersupply may not continue indefinitely

(iv) the income is secure in real terms as there will usually be annual reviews

(v)   the reduced rights of succession and of security of tenure must be evaluated in the circumstances of each tenancy

(vi) a sale to the sitting tenant should always be considered as there will be circumstances where the property will be worth more to him than to the general market. This will always be the case where there is security of tenure, and an investment valuation produces a figure which is lower than the vacant possession value.

Example 2.1

A 1930s semi-detached house is let on a periodic assured tenancy with the tenant responsible for internal decorations only. The rent has recently been assessed by the rent assessment committee at £150 per week. The vacant possession value is £200,000.

  The valuation of the freehold interest in this property may take two forms:

(a)	Rent received		£7,800 pa	
	Less outgoings:			
	External repairs, say	£1,000		
	Insurance, say	£400		
	Management @ 5% rent	£390		
			£1,790 pa	
	Net income			£6,010 pa
	YP perp. @ 6.5%		15.38	
	Say			£92,000
(b)	Vacant possession value		£200,000	
	Take 75%		0.75	
	Say			£150,000

Valuation (b) would be possible only if the investment market for tenanted residential property reflects a consistent pattern of values in relation to the value of owner-occupied property. The example demonstrates that even using a very low yield, an investment valuation produces a figure far below that which an investor is likely to have to pay in the open market. This reflects the prospects of obtaining vacant possession which could happen in a number of ways. The landlord could seek to obtain vacant possession using the grounds available under the act, such as providing alternative accommodation. He could simply wait for the tenant to vacate, or he could offer the tenant a share of the marriage value (the difference between vacant possession value of £200,000, and tenanted value of £150,000) as an incentive to vacate.

2.4 Protected tenancies

2.4.1 The law: general application

While, with a few minor exceptions, there will be no new regulated tenancies under the 1977 Act as a result of the 1988 Act, existing regulated tenancies will continue for some time. The 1977 Act applied to residential lettings subject to three requirements, as follows:

(a) There must be a tenancy whereby a dwelling house is let as a separate dwelling. Under section 1, "a tenancy under which a dwelling house (which may be a house or part of a house) is let as a separate dwelling is a protected tenancy for the purposes of this Act".

All forms of tenancy are liable to be protected under the Act, but it is apparent that licensees are not protected tenants.

Exactly what constitutes a licence as opposed to a protected tenancy has often been the cause of dispute but considerable guidance was given by the House of Lords in Street v Mountford (1985). In this case the agreement was described as a licence and included a declaration by the tenant accepting that the agreement did not give a protected tenancy under the Rent Acts. However, it was held that as exclusive possession was granted the agreement constituted a protected tenancy. The "exclusive possession" test is, therefore, a useful pointer when distinguishing a licence from a protected tenancy and care should be taken not to rely too heavily on the title of the document.

Further, and hopefully final, guidelines in this area were given by the House of Lords' decision in *AG Securities* v *Vaughan and Antoniades* v *Villiers* (1988). Lord Templeman discussed in detail documents which purported to grant a licence, but in fact created a tenancy, stating:

> "*Street* v *Mountford* reasserted three principles. First, parties to an agreement cannot contract out of the Rent Acts. Second, in the absence of special circumstances, not here relevant, the enjoyment of exclusive occupation for a term in consideration of periodic payments creates a tenancy. Third, where the language of licence contradicts the reality of lease, the facts must prevail. The facts must prevail over the language in order that the parties may not contract out of the Rents Acts. In the present case clause 16 (under which the landlord reserved a right to come into and share the premises with the occupiers) was a pretence."

It is clear from *Langford Property Co Ltd* v *Goldrich* (1949) that two separate flats let together as a single dwelling can constitute a house for the purposes of protection. The purpose of the letting is persuasive: consequently, where a building is let for business purposes protection is to be sought in the 1954 Landlord and Tenant Act Part II (see Chapter 1) rather than in the Rent Act, even if the tenant lives on the premises.

Where accommodation is shared, section 22 of the 1977 Act provides that no order may be made for possession of the shared accommodation unless an order can be made for possession of the separate accommodation let with it.

For premises to be regarded as a dwelling, it is apparent that more than one of the residential functions of sleeping, cooking, and eating and so on must be carried on. Hence in *Curl* v *Angelo* (1948) the letting of two rooms to an hotelier who used them as sleeping accommodation for guests and employees created no protected tenancy.

(b) The dwelling house must have a rateable value falling within the prescribed limits. In order that a tenancy is protected, its rateable value must not exceed certain limits on the appropriate day. The "appropriate day" is 23 March 1965 for any property which appeared on the valuation list on that date. For other properties, the appropriate day is the date on which the rateable value is first shown on the valuation list. Following the abolition of domestic rating, the limits, set out in section 4, have been considerably amended by the References to Rating (Housing) Regulations 1990 (SI 1990, No 424), and the Local Government Finance (Housing) (Consequential Amendments) Order 1993 (SI 1993 No 651). They are now as follows:

A tenancy which is entered into after 1 April 1990 is not a protected tenancy where:

Class A. The appropriate day fell on or after 1 April 1973 and the dwelling house had a rateable value (RV) on that day exceeding £1,500 in London, or £750 elsewhere.

Class B. The appropriate day fell on or after 22 March 1973 but before 1 April 1973 and the house had an RV exceeding £600, or £800 in London, and on 1 April 1973 had an RV exceeding £750, or £1,500 in London.

Class C. The appropriate day fell before 22 March 1973 and the house:
(a) on the appropriate day had an RV exceeding £200, or £400 in London and
(b) on 22 March 1973 had an RV exceeding £300, or £600 in London and
(c) on 1 April 1973 had an RV exceeding £750, or £1,500 in London.

There is a presumption in any case of doubt, that the tenancy is outside any class.

(c) The tenancy must not be exempt. Various categories of tenancy are exempted from protection under the 1977 Act and it is interesting to compare these exemptions with those applicable to the 1988 Act. One of the most important differences is that housing association lettings are excluded under the 1977 Act but are assured tenancies under the 1988 Act. The major exceptions to the 1977 Act are as follows:

(i) Where the dwelling house is let with land other than the site of the house (section 6). Note, however, that under section 26 any land let with a house shall, unless it consists of agricultural land exceeding two acres in extent, be treated as part of the dwelling house!

This appears to be a classic example of statutory nonsense, best resolved by an objective study of the main purpose of the letting. If the land is let with the house, residence being the main purpose of the letting, the tenancy will be protected

Agricultural tenancies are exempt (see Agricultural Holdings Act 1948, outside the scope of this book).

(ii) Other exempt properties include public houses (section 11), parsonage houses, and overcrowded dwellings, where the tenant is guilty of the offence (section 101).

(iii) Tenancies granted by the Crown or by an exempt body are not protected (section 13, as amended by section 73. 1980 Act and sections.14–16).

(iv) A tenancy granted by a resident landlord will not be protected if it was granted on or after 14 August 1974. The dwelling house must form part of a building but the building must not be a purpose-built block of flats (section12(1)) (see *Barnes v Gorsuch* (1982)). This type of tenancy is known as a restricted contract (see p 71).

Tenants sharing accommodation with their landlord, eg lodgers, are exempted (section 21), although this type of tenancy is also known as a restricted contract (see p 71).

(v) Tenancies granted by certain educational establishments are exempt under section 8.

(vi) A tenancy is not protected if the rent payable is less than two-thirds of the rateable value on the appropriate day (section 5(1)). In such cases either or both of Part I of the 1954

Landlord and Tenant Act (see 2.7 on p 71 below) or the 1967 Leasehold Reform Act (see Chapter 3) might apply to the tenancy.

(vii)   Under section 7, a tenancy is not protected if the dwelling is let at a rent which includes a payment in respect of board or attendance. Sham transactions are excluded and the amount of rent fairly attributable to attendance must form a substantial part of the whole rent.

(viii)   A dwelling let for holiday occupation is exempt under section 9.

(ix)   The Secretary of State is empowered to exempt certain dwellings from rent regulation where he is satisfied that there is no scarcity in the area. This power, given by section 143, has never been exercised.

(x)   Shorthold lettings and assured tenancies, creatures of the 1980 Act, are exempt from the full control and protection imposed by the 1977 Act (1980 Act, sections.55 and 56, see p 52).

(xi)   Lettings after 15 January 1989 are exempt as a result of Part I of the 1988 Act, subject to the exceptions set out in section 34 of that Act. These exceptions include the situation where suitable alternative accommodation is provided to a protected tenant and the court considers that an assured tenancy would not afford the required security. In any event, tenancies entered into after 1 April 1990 are exempt if the rent exceeds £25,000 pa.

Subject to these 11 exemptions, section 2 of the 1977 Act provides that a tenant protected by the Act shall, after the termination of his contractual tenancy, become the statutory tenant of the house and shall continue to be so as long as he occupies the house as his residence.

Under section 3, the terms and conditions of the statutory tenancy are simply a continuation of the terms of the contractual tenancy so long as they are consistent with the Act. A statutory tenant is liable to pay only the rent provided by the Act.

The tenancy cannot be assigned or sub-let, so any provisions for this in the original tenancy agreement are inoperative.

There are only two ways in which the statutory tenancy may be determined: either the court will give an order for possession (section 98) or the tenant will voluntarily give up possession. The tenant may request a payment from his landlord as a condition of giving up possession, but from no one else. A statutory tenancy cannot be alienated.

The only form of statutory tenancy intended to continue as a result of section 64 of the 1980 Act (which provides for the abolition of controlled tenancies) was the regulated tenancy, and controlled tenancies are not considered in this book due to their demise.

2.4.2 *Security of tenure*

The 1977 Rent Act, as amended, has twin intentions: to award security of tenure to protected tenants and to regulate the rents they pay. Security of tenure is ensured by section 3, under which the tenant retains the benefits of his original contract as long as he remains in possession and cannot be evicted unless a court order for possession is granted (section 98).

If the term of the contractual tenancy is current, the tenant is further protected by the terms of that tenancy and the Protection from Eviction Act of 1977. If the landlord is able to terminate that tenancy, the tenant becomes a statutory tenant and is protected by the 1977 Rent Act as above.

Certain of the grounds upon which the landlord can claim possession are discretionary, as the court will not order possession unless it "considers it reasonable to make such an order" (section 98). The court will not make the order if it is satisfied that greater hardship would be caused by granting the order than by refusing it. In *Hodges* v *Blee* (1987) the rather elderly tenant was in poor financial circumstances and would have problems with rehousing. The landlord had reasonable grounds for obtaining possession of the residence for his sons under case (ix) below. Perhaps surprisingly, the county court judge granted the landlord his possession order. The Court of Appeal held that it was not for them to review the earlier decision on the facts and that, as there was no error in law, they must dismiss the appeal.

Other grounds for possession are mandatory, where the court must order possession, giving a total of 10 discretionary grounds, six general mandatory grounds and three mandatory grounds confined to agricultural property and not considered here.

As with the exemptions of the Act, it is interesting to compare the grounds of possession with those applying under the 1988 Act. Given the intention of the 1988 Act to free the privately rented housing sector. It is not surprising that the grounds of possession are considerably broadened under the later enactment.

(a) Discretionary grounds:

(i)   Suitable alternative accommodation is available to the tenant. Such accommodation may be provided by the local housing authority or by the landlord himself (schedule 15, Part IV): see, for example, *Yoland Ltd* v *Reddington* (1981)

(ii)   Non-payment by the tenant of rent, or other breaches of covenant (schedule 15, Part I, case 1)

(iii)   Nuisance, or immoral or illegal use by the tenant (schedule 15, Part I, case 2)

(iv)   Deterioration of the dwelling house owing to neglect by the tenant or acts of waste (schedule 15, Part I, case 3)

(v)   Deterioration of furniture due to ill-treatment by the tenant (schedule 15, Part I, case 4)

(vi)   Where the tenant has given notice to quit and later changes his mind the landlord may be able to gain possession if, in the meantime, he has contracted to sell or let the house (schedule 15, Part I, case 5)

(vii)   Where the tenant has, without consent, assigned or sub-let the whole. As any assignment or sub-letting of the whole will put the tenancy outside the protection of the Act this case will only be of relevance during a contractual tenancy (schedule 15, Part I, case 6)

(viii) Where the premises are occupied by a service tenant and the premises are required for a new employee (schedule 15, Part I, case 8)

(ix)   Where the premises are reasonably required for occupation by the landlord, his son or daughter over 18, his father or mother, or the father or mother of his spouse, provided the landlord acquired the interest after a certain date (schedule 15, Part I, case 9)

(x)   Where the tenant sub-lets part of the house at a rent which exceeds the maximum rent recoverable under the Act (schedule 15, Part I, case 10).

(b) Mandatory grounds:

(i)   As an alternative to (ix) above, there is a mandatory right (schedule 15, Part II, Case 11, as amended by the 1980 Act, section 66 and schedule 7, and by the 1984 Rent (Amendment) Act) to possession by an owner-occupier who, at any time before the letting, occupied the house as his residence and let it on a regulated tenancy, giving notice that possession might be recovered on this ground, provided the court is satisfied that:

(1)   the dwelling is required as a residence for the owner-occupier or any member of his family who resided with him when he last occupied the house as his residence or

(2)   the owner has died, and the dwelling is required as a residence for a member of his family or

(3)   the owner has died and the house is required by a successor in title as his residence or for the purpose of disposing of it with vacant possession or

(4)   the dwelling is subject to a mortgage by deed granted before the tenancy, and the mortgagee is entitled to sell under a contractual or statutory power, needing possession to dispose of the property with vacant possession or

(5)   the dwelling is not reasonably suited to the landlord's needs in relation to his place of work and he requires possession in order to sell with vacant possession, using the proceeds to acquire a more suitable property.

(ii)   Where the landlord intends to occupy the house as his residence upon retirement, subject to certain conditions including the landlord having given notice that possession might be required under this ground, possession must be ordered (schedule 15, Part II, case 12).

(iii)   Where a holiday home is let out of season, the landlord may obtain possession where the dwelling is let under a tenancy not exceeding eight months and where he gave notice that possession might be required under this ground and at some time in the last 12 months the dwelling had been occupied for a holiday. This is a corollary ground to section 9 of the Act (which exempts holiday homes from protection) and enables a landlord to combine the letting of a holiday home with an out-of-season letting of the same property, without creating a protected tenancy (schedule 15, Part II, case 13).

(iv)   Vacation lettings by institutions of further education for a term not exceeding 12 months will not lead to protection provided the tenant was notified that possession might be recovered under this ground (schedule 15, Part II, case 14).

(v)   Where a house is required for occupation by a minister of religion as his residence and any current tenant is given notice not later than at the commencement of the tenancy that possession might be recovered on this ground (schedule 15, Part II, case 15), possession must be ordered by the court.

(vi)   Possession will be given where the landlord is a member of the armed forces and let a house after the Housing Act 1980, section 67, came into force, giving notice that possession might be recovered under this ground, where the landlord requires the house as a residence for himself or for other reasons similar to those stated under (i) above (schedule 15, Part II, case 20, introduced by section 67 of the 1980 Act).

2.4.3 Succession

Under schedule 1 para 2 of the 1977 Act, as amended by section 76 of the 1980 Act and section 39 of the 1988 Act, the statutory tenancy may be transmitted on the death of the original statutory tenant to the tenant's spouse. Under schedule 4 of the 1988 Act "a person who was living with the original tenant as his or her wife or husband shall be treated as a spouse of the original tenant".

Other members of the tenant's family who have been living with the tenant for at least two years prior to his death will also be entitled to succeed but will only obtain an assured tenancy.

2.4.4 Rent control

(a) Introduction

Coupled with the measures designed to protect a tenant from eviction are the parallel arrangements to restrict the rent paid, without which security of tenure would be a worthless gesture.

On the grant of a new tenancy of residential property not previously let, any rent can be agreed between the parties, but either party can then ask the rent officer to determine a fair rent for that property. The fair rent will be registered and becomes the maximum recoverable rent for the property under the existing, or any new, tenancy, unless and until it is properly changed, or unless the property is substantially altered. A registered rent can be altered by further application to the rent officer, such applications being considered at a minimum interval of two years, unless there has been a significant change in circumstances.

If the fair rent turns out to be higher than the rent agreed between the parties under their contract, the contractual rent will nonetheless continue to be paid for the agreed term.

Appeals against the rent officer's assessment are considered by a rent assessment committee, which includes a surveyor and which will either confirm the rent officer's figure or fix a new fair rent.

(b) Fair rent

(i) In determining, for the purpose of this Part of this Act, what rent is or would be a fair rent under a regulated tenancy of a dwelling house, regard shall be had to all the circumstances (other than personal circumstances) and in particular to
  (a) the age, character, locality and state of repair of the dwelling house and
  (b) if any furniture is provided for use under the tenancy, the quantity, quality and condition of the furniture.
(ii) For the purpose of the determination it shall be assumed that the number of persons seeking to become tenants of similar dwelling-houses in the locality on the terms (other than those relating to rent) of the regulated tenancy is not substantially greater than the number of such dwelling-houses in the locality which are available for letting on such terms.
(iii) There shall be disregarded:
  (a) any disrepair or other defect attributable to a failure by the tenant under the regulated tenancy or any predecessor in title of his to comply with any terms thereof
  (b) any improvement carried out, otherwise than in pursuance of the terms of the tenancy, by the tenant under the regulated tenancy or any predecessor in title of his: see *Smith's (Henry) Charity Trustees* v *Hemmings* (1981)
  (c) if any furniture is provided for use under the regulated tenancy, any improvement to the furniture by the tenant under the regulated tenancy or any predecessor in title of his or, as the case may be, any deterioration in the condition of the furniture due to any ill-treatment by the tenant, any person residing or lodging with him, or any sub-tenant of his.
(iv) In this section, "improvement" includes the replacement of any fixture or fitting.

(Rent Act 1977, section 70, as amended by the Housing Act 1980, section 26.)

This instruction to valuers appears to be largely based upon a commonsense interpretation of market rental value restricted by acts of the tenant (in a similar manner to parts of section 34 of the 1954 Landlord and Tenant Act in its application to business tenancies) with the single proviso that scarcity should be ignored.

This places the valuer in an impossible position, as it can be argued that without scarcity a product has no value. Following the introduction of this valuation basis it was the attitude of rent officers and the results of appeals against their decisions which have set the standard for the fixing of fair rents, often at what would appear to be a market rent less a deduction for scarcity of 10–25%.

Assessing the market rental value as a part of this exercise is aided by the increasingly large number of assured and assured shorthold tenancies, which provide good comparable evidence of market rents. Traditionally, however, precedent took the place of rationale and the approved method of determining a fair rent was simply by reference to registered fair rents for comparable properties (see *Mason* v *Skilling* (1974)). Where such evidence is hard to find a contractor's approach is sometimes adopted, in which case a percentage of the inclusive costs of construction is taken as a reasonable rental figure and discounted to allow for scarcity (see *Tormes Property Co Ltd* v *Landau* (1971)), but this is not to be relied upon.

A limit to the amount of deduction that can be made to take account of the poor state of repair or poor character of the house is indicated by the decision in *Williams* v *Khan* (1980) in which it was held that a house subject to a closing order due to its unfitness for human habitation was still worthy of a fair rent of £5.75 per week!

There is some evidence that the reduction in the re-registration interval from three years to two by the Housing Act 1980, and the abolition of the phasing of increases by the Rent (Relief from Phasing) Order 1987 had some effect in bringing rents closer to market levels, thereby encouraging private renting. During the period from 1980 to 1985 Department of Environment statistics show that fair rents increased at a greater rate than retail prices and house prices (Kemp 1988). The movement of fair rents closer to market rents was given a considerable boost by the Court of Appeal in *Spath Holme Ltd* v *Greater Manchester and Lancashire Rent Assessment Committee* (1995). At the original rent assessment committee (RAC), evidence of market rents were rejected, as had become common practice, as their level was assumed to reflect an element of scarcity. The Court of Appeal made it clear that the starting point for any fair rent must be market evidence, which should then be adjusted, for (a) any personal circumstances and (b) scarcity. The court instructed RAC's that a fair rent should be the same as a market rent in the absence of those two areas of influence. If it decided to depart from the market evidence, the RAC should explain its reasons and provide calculations. This precedent had the effect of increasing fair rents fixed by RAC's to a level at or close to open market rental value. The increase in the level of fair rents caused some concern for the position of tenants who could be faced with large increases in their rents, which may previously have been fixed at well below market value for many years. To address this problem the government brought in a system of transition, similar to that which applies to rates increases. The Rent Acts (Maximum Fair Rent) Order 1999 limits the amount of rent that can be charged by linking increases to the Retail Prices Index, plus an additional fixed percentage.

(c) Premiums

It would be facile for landlords to defeat the aims of rent control by agreeing to charge a fair rent while requiring that a premium is paid by the tenant. Consequently, section 119 of the 1977 Act makes it an offence to require a premium or loan as a condition of the granting, continuance, or renewal of a protected tenancy: see *Adair* v *Murrell* (1981). It is also normally an offence to receive a premium. This provision applies whether the premium is payable to a landlord or to an outgoing tenant, effectively killing any open market for tenancies.

Associated payments, such as deposits, excessive prices for furniture, advance payments of rent, and so on are also prohibited.

2.4.5 *Valuation*

The valuation of residential property which is subject to protection under the Rent Acts, should arguably be the preserve of the investment method of valuation in which the net income produced by the investment is capitalised at a rate of return which reflects the advantages and disadvantages of the investment. As with assured tenancies, this is not always the preferred method however, due to the vast difference in capital values between a vacant residential property and a tenanted one, which in the past may have attracted a value as low as 25%. Again, the percentage deduction will reflect the prospect of obtaining vacant possession. With no new regulated tenancies since the 1988 Act, tenants are generally ageing; the likely time before obtaining vacant possession is reducing; and sale prices are reflecting much lower deductions from vacant possession value

This difference — arguably a major reason for the undersupply of privately rented residential property which prompted the phasing out of regulated tenancies — is the result of the fixing of below-market rents and the reluctance of investors to accept low returns on an investment which is subject to increasing expenses of ownership. The effect of the disparity in values is that the valuer must concern himself with the prospect of the landlord obtaining vacant possession, thereby realising a capital gain, so that vacant possession value may have to be taken into account in the valuation.

The points to be borne in mind by the valuer concerned with residential property let on a regulated tenancy are as follows:

(a)   the income is relatively secure in money terms, due to the undersupply of residential property
(b)   the usual terms of a residential letting reserve the responsibility for external and structural repairs and insurance to the landlord. In older property this is likely to be of particular relevance
(c)   the income is subject to regular increases, although the quantum of such an increase is likely to be small and is subject to rent regulation
(d)   the prospect of vacant possession or conversion to an assured tenancy at a market rent must be evaluated with regard to the succession rules
(e)   a sale to the tenant should be considered as a possibility
(f)   the risk of voids is not relevant to the valuation of single units, as the prospect of vacant possession is likely to be a benefit
(g)   the valuation of multi-unit residential property (eg a block of flats) is subject to a different approach as the landlord may have no desire to obtain vacant possession due to the unattractiveness of such a property to a potential purchaser. Such valuations are also affected by the provisions of the 1987 Landlord and Tenant Act (see p 75). A realistic approach would be to carry out an investment valuation based on current rental income, which is likely to be a mixture of market and fair rents, with a reversion to market rents in respect of the regulated tenancies, based on the prospects of obtaining vacant possession, and re-letting at market rents, in each case.

Example 2.2

A 1930s semi-detached house is let on a regulated tenancy on the usual terms (tenant responsible for internal decorations only) to a fairly young tenant at a recently assessed fair rent of £80 per week. The vacant possession value is £180,000.

The valuation of the freehold interest in this type of property might take two forms:

(a)	Rent received		£4,160 pa	
	Less outgoings:			
	External repairs, say	£800		
	Insurance, say	£250		
	Management @ 10% rent	£416		
			£1,466 pa	
	Net income			£2,694 pa
	YP perp. @ 4%			25
	Capital value			£67,350
	Say			£67,500
(b)	Vacant possession value		£180,000	
	Take two-thirds		0.667	
				£120,000

Valuation (a) demonstrates that, partly because of the low rent, an investment valuation is likely to produce a very low figure, even though yields will be very low to reflect hope the prospect of obtaining vacant possession at some time in the future. Valuation (b) is simply an extension of the comparative method of valuation. It tends to be preferred in the market place, though it does depend on a consistent pattern of values being established, in relation to the value of owner-occupied property. In some markets, in fact, a remarkably uniform relationship might be discernible, justifying such an approach.

In theory, the valuer should depend upon an investment valuation, employing comparison as a check. In practice valuation (b) is almost invariably preferred.

Example 2.3

Two houses identical to the house in Example 2.2, both let on regulated tenancies, are to be valued. The first (No 1) is let to a man aged 94 who lives alone, while the tenant of No 2 is keen to purchase the freehold interest.

No 1: The prospect of obtaining vacant possession is of particular relevance in this valuation. The life expectancy of the tenant is (in probabilistic terms) between two and three years. After his death the possibility of succession should be considered. Does he have an estranged wife who might succeed him, or might a member of his family join him and live with him for a period of six months before his death?

It is suggested that should either of these be a possibility, the valuer would be over-optimistic in his valuation should he place an estimate far in excess of £120,000 upon the freehold interest.

On the other hand, assuming that the tenant has no qualifying family, or is a second successor himself, the value of the freehold interest must lie somewhere between £120,000 and £180,000 depending upon the attitude of a potential purchasing speculator. The valuer can apply a term and reversion approach as follows, increasing the deferment yield to reflect the possibility of longer occupation by the tenant. Again the market practice may well be for an arbitrary deduction from vacant possession value.

Rent received (net)	£2,694 pa		
YP 3 years @ 5%	2.72		
		£7,328	
Reversion to			
vacant possession value	£180,000		
PV 3 years @ 7%	0.816		
		£146,880	
Capital value			£154,208
Say			£154,000

No 2: The interest of the tenant in purchasing the freehold introduces the possibility of considerable marriage value being exploited. The purchase of an interest worth £120,000 on the open market will present the tenant with an asset worth £180,000.

The distribution of marriage value between the two parties to a transaction such as this is usually decided by the relative bargaining strengths of the parties. In this case, the tenant is in a strong position, as his interest in purchase probably reflects his long-term desire to remain in occupation, which he would otherwise achieve by paying a fair rent.

Consequently, a small increase over the market value of the freehold interest is likely to be agreed between the parties.

Capital value, say   £140,000

Example 2.4

The long leaseholder of a purpose-built block of 16 identical flats pays a nominal ground rent for the remaining 106 years of a 125-year lease. There are two vacant flats. Ten flats are let at the recently fixed fair rent of £90 per week, while the remaining four are let at a rent of £70 per week, registered two years ago. Tenants are liable for internal decorations only.

The new fair rent includes a service charge of £10 per week to cover the cost of heating, lighting, and cleaning the common parts. £7 was the corresponding charge in the old rent.

Value the interest:

(a) on the assumption that the block is incapable of being broken up

(b) on the assumption that the individual flats are saleable at a premium of £100,000 each if let on long sub-leases at nominal ground rents.

(a) Current total outgoings:

Services, say	£6,000	pa
Repairs, say	£5,000	pa
Insurance, say	£4,000	pa
	£15,000	pa

Current rental income if fully let:
$(12 \times £90 \times 52) + (4 \times £70 \times 52) = £70,720$ pa

Total rent upon re-registration of flats currently let at £30 per week
$(16 \times £90 \times 52) = £74,880$ pa

Valuation			
Rent received		£70,720	
Less			
Outgoings	£15,000		
Voids, say 5%	£3,536		
		£18,536	
Net income		£52,184	
YP perp. @ 7%		14.286	
			£745,500
Say			£750.000

(b) Two flats currently vacant:
     2 × £100,000 each   £200,000

     14 remaining flats:
     Either:
     Rent received upon re-registration of flats
     currently let at £30 per week
     (14 × £90 × 52) =   £65,520 pa
     Less:
     Outgoings (reduced pro rata)   £13,125
     (voids are now an advantage)
                                                                      £52,395 pa
     YP perp. @ 7%   14.286
                                                                     £748,515

     Or:
     14 flats if vacant @ £100,000   £1,400,000
     take, say, two-thirds to reflect risk and time to
     obtain vacant possession   £933,880
     take second option as it is higher   £933,880

     Total   £1,133,880

     Say   £1,135,000

While the receipt of a premium by a landlord or assignor is forbidden by the 1977 Act, a landlord is not prevented from offering a financial inducement to a tenant to give up possession (eg by paying for the deposit on the tenant's new home). Should the tenant be interested in such an arrangement, the valuation of the freehold interest must reflect that possibility.

Where the quantum of such an inducement has been agreed or is close to agreement, the value becomes:

Value of freehold in possession	x
Less:	
Inducement to tenant	y
Allowance for risk and delay	z
Value of freehold interest	$\dfrac{(y + z)}{x - (y + z)}$

The market for protected tenancies in residential property would no doubt flourish were it not for the measures preventing the payment of premiums upon assignment (section 120, Rent Act 1977). The discrepancy between the value of freehold interests subject to tenancies (£120,000 in Example 2.3) and those with vacant possession (£180,000 in the example) would suggest that the tenancy of such a property has a value equal to the difference.

However, because there is no legal market for such interests, the only relevance of this figure is to fix a ceiling to the inducements a landlord might offer to his tenant to vacate the property.

The attitude of the courts in preventing a market for tenancies becoming established is illustrated in *Farrell* v *Alexander* (1976). In this case a tenant (A) wished to assign her lease to B. Her tenancy, being protected rather than statutory, was subject to a condition that upon any assignment she must first offer the tenancy, without consideration, to the landlords, who were willing to grant a new tenancy to B. B agreed to pay £4,000 to A for fixtures and fittings on condition that A surrendered her lease to the landlord. B, on discovering that the fixtures and fittings were worth only £1,002, claimed the balance from A as an illegal premium, and the House of Lords confirmed that he was entitled to do so.

2.5 Restricted contracts

Before the plethora of legislation affecting residential property which was enacted in the early 1970s, consolidated in 1977, and amended in the 1980s, furnished lettings were treated by the legislation in a particular way and tenants of furnished property were afforded less protection than the tenants of unfurnished flats and houses. The Rent Act of 1974 converted most furnished tenancies into regulated tenancies, leaving a remainder of tenancies which became known as restricted contracts.

As a result of section 36 of the 1988 Act there will be no new restricted contracts and existing restricted contracts have been converted to assured tenancies as soon as the rent payable under the agreement was varied. This has had the effect of bringing about the demise of the restricted contract and, in view of this fact, the valuation of properties let subject to such tenancies is not considered in this book.

2.6 Shorthold tenancies

The shorthold tenancy is an innovation of the 1980 Housing Act and was designed to increase the supply of residential accommodation available to rent by overruling the tenant's right to remain in occupation as a regulated tenant. Its replacement by assured shorthold tenancies in the 1988 Act may be considered the result of its failure to achieve this objective, or a compliment to the success of this experiment, depending on personal opinion.

As with other tenancies subject to the protection of the 1977 Act, there will be no new shorthold tenancies but existing tenancies will be allowed to run their course. As a shorthold tenancy must be for a term certain of not more than five years, such lettings have died of natural causes.

2.7 Long tenancies

2.7.1 The law

Before the introduction of Part I of the 1954 Landlord and Tenant Act, long tenancies granted for more than 21 years at low rents fell outside the protection of the Rent Acts. The 1954 provisions brought them within this net.

The 1967 Leasehold Reform Act and the Leasehold Reform, Housing and Urban Development Act 1993 apply to largely identical tenancies and give much more valuable protection. This is dealt with in Chapters 3 and 4, where this type of tenancy is introduced in detail. The 1954 provisions continue to be of (minor) importance because the 1977 and 1988 Acts specifically exclude tenancies let at a low rent and because the 1967 and 1993 Acts are not always applicable. Section I of the 1954 Act provided for the application of full Rent Act protection on the termination of such tenancies providing they fell within the rateable value limits of the Rent Acts, otherwise there was no protection for the tenant. Provisions in the 1989 Local Government and Housing Act require that for long tenancies terminating after 15 January 1999, protection will instead be provided by the 1988 Housing Act, ie an assured tenancy at a market rent, and not an assured shorthold tenancy.

Section 2 of the 1954 Act describes the tenancies to which Part I applies. These are any long tenancies at a low rent where the "qualifying condition" is fulfilled. This condition is that "on the coming to an end of the tenancy at that time the tenant would, if the tenancy had not been one at a low rent, be entitled by virtue of the Rent Acts to retain possession of the whole or part of the property comprised in the tenancy". Applying section 22(3) of the Landlord and Tenant Act 1954 as an elucidation of this condition, it is clear that it is the circumstances subsisting at the end of the long lease that should be considered in order to determine whether the tenancy would receive the benefit of full Rent Act protection or not.

The 1954 provisions are of particular use to the qualifying tenant where, for some reason, he does not qualify for protection under the 1993 Leasehold Reform, Housing and Urban Development Act or the 1967 Leasehold Reform Act (where, for example he does not meet the residence requirements).

In such cases, the extent of the property covered by the 1954 Act might be the subject of dispute.

In *Herbert* v *Bryne* (1964) a complete house had been let in 1863 at a ground rent of £2 pa for 99 years. It was later divided into flats. Immediately before the lease was due to expire, the remainder of the term was acquired with vacant possession of the lower two floors, which the new tenant occupied. It was held that the whole property was covered by the Act and the tenants of the upper floors became sub-tenants of the litigant, who occupied the property as a regulated tenant.

In order to engineer the conversion of the long tenancy into an assured tenancy the landlord must serve a notice specifying that the tenancy will come to an end at the term date or a later date: section 4(1), Landlord and Tenant Act 1954. It will be in his interests to achieve this change at the term date as even a fair rent will exceed the original low rent. The landlord can serve a schedule of dilapidations at the end of the long tenancy and the tenant has a requirement to carry out initial repairs prior to commencement of the regulated or assured tenancy.

There are, however, grounds which the landlord can employ to recover possession at the end of the long tenancy, laid down in schedule 3 to the 1954 Act.

2.7.2 *Valuation*

The valuation of a freehold interest subject to a long tenancy at a low rent is primarily influenced by the potential application of the Leasehold Reform Acts which are described in Chapters 3 and 4. As those Acts only apply to lessees who have been resident for three years, the 1954 Act may be of greater concern where the tenant does not reside, cannot build up a sufficient period of residency prior to the termination of the lease, or could not finance the purchase of the freehold or a statutory lease extension. In these circumstances the 1954 Act can still play a material part in the valuation under the Leasehold Reform Acts.

Apart from the effects of the Leasehold Reform Acts, the valuation will fall into two parts. The term rents will continue for the remainder of he "long leases" and there will be a reversion to fair or market rents thereafter (see Example 2.4), taking into account any prospect of vacant possession at the end of the long leases.

2.8 Secure tenancies (public sector housing)

2.8.1 The law

The concept of the secure tenant was introduced by the 1980 Act and has been amended by the 1985 Housing Act and the 1986 Housing and Planning Act. A secure tenancy depends to a large extent upon the nature of the landlord, subject to the proviso that the tenant occupies the dwelling house as his only or principal home. The landlord must be a local authority, a new town corporation, an urban development corporation, the Development Board for Rural Wales, or a housing co-operative (1985 Act, section 80). The Act originally applied also to housing associations but the 1988 Act shifts housing associations across the barrier between public and private sector tenancies and any new lettings by such bodies will be assured tenancies. Pre-1989 housing association secure tenancies will be unaffected and a subsequent letting will be a secure tenancy where it is granted to a person who, immediately before it was entered into, was a secure tenant, and where the landlord also remains unchanged.

The most important type of tenancy falling within this definition is the local authority or new town letting: the letting of a "council house". Under the Housing Act of 1985 local authorities have a duty to review the housing conditions in their areas and a power to provide housing to ameliorate those conditions. Where they provide houses, they are empowered by section 24(1) of that Act to "make such reasonable charges for the tenancy or occupation of the houses as they may determine". In practice, "reasonable rents" have been loosely derived from private sector rentals, but are considerably below what might be expected as the open market rental value.

No market for secure tenancies is permitted as it is a condition of every such tenancy that the tenant shall not sub-let or part with possession of his council house (1985 Act, section 93) and if a secure tenancy is assigned it ceases to be a secure tenancy (1985 Act, section 91).

A major aim of the 1980 and 1985 Acts and the moving force behind the secure tenancy concept was the desire on the part of the Conservative government to give local authority tenants the right to buy their homes at a discounted price (see Part V, 1985 Act). It is this innovation which has had the greatest impact upon valuation in the public sector. This right to buy was originally available to a public sector tenant of least two years. In order to reduce the potential for profiteering the 2004 Housing Act introduced some restrictions on the right to buy provisions. These include a maximum limit on the discount available, an increase on the residency qualification limit from two years to five years, and an increase from three to five years on the period during which the discount must be repaid if the property is resold. Therefore if the tenancy began on or after 18 January 2005, the qualifying period is five years.

2.8.2 Valuation

Under section 126 and 127 of the 1985 Act, the price payable by a secure tenant for his home is the market value disregarding the tenant's improvements and the neglect of internal redecoration, less a discount.

The market value on the sale of the freehold is based on the assumption that the council sells an estate in fee simple with vacant possession, so that the presence of the sitting tenant is to be ignored, and discounting the fact that the tenant, or a member of his family residing with him, wants to buy the house. The result of these instructions is that the valuer must ignore the inflationary effect of the tenant's bid and any marriage value that might be exploited. These assumptions considerably ease the valuer's task: he is simply to estimate the open market value of the property with vacant possession, advisedly by reference to comparable evidence of sales of similar property in the open market.

This was difficult in the early days of the right to buy legislation, as it was difficult to compare the value of houses in a council estate with similar property on a private estate. There is, however, now a well-established resale market for former council houses which provides sound comparable evidence of open market values.

The discount to be deducted is governed by section 129 of the 1985 Act and depends upon the tenant's (or the tenant's spouse's) period of occupation of the house as a secure tenant. Where this is less than three years, the discount for houses is 32%: where it is three years or more, the discount is 32% plus 1% for each complete year by which that period exceeds two years, subject to a maximum deduction of 60%. For flats, the range of discounts is from 44% to 70%. The discount cannot exceed the regional upper limits, which, in England and Wales, range from £16,000 to £38,000.

A secure or assured tenant of a registered social landlord, for example, a housing association or a local housing company, may have the right to buy their home under a different scheme called the right to acquire. The right to acquire only applies to a limited number of properties, for example, homes built with public funds on or after 1 April 1997.

Example 2.8

A secure council tenant wishes to buy his home, which is a three-bedroomed semi-detached house on a council estate. He has occupied the property as a secure tenant for the last 11 years.

Similar houses not on council estates sell for around £175,000 and there is some evidence that upon the resale of acquired council houses somewhat lower prices are obtained.

The tenant has extended his home to the rear to provide a utility room.

Advise him as to the price he will be asked to pay.

Open market value, say	£150,000	
Less value of tenant's improvements, say	£7,500	
		£142,500
Less discount (32% + 9% = 41%)		£58,425
But maximum discount for the area will apply, say		£38,000
Price		£104,500

If the dwelling house is a flat, the tenant has the right to be granted a long lease, probably of 125 years, the ground rent not exceeding £10 pa. The estimation of the price to be paid upon the grant of such a lease follows similar lines to Example 2.8.

2.9 Housing Act 2004

The 2004 Housing Act introduced a number of changes, particularly in relation to houses in multiple occupation (HMOs). A mandatory licensing scheme is introduced for large HMOs, which are defined as properties over three or more stories with five or more occupants in at least two households. These include houses flats and hostels where there is a degree of sharing of facilities. There are a number of exemptions, including buildings owned by registered social landlords, and local housing authorities. (LHAs) have some flexibility to extend the licensing scheme within their area. The LHA will only issue a licence to a person considered to be a fit and proper person. It will last for up to five years and will impose conditions in respect of matters such as the maximum number of occupants, as well as safety measures. The licence can be revoked if there is a serious breach, or the licence holder is no longer considered fit and proper. An appeal against the conditions of a licence, or a revocation, may be made to the Residential Property Tribunal.

Some concern has been expressed that the licensing provisions of the Act could deter landlords, and thereby reduce the availability of accommodation in hostels and other large shared buildings. It is too early to assess whether this is the case, but it does appear that the market for such properties was adversely affected during the period immediately prior to, and after, the introduction of the new licensing regime. It may be that there will be long term fall in demand, and therefore value, in respect of certain classes of property which are generally valued by the investment approach eg hostels catering for asylum seekers or tenants receiving housing benefit. The impact on the value of many licensed HMOs is, however, likely to be less severe as tenants are unlikely to have long term security of tenure, and the highest value may well be based on vacant possession value, less an allowance for the time and risk involved in obtaining vacant possession

2.10 Introductory Tenancies

Introductory tenancies are a new form of tenancy introduced by section 124 of the 1996 Act. A local housing authority or housing action trust may opt to operate an introductory tenancy regime under which all lettings by that body will be introductory tenancies, unless the tenant was previously a secure tenant or an assured tenant of a registered social landlord, other than under an assured shorthold tenancy. Under an introductory tenancy there is a one-year trial period during which there are strict procedural rules which the landlord must comply with if he wishes to seek possession.

As introductory tenancies belong in the public, rather than the private sector, they are not considered in detail in this book.

2.11 Tenants' rights of first refusal

The Landlord and Tenant Act 1987, as amended by the Housing Act 1996 Act, gives tenants of privately owned flats rights which considerably complicate the sale of this type of property and may well have a significant effect on values.

Part I of the Act gives tenants a right of pre-emption should the landlord decide to sell his interest. Where a landlord wishes to dispose of a building or part of a building which is let in flats, he must first serve a notice on the tenants setting out the terms of the sale and offering to sell to the tenants on those terms. The tenants then have several courses open to them:

(a) where the requisite majority of tenants wish to accept the offer they may serve a notice to that effect on the landlord who then cannot sell the property except to a person or body nominated by the tenants

(b) where the tenants do not serve a notice accepting the offer the landlord may proceed to sell the property privately at any time during the next 12 months providing the sale price is not less than that specified in the notice

(c) the tenants may serve a counter-notice on the landlord outlining terms acceptable to them. This may lead to further negotiations or the landlord may reject the offer and proceed to sell the property privately as if his offer had been refused.

If the correct procedure is not followed the new owner may find himself forced to sell to the tenants at the price he paid.

The provisions apply to buildings or part of a building which is no more than 50% in non-residential use, contains two or more flats and more than 50% of the flats are held by qualifying tenants.

Qualifying tenants include regulated tenants under the Rent Acts and tenants under long leases at a low rent. Shorthold tenants, assured tenants, business tenants, and service tenants are excluded.

Some landlords are exempt including local authorities, the Crown, housing associations, and resident landlords.

The 1987 Landlord and Tenant Act gives other important rights to tenants including the appointment of a manager or the compulsory acquisition of their landlord's interest where the landlord is in breach of his obligations.

Taken overall, the Act imposes an increased administrative burden on private landlords of qualifying properties, particularly on disposal, and is a further disincentive to investment in this sector. There will be circumstances where, but for the Act, it would have been possible to obtain a special value on a sale to the sitting tenants by exploiting the marriage value released by merging the interests of landlord and tenant. This opportunity is more difficult to exploit as a result of the procedures required under the 1987 Act. However, as assured and assured shortholds have become the norm, there are fewer qualifying tenants and the right of first refusal will rarely apply.

2.12 Residential property outside the Rent Acts

Tenanted residential property might fall outside the protection of the Rent Acts for several reasons, the major exception being properties which have had rateable values too high to bring the property within protection. A common example of this class of excepted property is the flat let in a fashionable area of central London. The valuation of such a property is likely to consist of two parts: first, the existing rent is capitalised for the remainder of the lease; second, there is a reversion to vacant possession value.

Other unprotected lettings include licences and holiday lettings. The valuation of such property is carried out by direct capital comparison with similar properties sold with vacant possession, perhaps with a deduction to allow for the risk, expense, and/or time delay involved where a tenant or licensee might not agree to immediate vacation of the property.

A final category is owner-occupied residential property. By contrast, this staple diet of estate agents is valued in splendid isolation (albeit by comparison) without an explicit allowance for the normally seismic effect of statutory interference.

Residential Property: Leasehold Reform — The Rights

3.1 Introduction: the background to the legislation

The leasehold reform legislation was first applied to houses and grew out of the unease amongst residential long leaseholders in the 1960s, when the building leases granted during the period of expansion of towns in the mid- to late-19th century, especially in London and certain other regions such as South Wales, were about the expire. The intention of the freeholder when granting the 99-year lease had been that the land would revert to him to arrange for the anticipated redevelopment to take place.

However several factors undermined the original plan. First, most buildings, if well maintained and periodically refurbished, have a useful life well beyond that anticipated for their component parts. Second, a residential property is a home as well as an investment and the lessee might not want to leave at the lease end. Third, although the tenants were afforded protection from eviction, their rents were poised to increase from Victorian ground rent levels (not reviewable) to 1960s fair rents, a multifold rise in the majority of cases.

The legislation which ensued, the Leasehold Reform Act 1967 (the 1967 Act), gave resident leaseholders a right to compulsorily acquire the house they occupied and the land on which it stands. The principle underlying the legislation was challenged at the European Court in 1979. When the Grosvenor Estate challenged the validity of the Act[1] saying that it was in breach of the European Convention of Human Rights.[2] However although the European Court held that Grosvenor had indeed "been deprived of their possessions", they decided it was "in the public interest" and therefore not a breach. The court also held that article 1 did not guarantee the right to full compensation in all circumstances and since it was Parliament's view that "in equity the bricks and mortar belong to the leaseholder", the Act afforded a fair balance between the interest of society and the landlord's right of property, so the level of compensation payable in valuations under section 9(1) of the 1967 Act was not unjust.

---

1   (*Petition application 8793/79, James v United Kingdom*, decision dated 21 February 1986)
2   Article 1 of the First Protocol: "Every natural or legal person is entitled to the peaceful enjoyment of his possessions. No-one shall be deprived of his possessions except in the public interest and subject to the conditions provided for by law and by the general principles of international law."

Soon after 1967 the mission started to extend the enfranchisement principle to flats. Prior to the Second World War the sale of flats for owner occupation was virtually unknown. Bomb damage in the War meant that housing stocks were low. Incomes were increasing as reconstruction of the economy meant that there was full employment, and consumers' spending expectations were high. Conservative governments actively pursued policies encouraging home ownership, but this put severe pressure on housing stocks and flat dwelling became an acceptable lifestyle for all classes. The sale of flats blossomed; first in purpose built blocks then, as private investment in accommodation for rental dwindled, in what became known as break-ups of previously rented mansions and other types of flats. In the early 1980s, the rise in flat ownership was further fuelled by Mrs Thatcher's right-to-buy policies, in which she realised her research-based belief that homeowners would not vote Labour.

Prior to 2002, in the absence of a commonhold or condominium system, there was no truly satisfactory alternative to leasehold as a form of tenure for the ownership of flats. This is because positive covenants are unenforceable under freehold tenure, except against the original purchaser. It is fundamental to the ownership of a flat that the others in a block are positively repaired and maintained and that they provide support. Therefore, virtually all flats in England and Wales are held leasehold. Freehold flats are regarded with great suspicion, and do not normally provide satisfactory security for a bank or building society loan. While house ownership is usually on a freehold basis and is of the property itself, ie the bricks and mortar and the land, with a leasehold flat you only buy a right to live in it for the unexpired term of the lease.

Problems also arose associated with the personalities in a block. Inevitably most landlords are not resident in their blocks and they are often perceived as being like the notorious absentee landlords of Ireland, with no great interest in the property other than what they can get out of it. The leaseholder rarely appreciates that in a block of flats you cannot do just as you would like, but have to conform to standards and patterns set by others. You live in close proximity to other people and, while you should all share a common interest in the block just as in a co-operative, that is not always the case. You must contribute to repair and redecoration costs when they arise, put up with the decorations in the common parts that are generally acceptable, and live with the quirks, visitors and children of neighbours.

During the 1970s Members of Parliament found an increasing percentage of their post bag related to badly managed flats which were also wasting assets. To the non-professionals the main culprit appeared to be the leasehold system itself. Practitioners know that the reasons for bad management are manifold and include badly drawn leases as well as incompetent agents and unscrupulous landlords. However, it appeared to the non-professionals that if there were no leases that would solve the wasting asset problem and there would be no landlords; wicked, incompetent or otherwise. If flat owners could manage their block themselves, or choose their managing agents, they could make sure that they got the standard of service they required.

Over the years there were several attempts to make landlords more accountable to tenants and in particular, to control service charges to a reasonable level. Theses were consolidated in the Landlord and Tenant Act 1985. Part I of the Landlord and Tenant Act 1987 gave residents a right of first refusal if the landlord's interest was sold, and a right of compulsory purchase under Part III if the management was truly awful.

In 1987 the Law Commission recommended a new form of tenure: commonhold and, in November 1990, the Lord Chancellor's Department issued a consultation paper and draft Bill, though commonhold was then sidelined in preference to reforms of the legal system.

In 1991 the Department of the Environment put out proposals for lessees of flats to have rights to enfranchise blocks of flats, and for lessees who could not take part in an enfranchisement, to be able to buy extended leases. In the run-up to the 1992 general election, the Conservative government promised the leasehold reform proposals and after a stormy reception from some Lords, and some fairly radical revisions to the original proposals, the Leasehold Reform, Housing and Urban Development Act 1993 was passed and came into effect on 1 November that year.

The first amendments to the Act were made by the Housing Act 1996. These were generally to close up a loophole that had been discovered, whereby a block held under more than one freehold interest was not enfranchiseable, and to clarify some vagaries. In line with the alterations to the 1967 Act, the low rent test was abolished for leases of over 35 years.

Commonhold was introduced by the Commonhold and Leasehold Reform Act 2002, which also made amendments to the 1993 Act to bring commonality to the processes for houses and flats.

The rights, and the process of exercising the rights, are complex. A brief summary follows. For a full explanation see *Hague on Enfranchisement* Radevsky & Greenish. There are also useful publications on the web-site of LEASE, the government funded advice service (*www.lease-advice.org*).

3.2 The legislative framework

3.2.1 1967 Act

Since 1967, the principles of the original Act have been amended to widen its application or to nullify the effect of loopholes upheld by the courts. The subsequent legislation and its principal effects is as follows:

- *Housing Act 1969*
  Excluded the tenant's overbid from valuations under section 9(1).

- *Housing Act 1974 (the 1974 Act)*
  Extended the application of the Act to houses having a rateable value of between £1,000 and £1,500 in Greater London or £500 to £750 elsewhere, with new valuation rules in section 9(1A).

- *Leasehold Reform Act 1979*
  Prevented the enfranchisement price being increased by the creation of a superior interest.

- *Housing Act 1980*
  Reduced the period of occupation for residency from five years to three years, introduced the formula for the valuation of minor intermediate interests, set up Leasehold Valuation Tribunals (LVTs) as valuation courts of first instance with only appeals being heard by the Lands Tribunal.

- *Housing Act 1985*
  Extended the Act to former secure tenants who had exercised the right-to-buy

- *Housing & Planning Act 1986*
  Extended the Act to leaseholders under shared ownership leases provided their ownership was 100% and plugged the 'Hickman' loophole whereby lessees had been able to secure large reductions in the enfranchisement price by opting first for a statutory 50-year lease extension, and then enfranchising.

- *References to Rating (Housing) Regulations 1990* (SI 90/434)
  Made amendments to the rateable value and low rent tests consequent upon the abolition of rateable values

- *Leasehold Reform, Housing & Urban Development Act 1993* (the 1993 Act)
  Abolished the rateable value test, amendments to low rent and long tenancy tests (providing the sections 1A and 1B rights to enfranchise) and introduction of section 9(1C) valuations.

- *Housing Act 1996* (the 1996 Act)
  Abolished the low rent test for houses held on leases granted for terms in excess of 35 years (providing the section 1AA right to enfranchise).

- *Commonhold and Leasehold Reform Act 2002* (the 2002 Act)
  Removed the residency qualification, allowed holders of section 14 extended leases to enfranchise, fixed the share of marriage value payable at 50% where the unexpired term is below 80 years and at nil above 80 years.

3.2.2 1993 Act

Since 1993 the principles of the original Act have been amended to widen its application or to nullify the effect of loopholes upheld by the courts. The subsequent legislation and its principal effects is as follows:

- *Housing Act 1996* (the 1996 Act)
  Removed the requirement for a professional valuation to be carried out before an enfranchisement claim is made, enabled blocks held under more than one freehold to be enfranchised, abolished the low rent test for flats held on leases granted for terms in excess of 35 years, provided for the claimant's interests to valued on similar assumption to those adopted in valuing the freeholder's interests

- *Commonhold and Leasehold Reform Act 2002* (the 2002 Act)
  Abolished the residency qualification and replaced it only for lease extensions with an ownership qualification, abolished the low rent test for leases granted for over 35 years, fixed the share of marriage value payable at 50% where the unexpired term is below 80 years and at nil above 80 years, fixed the date of valuation in all cases at the date of claim, extended the right to a new lease to the personal representatives of deceased lessees, removed the right to enfranchise for premises including railway track, reduced the proportion of qualifying lessees required to participate to 50% of all the flats, increased the permissible non-residential element of a block to 25%.

3.3 The rights available

3.3.1 Houses

Lessees who qualifying by virtue of sections 1A, 1AA or 1B (the 1993 and 1996 Act amendments), only have the right to purchase the freehold interest and any intermediate leasehold interests.

Claimants under previous legislation have the option of either enfranchising, or taking a 50 year lease commencing with expiry of the current lease at a then current market rent with a review after 25 years.

The extended lease is very much a poor relation to the right to enfranchise for several reasons:

(a) The ground rent during the 50 years of the term of the extended lease is likely to be very high in relation to the ground rent under the original lease: in some cases the ground rent will be higher than the fair rent. This will depress the value of the leasehold interest.

(b) There is no right to remain in possession under Part I of the Landlord and Tenant Act 1954 or schedule 10 of the Housing Act 1988 at the end of the extended term.

(c) The lessee under the extended lease has a right to enfranchise, but since nothing was paid for the new lease, it is at a price based on the assumption that the section 14 lease had not been granted.

3.2.2 Flats

Under Chapter II of the 1993 Act, a qualifying lessee has an individual right to buy a new lease which is 90 years longer than the existing lease at a peppercorn ground rent throughout the term. The existing lease is surrendered. There is nothing any of his neighbours can do to curtail or restrain that individual right.

The new lease is granted by the landlord who has a sufficient reversion (the competent landlord) direct to the claimant. The interests of any intermediate landlords are preserved by a deemed surrender and regrant of their intermediate leases. The terms of the new lease should follow the existing lease, but there is provision for the lease to be brought up to date, for example by including a comprehensive service charge from the date of expiry of the original lease. In other words if the lessee benefits at present from not having to pay a proper service charge, they will not lose that benefit by buying the lease extension. It is one way in which a lessee with a lease which is so defective that it does not provide suitable security for a loan, can rectify the situation. However, the low value of the current defective lease, and the improved value after the grant of the new lease will be fully reflected in the price payable.

The right to a lease extension can be exercised repeatedly, and thus can be used to obtain a very long leasehold interest.

If a claim for a longer lease is made, and that is followed by a claim to enfranchise the building in which the flat is located, the lease extension claim is frozen until the enfranchisement has taken place. It is then reactivated with the new freeholder.

Lease extensions are available to qualifying residents of flats (see 4.3) in any type of building.

The lessee has to meet the vendors' legal and valuation costs in the purchase, and pay all amounts owing on completion of the lease.

3.3.3 Blocks of flats

Under Chapter I of the 1993 Act, each qualifying lessee also has a collective right, exercisable in conjunction with his neighbours, to take part in an enfranchisement, ie the purchase, of the freehold and any intermediate leasehold interests in the building of which his flat forms part. The right may also extend to the gardens, grounds and appurtenant property.

In an enfranchisement, the leaseholder residents remain lessees, simply swapping landlords. Participants in the enfranchisement are usually also joint owners of the freehold, and would normally grant themselves longer leases at peppercorn ground rents "back-to-back" with the purchase of the freehold.

Where there are non-qualifying units, the participants must offer to buy those units even though the freeholder can claim leasebacks of them (for 999 years at peppercorn ground rents).

3.4 Qualification

3.4.1 To enfranchise a house

There are three tests to qualify to exercise the right of compulsory purchase. They are:

(a) the property leased must be a house: see 3.6 below
(b) the tenancy must be a long tenancy: see 3.6 below
(c) the tenant must have owned the lease for more than two years.

If the lease of the house was granted for a term of 35 years or less, there is also a low rent test: see 3.6 below.

3.4.2 To extend the lease of an individual flat

There are three tests to qualify to exercise the right of purchase of an extended lease and a fourth if the lease of the flat was granted for a term of 35 years or less. They are:

(a) the property leased must be a flat
(b) the tenancy must be a long tenancy
(c) the tenant must have owned the lease for more than two years.

If the lease of the flat was granted for a term of 35 years or less, there is also a low rent test.

3.4.3 To enfranchise a block of flats

The following are the tests as to whether the block itself can be enfranchised.

(a) It must be a self-contained building with vertical separation from other buildings and capable of being redeveloped independently. If the building shares services with other premises, they must be capable of being provided independently "without involving the carrying out of any works likely to result in significant interruption in the provision of such services for the occupiers of the remainder of the building".[3] This means that an enfranchiseable block need not be detached; semi-detached, or terraced, and even parts of blocks may be enfranchiseable.

Since a flat includes a maisonette and other properties which do not satisfy the definition of a house under the 1967 Act, two or more abutting houses from which garages, for example, have been excluded, may form a block of flats. The possibilities are endless.

---

3   Section 3(2) of the 1993 Act.

(b) There must be two or more flats in the block. Therefore a mews house (from which a garage has been excluded) and the excluded garage are not a block of flats, but two mews houses and two garages could be.

(c) At least 75% of the internal floor area of the building must be in residential use. This includes all areas used or intended for use in conjunction with a dwelling and also dwellings used in conjunction with commercial premises.[4] Common parts are excluded from the calculation of internal floor area, as are areas that are not wholly enclosed.[5] No method of measurement is prescribed but the RICS Code of Practice has been accepted as suitable. Residential premises are measured to gross internal area.

(d) Blocks of up to four flats where the landlord is resident are excluded from the Act.

(e) Two thirds of flats in the block must be let to qualifying tenants.

Blocks which are physically separate from one another and capable of being redeveloped individually must be enfranchised separately, even though they share common facilities or are situated in a private estate.

The number of participants in the claim must be not less than half of the total number of flats, and they must all be qualifying tenants. Participation involves signing the initial notice of claim and thus sharing responsibility for the vendors' costs. There is no minimum period of ownership required before a qualifying tenant can participate in an enfranchisement.

Example 3.1: How do the tests work in practice in relation to a block of 14 flats with a resident caretaker's flat?

Test 1 (qualification): Are two thirds let to qualifying tenants?
There are 15 flats in total. Two thirds of 15 = 10. So 10 flats must be let to qualifying tenants. Let us assume that only 10 of the units are let to qualifying tenants.

Test 2 (participation): The minimum number of participants is 50% of 15 = 7.5, rounded up to 8. Therefore 8 qualifying tenants must be willing to participate.
So as long as 8 of the 10 qualifying tenants can be found to sign the claim, it can proceed.

3.5 Exemptions

3.5.1 House

The right to enfranchise cannot be exercised where the freehold is owned by:

(i) the Crown
(ii) the National Trust
(iii) some public authorities (if a minister of the Crown certifies that they will require the house for redevelopment) or
(iv) a charitable housing trust.

---

4   *WHRA RTM Co Ltd* v *Gaingold Ltd* [2004] an unreported LVT case.
5   *Indiana Investments* v *Taylor* [2004] 3 EGLR 63.

A lessor can resist enfranchisement by showing that he requires the house for his own, or his close family's occupation, but not that he intends to redevelop it. However, redevelopment can be used as a means of preventing the tenant from gaining a 50 year extension to his lease.

3.5.2 Block of flats

The right to enfranchise cannot be exercised where the freehold:

(i)   belongs to the Crown
(ii) is in a cathedral precinct
(iii) is one of the inalienable properties of the National Trust or
(iv) includes operational railway track. Since premises have to consist of a building, the exclusion only applies if part of the building (including the foundations) includes a railway track. Track in a tunnel running through the ground beneath a building, does not exclude the building.

3.5.3 Flat

The right to purchase an extended lease cannot be exercised if the freehold:

(i)   belongs to the Crown
(ii) is in a cathedral precinct or
(iii) is one of the inalienable properties of the National Trust.

3.6 Definitions

3.6.1 House

The use of the property need not be exclusively residential but the property must be "reasonably so called". It need not have been purpose-built as a residence, but the mere fact that it is being used as a residence is not, of itself, sufficient. The definition of a house is undergoing detailed testing at the time of writing, following the removal of the residency qualification by the 2002 Act. However, under the pre-2002 Act legislation, a property was held to be a house even though the ground floor was used as a shop (but otherwise the accommodation was residential) (*Lake* v *Bennett* (1970)). Even a purpose-built shop with living accommodation above has been held to be a house (*Tandon* v *Trustees of Spurgeons Homes* (1982).

A property which was so lacking in facilities as to be incapable of occupation as a house was held not to be a house (*Boss Holdings* v *Grosvenor* (2006)), and a former house converted for a commercial use was also held not to satisfy the definition (*Mallett* v *Grosvenor* (2006) and *Colliers CRE* v *Portman* (2006)).

It does not matter whether the house has been sub-divided into flats and is also subject to the 1993 Act right to enfranchise, provided that the sub-divisions are not in themselves "houses" because houses cannot be grouped together in an enfranchisement (*Malekshad* v *Howard de Walden Estates Ltd* (2003)).

The separation between the house and other property must be vertical. If a substantial part lies over or under another property it will not be a house (*Parsons* v *Trustees of Henry Smith's Charity*, 1974,

Birrane v *Grosvenor Estate* (1995) and *Malekshad* v *Howard de Walden Estates Ltd* (2003)). Thus a mews house where a garage is not included in the lease of the house, or which is built over the cellar of another house, is not a house within the meaning of the Act.

3.6.2 Flat

A separate set of premises, whether or not on the same floor, constructed or adapted for use as a dwelling, where either the whole or a material part lies above or below some other part of the building.

3.6.3 Qualifying tenant

A qualifying tenant must have a particularly long term or a long lease at a low rent.

3.6.4 Long lease

A long tenancy is one granted for 21 years or more which is not determinable within that time. The 21 years runs from the date the lease was granted if that post-dates the date on which the term commenced. References to the death or marriage of the Prince of Wales are no longer relevant termination provisions.

Where a long tenancy is surrendered in return for a new long tenancy, the two terms should be treated as if there had been a single tenancy starting with the commencement of the earlier of the two tenancies.

3.6.5 Particularly long term

Broadly a lease granted for more than 35 years.

3.6.6 Low rents for houses

There are now only a few cases where a test remains applicable. They are:

1   to determine whether the tenancy is excluded under section 1AA
2   to determine whether section 3(2) applies (ie whether an extended tenancy is a long tenancy at a low rent)
3   where a claim is to be made for a section 14 extended lease
4   to determine whether the price is to be assessed under section 9(1): see also section 4.25.

The various tests of low rent are in sections 4 and 4A, the latter introduced by the 1993 Act.

3.7 Ownership

All three rights were originally granted only to lessees who resided in the property concerned. Not being persons, companies could not reside so it was easy to avoid enfranchisebility by insisting that leases were taken in the name of companies. In 2002 the residency test was replaced, in respect of enfranchisement of houses and the purchase of lease extensions, by an ownership test. There was no replacement in respect of enfranchisement, where now neither residency nor a period of ownership is a requirement. Since the introduction of the 2002 Act, this means that company lessees can enfranchise.

For houses and flats, the tenant must have been the registered owner for two years prior to the date of claim. The ownership runs from formal completion of the purchase, ie from registration at the Land Registry, not from completion of the sale.

The residency test now only relates in two circumstances, both of which relate to houses

(a) when a flat forming part of a house is let to a qualifying tenant, the tenant of the house must satisfy a residency test
(b) when Part 2 of the Landlord and Tenant Act 1954 applies to the tenancy.

The test is that the tenant must have been in occupation of at least part of the house as his main residence for the last two years, or for periods amounting to two years in the last ten. That occupation can have been on any basis, for example as a long lessee, as a regulated tenant or as a member of the family of a former lessee.

3.8 Rateable Value

The only relevance of the former rateable value is in relation to houses and is:

(i)   as a qualifying condition to extend the lease: see 3.3
(ii)   in relation to the low rent test where that still applies: see 3.6 and
(iii) to establish under which set of valuation rules, the enfranchisement price is to be assessed: see 3.9.

Since the claimant is to have the benefit of any tenant's improvements (per section 9 (1A)(d)), the valuation rules are assessed on the notional (reduced) rateable value of the unimproved property. Schedule 8 to the 1974 Act sets out the procedure to obtain the notional RV from the district valuer if it cannot be agreed with the freeholder (see *Pearlman* v *Keepers and Governors of Harrow School*). This is useful if the notional rateable value can be reduced below £1000/£500, so that section 9(1) applies rather than section 9(1A).

3.9 Valuation Rules

The valuations are considered in detail in the next chapter, but the rules governing those valuations are as follows.

3.9.1 Houses

There are four sets of valuation rules to establish the enfranchisement price. They apply as follows:

section 9(1):   if the house could be enfranchised under the 1967 Act unamended ie had a rateable value under £1,000 in Greater London or £500 elsewhere (see 3.10) and satisfied the original tests of low rent which were, put simply:

1: if the rateable value was less than £1,500/£750 or if the tenancy was granted between 1 September 1939 and 31 August 1963 on a building lease and the rateable value was less than £1,500/£750, the test is less than two thirds of the rateable value on the "appropriate day" or the first day of the term

2: if the tenancy was granted between 1 September 1939 and 31 August 1963 on a building lease and the rateable value was more than £1,500/£750 or if the tenancy was granted before 1 April 1964 and the rateable value was above £1,500/£750, the test is equal to or less than two thirds of the letting value of the property (on the same terms) at commencement of the tenancy

3: if the tenancy was granted between 1 April 1963 and 31 March 1990 and the property had a rateable value before 1 April 1990, the test is equal to or less than two thirds of the rateable value at commencement or the date when it first had a rateable value

4: if the tenancy was granted between 1 April 1963 and 31 March 1990 and the property never had a rateable value or if the tenancy was granted after 1 April 1990, the test is equal to or less than £1,000 pa in Greater London or £250 elsewhere.

section 9(1A): if the house could only be enfranchised by virtue of the amendments introduced by the 1974 Act ie had a rateable value of between £1001 and £1500 in Greater London or £501 and £750 elsewhere (see 3.11) and satisfied the original four tests.

section 9(1C): if the house could only be enfranchised by virtue of the amendments introduced by the 1993 and 1996 Acts ie under sections 1A, 1AA or 1B (see 3.13), eg had a high rateable value or did not have a low rent.

section 9(1AA): if the right to a section 14 lease extension has been taken up, the new lease has started and the rateable value is such that the valuation is to be under section 9(1A).

See also section 3.8 as to the effect of tenant's improvements on rateable value and, therefore, on the applicable rules.

3.9.2 Flats

The basis of assessment of the premium payable for the new lease is set out in schedule 13 to the 1993 Act.

3.9.3 Blocks of Flats

The basis of assessment of the price payable for the freehold is set out in schedule 6 to the 1993 Act. In all cases the freeholder's interest and that of each intermediate leaseholder has to be purchased separately. The price for each comprises the aggregate of:

1. the open market value of interest
2. a share of the marriage value
3. any compensation for losses to other properties retained resulting from the enfranchisement.

This formula is similar to the elements which make up the enfranchisement price under section 9(1C) of the Leasehold Reform Act 1967 (see section 4.2) and echo the way in which the price was set under section 9(1A).

3.9.4 Valuation Date

The valuation date in all cases has been the date of claim since February 2005. Generally speaking, this is to the landlord's disadvantage, particularly if the market is rising or the lease is short. By the time the purchase is completed, the purchaser could have made a substantial profit simply from the operation of a rising market. No interest is payable on the purchase price, but the rent continues to be payable to the date of completion, even though the landlord has been compensated for the loss of the rent from the date of claim.

3.10 Mechanics of purchase in enfranchisement

Because there may be many parties involved, both on the purchasers' side and on that of the vendor(s), the Act provides in effect, for each side to have a single voice. The residents operate through a nominee purchaser which is the party that will acquire the freehold. The vendors are represented by the reversioner who is the lessor with the longest interest and is usually the freeholder.

Up to four persons can hold an interest in land so, if more people are involved, eg when there are more than four units in the block, or if the units have joint owners and the aggregate is more than four, then another vehicle must be used. The most common is a company, limited by share or by guarantee, but it is possible to use a form of trust.

An intermediate landlord can give notice to act independently, but not until after the counter-notice has been served. If no such notice is given, the reversioner owes a duty of care to the intermediate landlords.

3.11 Intermediate interests

In many cases, there is a series of tenancies between the tenants in occupation and the freeholder. In enfranchisement each interest is compulsorily acquired.

In lease extension, the competent landlord who grants the new lease is that landlord who is able to grant a lease which is 90 years longer than the current lease. There are deemed surrenders and re-grants of all intermediate leasehold interests.

Each intermediate landlord is entitled to be represented individually, and to have the open market value of his interest and any compensation due to him assessed separately. However, when it comes to the vendors' share of marriage value, there is only one assessment made and the vendors' share is apportioned amongst them in proportion to the value of their interests.

Where an intermediate interest has an expectation of possession of not more than one month and a profit rent of not more than £5 pa, it is termed a minor superior tenancy under the 1967 Act and a minor intermediate leasehold under the 1993 Act (commonly known as a MILI). The price payable for a MILI is calculated in accordance with the formula set out in the Acts which is considered in detail in section 4.20.

Intermediate leaseholds which are not MILIs are valued in accordance with the rules for valuing the freehold.

3.12 Rentcharges

A rentcharge is a sum of money payable annually to the former owner of an interest in land. It can be secured against a freehold or a leasehold interest, and it can be imposed in perpetuity or for a fixed period of time. The 1977 Rentcharges Act provides for the eventual expungement of rentcharges

Following a claim under the 1967 Act, it is for the landlord to decide whether to convey the freehold subject to the rentcharge or not, but if the rentcharge exceeds the rent payable by the claimant, he must discharge the excess so that, after enfranchisement, the tenant is subject to no great burden than before: see 4.21.

3.13 Third party determinations

Both the 1967 and the 1993 Acts contain provision for matters in dispute to be determined by a third party. The courts deal with such matters as the validity of notices, but valuation issues are determined in the first instance by an LVT. Their jurisdiction also covers the terms of the transfer or the new lease and the reasonableness of the costs the vendor seeks to recover.

Originally, the third party was the Land's Tribunal. However, the Housing Act 1980 provided that, in the event of non-agreement, a determination was first to be sought from a leasehold valuation tribunal (LVT).

3.13.1 LVTs

LVTs are organised by one of the five Rent Assessment Panels in England which collectively form the Residential Property Tribunal Service or the Panel in Wales. The RTPS publishes information about the service and there is also a comprehensive reference book.[6]

LVTs are intended to be an economic, accessible and informed forum for the resolution of disputes. They are intended to be less formal than the Land's Tribunal and they set their own rules of evidence. Cases can be dealt with on paper or at a hearing. The parties can appear in person or be represented by a valuer, solicitor, barrister or any other person and, whilst written proofs of evidence are often produced, the parties can give oral evidence on an informal basis. Because the strict rules of evidence do not apply, hearsay can be acceptable and comparables do not need to be proved.

LVTs are expert tribunals and are not bound to decide the case on the evidence presented; they can use their own knowledge and expertise. For this reason tribunal members usually inspect the property and the comparables. LVTs must reach a conclusion, and they do not have the option of making a finding that there is insufficient evidence before them on which to make a determination.

The role of the LVT has been considered in several recent Lands Tribunal cases. In *Arrowdell Ltd* v *Coniston Court (North) Hove Ltd* [2006] LRA/72/2005, the Members said:

> ... It is entirely appropriate that, as an expert tribunal, an LVT should use its knowledge and experience to test, and if necessary to reject, evidence that is before it. But there are three in escapable requirements. Firstly, as a Tribunal deciding issues between the parties, it must reach its decision on the basis of evidence that is before it. Secondly, it must not reach a conclusion on the basis of evidence that has not been exposed to the parties for comment. Thirdly, it must give reasons for its decision. In the present case the Tribunal reject the evidence of

---

6   *Leasehold Valuation Tribunals* by A Dymond, A Cafferkey and S Gallagher 2004.

both experts on relativity, and it was entitled to do this provided reasons for doing so were explained. But in basing it decision on "its own knowledge and experience, particularly in relation to relativities which had been agreed between parties or their valuers in other similar cases" it was in error because those agreements on relativity had not been identified nor had the parties had the opportunity to comment on them. As expressed, the decision contravened the second requirement. In refusing permission to appeal, the Tribunal said that it did not rely on any specific case or cases but, on this basis the first requirement was contravened. As for the third requirement, reasons that state that a decision was based on no evidence or on evidence that was not disclosed to the parties are adequate in one sense: that they enable the invalidity of the decision to be established. But to support a valid decision the reasons must enable the parties to understand why it was that the Tribunal reached the conclusion that it did rather than some other conclusion, so as to show that the conclusion was one to which the Tribunal was entitled to come on the basis of the evidence before it.

This confirms what had long been thought to be the case — if evidence is given by experienced valuers on both sides, it is to be expected that an LVT will be slow to depart from the limits presented by the valuers or to arrive at a decision by a method not adopted by either valuer. If it proposes to do so, the LVT should give the parties an opportunity to comment before giving a decision.

Then in *Elmbirch Properties plc* [2007] LRA/28/2006, the Lands Tribunal said:

> The LVT is a Tribunal which will have at least one, and sometimes more, specialist members. It is not bound to accept the evidence given by witnesses who appear before it, including expert witnesses. That does not imply that it can simply substitute its own view without giving the parties a chance to comment upon it. It is axiomatic in this area of the law but if points are to be taken or conclusions reached by a specialist tribunal, which do not arise from the evidence before it, or which represent a conclusion or judgement which have not been put forward by the parties (including those matters which arise from a site inspection), then it must invite the parties to make representations before them before it reaches a settled conclusion. A failure to do so is a breach of natural justice, see *Fairmount Investments Ltd* v *The Secretary of State for the Environment* [1976] 1WLR1255.
>
> We conclude from the above that the LVT based its decision upon evidence that either has not been submitted by the parties or upon which those parties were not given an opportunity to comment. It was wrong to have done so.

3.13.2 *The Lands Tribunal*

Until 1980, the LT was the court of first instance in enfranchisement matters and it has had a formative role in approving the format of valuations and the appropriate approach to matters of opinion and the interpretation of market evidence.

Since 1980, it is only after an LVT has made a determination that either party can apply to the Lands Tribunal for it to make a further determination. The Lands Tribunal is a branch of the High Court. There are both valuer and lawyer Members of the Tribunal. Usually they sit alone, but they can hear cases in panels of two or three Members.

Barristers and solicitors have a right of audience in the Land's Tribunal, but a valuer may represent a party only with the leave of the Tribunal. There is no appeal from the Land's Tribunal on matters of value, but an appeal on a point of law can be made to the Court of Appeal.

Wellcome Trust v *Romines* (1999), established that the appeal is by way of rehearing and that the appellant must show that the LVT decision was significantly wrong. The Lands Tribunal will not just tinker with an LVT decision.

The Tribunal will hear the case afresh and can take new evidence but new evidence will be treated "with, if not particular scepticism, at least with particular care."[7] It operates under the strict rules of

---

7   *Sinclair Gardens Investments (Kensington) Ltd* v *Franks* (1997) 76 P&CR 230.

evidence and has to make a decision on the evidence put before it. Thus it is perfectly possible for it to arrive at a different decision to that of the LVT on the same facts.

However, the valuation method and rates adopted by LVTs and the Lands Tribunal should not be followed slavishly. The President of the Lands Tribunal said in *W Clibbett Ltd* v *Avon County Council* (1976) that "previous decisions of the tribunal ... are relevant only to arguments on law or procedure. The assessment of compensation must be decided on, and only on, the evidence."

In *Arbib*,[8] the Lands Tribunal said:

> Decisions of LVTs and this Tribunal on questions of fact and opinion should not be treated as evidence of value in later cases. Such decisions do not establish any conventions or precedents (paras 112–166).

However they also said

> A decision of this Tribunal setting out general guidance on valuation principles or procedure, however, may be applied or referred to in subsequent cases (para 116).

---

8   *Arbib* v *Cadogan* [LRA/23/2004].

Residential Property: Leasehold Reform — Valuations

4.1 Valuer's Services

In the context of proceedings under the 1967 and 1993 Leasehold Reform Acts, a valuer may be asked to advise in a number of different situations such as:

- when a lessee or group of lessees are considering extending their leases, buying their freehold or exercising the right to manage
- to prepare the figures to be offered in a notice of claim
- when a landlord has received a claim and needs advice on whether the offer should be accepted and, if not, what counter-offer should be made
- in negotiating the premium or price
- in giving expert evidence to an LVT
- in considering whether to apply for leave to appeal against an LVT decision
- in advising on matter relating to negligence issues arising out of the failure of a claim.

4.1.1 Red Book

Valuers who are chartered surveyors must provide valuations in accordance with the RICS Valuation and Appraisals Manual (the Red Book) which contains guidance on valuations for enfranchisement (VIP 7). However the Red Book does not apply once a notice of claim has been served as, at that point, a dispute has arisen.

4.2 Basis of valuation

With the exception of low value houses where the valuation is to be carried out under section 9(1) of the 1967 Act (see 3.9 and 4.25), there is a common basis of valuation with three basic elements.

1   In every case, every landlord whose interest is taken in whole or in part, is to be compensated for the loss of his interest. In enfranchisements, the interest is being bought entirely, so the compensation is the value of the interest. In lease extensions, every landlord will retain an interest

after the grant of the new lease, so the compensation is the diminution in the value of the interest, found by valuing that interest both before and after the grant of the new lease.

2   In cases where section 9(1C) of the 1967 Act applies (see 3.9), the landlords are also entitled to compensation for losses caused to other property they retain after the sale.

3   In all cases other than where the lease has over 80 years unexpired, the landlords are entitled to 50% of the marriage value released by the transaction.

Since February 2005 the valuation date in all cases has been the date on which the claim was made.

4.3 Value of the interest to be acquired

The general valuation hypotheses under the Acts is rooted in section 9(1A) of the 1967 Act. This says that the enfranchisement price is the amount which the house and premises, if sold in the open market by a willing seller, might be expected to realise on the following assumptions:

(a) that the vendor was selling an estate in fee simple, subject to the tenancy, but ignoring any right to acquire the freehold

(b) that, at the end of the tenancy, the tenant has the right to remain in possession of the house under the provisions of Part I of the Landlord and Tenant Act 1954

(c) that the tenant has no liability to carry out any repairs, maintenance or redecorations under the terms of the tenancy or Part I of the Landlord and Tenant Act 1954

(d) that the price is diminished by the extent to which the value of the house and premises has been increased by any improvement carried out by the tenant or his predecessors in title at their own expense.

Sub-paragraphs (e) and (f) make further assumptions concerning rentcharges and other rights and burdens.

Those provisions have been repeated (with suitable adaptations) in schedule 6 of the 1993 Act for blocks of flats.

However, where lease extensions are concerned, the landlord's interest is not extinguished by the grant of the new lease. Therefore, schedule 13 of the 1993 Act refers to the compensation payable being the diminution in the value of the landlord's interest, which is to be found by valuing it twice, before and after the grant of the statutory lease. Those valuations are to be carried out on what are effectively the same provisions.

Simple term and reversion valuations are undertaken to value the landlord's interest. Ideally property interests should be valued by reference to market evidence, but then comparables would have to be found with similar characteristics to the interest being valued (eg fixed or stepped rents, similar length reversions). The Lands Tribunal[1] has said that all the transactional evidence of investment sales is of no assistance as it is tainted by the Act, so the likelihood of finding acceptable market comparables is remote.

The Tribunal also said that "... further thought [should be] given to the convention of valuing the rental income separately from the reversion ...", and that it is right to reject valuation by reference to a conventional rate established by reference to the settlement evidence which the Tribunal created.

---

1   In *Arbib*.

However, for the best part of 10 years, the general acceptance of a conventional rate enabled the deferment rate to be settled in the "6% area" for London and for house cases generally, without recourse to any tribunal, thus making savings on the public purse and the parties' costs.

4.4 Rent

The rent to be capitalised is not always the actual rent passing, or which will pass (where there are increases to fixed rents), or be assumed to pass (where the increases are on a formulaic basis such as a percentage of the capital value of the property).

If the rent passing was assessed at the last review by reference to the market or rental value of the property in its present condition, and if tenant's improvements had been carried out, the passing rent is to be reduced to that which would have applied had the rent review hypothesis included the statutory disregard of improvements. Similarly, any future reviews to a market value based rent are also to disregards tenant's improvements.

If the interest being valued is that of an intermediate landlord, it is the profit rent that is capitalised in an enfranchisement (see 4.20).

4.5 Vacant possession values

The 1993 Act gives three basic disregards to be made in the valuation of the vendors' interests. They are:

(a)   any rights under the leasehold reform Acts
(b)   the lessee's rights to remain in possession at the end of the term of the long lease
(c)   any increase in value due to improvements carried out and paid for by the lessee.

There are similar disregards to the 1967 Act, with another added: that the tenant has no liability to carry out any repairs, maintenance or redecorations under the terms of the tenancy, or Part I of the Landlord and Tenant Act 1954.

In block enfranchisements under the 1993 Act, by importation into para 4 of schedule 6, the same disregards apply to the vendors' interests when making valuations for the assessment of any marriage value. Whether the rights are to be excluded from the valuation of the tenant's interests is still uncertain. The only mention of similar disregards in the valuation of the tenants' interests is in valuations of tenants' interests for lease extension claims under schedule 13. In a house case, *Norfolk* v *Trinity College, Cambridge*, in the absence of any specific rules, the Land's Tribunal determined that the value of tenant's improvements should be deducted from both the landlord's and the tenant's interests, approximately in proportion to the value of their interests.

The point was raised but not determined in *Donath* v *Grosvenor re: 36/7 Eaton Mews South* (LON/ENF/6) and has not been taken since. If the point is taken again in the future, the arguments might be as follows:

(a)   for the vendors: the principle established in *Norfolk* v *Trinity College, Cambridge* should apply. Alternatively, that, although schedule 6 is worded differently to schedule 13, in the absence of any rules in schedule 6, the rules spelled out in schedule 13 should apply
(b)   for the purchasers: the position in schedule 6 contrasts with that in schedule 13 (see 4.15) where its is stated that the same assumptions are to be made in valuing both the landlords' and the

tenant's interest. On that basis, another part of the decision of the Land's Tribunal in *Norfolk* (relating to the tenant's over-bid) should prevail. There the Tribunal determined that the specific provision section 9(1) introduced by the 1969 Act, excluding the tenant from the market, could not be read as amending section 9(1A). In this case the 1996 Act amendments to schedule 13 cannot be read as amending schedule 6.

As there are tenant's interests both before and after marriage (because of the leasehold tenure, the nominee purchaser is merely a new landlord), the value of tenant's improvements will appear on both sides of the equation and be of relatively little significance.

4.6 Assumptions — the no Act world

In the 1967 Act this assumption is "that the vendor was selling an estate in fee simple, subject to the tenancy, but ignoring any right to acquire the freehold".

In the 1993 Act, in relation to blocks of flats, para 3(1)(b) of schedule 6 says "on the assumption that this Chapter and Chapter II confer no right to acquire an interest in the specified premises or to acquire any new lease (except that this shall not preclude the taking into account of a notice given under section 42 with respect to a flat contained in the specified premises where it is given by a person other than a participating tenant)".

In relation to flats, Schedule 13 of the 1993 Act says "on the assumption that Chapter 1 and this Chapter confer no right to acquire any interest in any premises containing the tenant's flat or to acquire any new lease".

It is commonly said of this assumption that the valuer is required to value in a no-Act world, ie a world in which the Act under which the right of compulsory purchase has been claimed does not exist. In fact none of the assumptions says that there is to be a no-Act world, but only that the Act does not apply to the house or block being enfranchised or the block where the flat whose lease is being extended, is situated. This point has not yet been aired at a tribunal.

In block enfranchisements, the alterative, individual right to a lease extension, is likely only to be of value before the marriage of the interests, and therefore it could significantly reduce the quantum of marriage value if it is to be excluded from the existing values of the participants' flats.

4.7 Assumptions — tenant's improvements

Section 9(1A)(d) of the 1967 Act provides that the value of the house is an open market value:

> On the assumption that the price be diminished to the extent to which the value of the house and premises has been increased by any improvement carried out by the tenant or his predecessors in title at their own expense.

This wording is unfortunate to say the least. The Act does not say that the improvements are to be ignored in carrying out the valuation, but that the price to be paid for the freehold is to be diminished by the increase in the value of the house occasioned by the improvements. However the effect of the assumption is generally held to be in diminishing the value of the house.

The 1993 Act has slightly different wording. In relation to blocks of flats, schedule 6 says:

> on the assumption that any increase in the value of any flat held by a participating tenant which is attributable to an improvement carried out at his own expense by the tenant or by any predecessor in title is to be disregarded;

This is more elegant in that it relates the value of improvements to the value of the flats. It appears however that improvements are not to be disregarded when valuing the flats of the non-participants. In relation to individual flats, schedule 13 says:

> on the assumption that any increase in the value of the flat which is attributable to an improvement carried out at his own expense by the tenant or by any predecessor in title is to be disregarded

Both Acts also say that any lease surrendered in consideration for the grant of the current lease, is part of the current lease, so the improvements to be discounted may predate the granting of the current lease.

The improvements can be disregarded both in assessing the capital values, and the rent on review to be adopted in the calculation under the Act, even though a higher rent (reflecting the value of the improvements) is actually payable from the review date to the date of completion of the purchase of the freehold.

The principles to be followed were laid down by the House of Lords in *Shalson*.[2] For works to be classed as improvements they must:

(a) have been carried out to the building and
(b) have been carried out during the term of the lease and
(c) have been paid for by the lessee at the time and
(d) have constituted an improvement at the time they were carried out.

Having established which works can be classed as an improvement, it is the amount by which those works increase the value of the property at the valuation date which is disregarded in the valuation. The basic premise is to start with the value of the property on the date of valuation and then to identify the improvements to establish the nature and gross internal area of the unimproved house when let. It is difficult to conceive of a way of valuing the improvements in isolation, and the general approach is to value the house improved and as though the works which add value had not been carried out; the difference between the two being the value of the improvements.

However the potential to carry out the valuable improvements should be included in the value of the property that will revert to the freeholder at the end of the lease, subject only to consideration as to the prospect of the improvements being able to be carried out in the context of changed planning policies.

4.8 Assumptions — rights to remain in possession

At one time, the prospect of the claimant remaining in possession at the end of the long lease played a significant part in the valuation process. In *Lloyd-Jones*, the tenant claimed and won a 10% deduction against the full vacant possession value of the freehold when assessing the reversionary value because of his entitlement to remain in possession at the end of the lease under the terms of Part I of the Landlord and Tenant Act 1954, ie on a regulated tenancy at a fair rent.

With the changes to Part I brought about by the Housing Act 1988, LVTs have tended to reduce considerably any discount previously allowed and, in early valuations of flats, made no allowance at all (please also see Chapter 2.)

---

2   *Trustees of John Lyon Estate* v *Shalson*.

However some cases where the leases were about to expire or had expired by the valuation date drew a tight focus on the value of the rights. In *Conley* v *Goldstein re: 27 Belsize Park Gardens* (LON/NL/108) where the term had only four months unexpired, the evidence from the tenant was that he intended to continue to live in the flat with his partner and ten year old child. The tribunal accepted the landlord's evidence that a regulated tenant reduced the value of the flat by 40%.

In *Horden* v *Cadogan re: 45 Cadogan Place* (LON/NL/95) the lease had expired by the date of the valuation. The LVT saw market evidence of a 52.5% discount for the risk and a landlord's submission that 10% would be appropriate. They also had regard to the Land's Tribunal's decision in *Vignaud* v *John Lyon School* (1996), a case under the 1967 Act where, in the face of evidence that the tenant could not afford to stay on at a market rent at the end of the lease, the Tribunal had applied the customary discount of 10%. They concluded in *Horden* that the risk of the tenant remaining in possession, able to undertake the dilapidations and to pay the rent was greater and applied a 25% discount.

However it is only lessees of properties with old RVs under £1,500 who could hold over on regulated tenancies, and none have been created since 1989. It is generally considered that an assured shorthold tenancy has no effect on the reversionary market value, there being no risk that vacant possession will be given.

In *Mutual Open Property* v *Cadogan* (LON/ENF/473/00) two of the lessees were holding over on leases which had expired. Neither was potentially subject to a continuation tenancy under the Local Government and Finance Act 1989. It was agreed that it would take nine and 12 months from the valuation date to obtain possession of the flats or to establish that no further tenancy would arise. The LVT valued the flats at the valuation date by deferring the VP value and allowing for finance costs, resale costs and profit.

4.9 Assumptions — lack of obligation to repair

The assumption "... that the tenant has no liability to carry out any repairs, maintenance or redecoration under the terms of the tenancy or part 1 of the Landlord and Tenant Act 1954" applies only in house cases. Its effect on the valuation was considered in *Arbib*, when the Tribunal said that the normal assumption is "that the purchaser [of the freehold interest] runs the risk that it might be returned in less good repair at the term date, without even the benefit of being able to seek damages for breach of covenant" but refers to "the improbability of [the properties] being seriously neglected even if the tenant had no rights under the Act".

The Tribunal say in *Arbib* that "what has to be judged, and, on the material which has been put before us it can be a matter of judgement only, is what reward the market will require for investing in an asset which may require management ... and which may be destroyed, and which may be expensive to realise at the end of the term." They received unanimous evidence that the assumption has no effect in relation to high value houses because of " ... the improbability of their being seriously neglected ...".

4.10 Right to break for redevelopment

Section 61 of the 1993 Act provides the landlord with a right to possession both on the date of expiry of the original term and of the extended term. The landlord must satisfy the court that he intends to demolish or reconstruct or carry out substantial works to the premises of which the flat forms part. If the court is so satisfied, the tenant is compensated by the open market value of the lease. The parties can contract out of the section 61 right by agreement.

If the unexpired term is short, the prospect of an uncertain but early break may adversely affect the value of that leasehold interest. In *McGirk* v *Viscount Chelsea*, the Lands Tribunal approved the principle of a deduction on account of the impending break (the lease had just under seven year unexpired) but found on the facts that no deduction was to be made. In *Sloane Stanley* v *Currie* [1996], the LVT made a deduction of 7.5% at 14.75 years unexpired. In *Watton* v *Ilchester Estate Trustees* [2002] LRA/21/2001, the Lands Tribunal approved the deduction of 10% where the lease had 40 years to run, in the context that the landlord had all the other flats in the block in hand and was seeking compensation for the loss of development value to reconvert the block back to the house it had originally been.

4.11 Leases subject to tenancies

Since 2002 there has no longer been a requirement that claimants must reside in the premises, so claims can be made even where claimant has sublet the premises on a short term basis. (If the premises are sublet on a long lease, it is only the long sub-lessee who will be able to exercise the rights.) Therefore allowance has to be made for the effect on the value of the flat of the sub-tenancy, particularly if that tenancy is regulated: see Chapter 2.

However, since no new regulated tenancies have been created since 1989 and most were created long before then, most regulated tenants are now elderly. The regulated tenancy can only be passed to a spouse who would usually be of more or less the same age, so most regulated tenancies can be expected to have ended by about 2020. Therefore, unless the lease will end before the regulated tenant can be expected to have vacated, the freehold reversionary value does not need to be discounted.

Another family member can succeed to an assured tenancy (see Chapter 2) and enjoy security of tenure, which also has the effect of depressing the VP value.

4.12 Onerous rent

Where the rent payable is above average, it has become known as an onerous rent. The Lands Tribunal[3] have determined that an onerous rent depresses the value of the leasehold interest. The Member said "I have not been shown any market evidence which leads me to conclude that a millionaire offered two otherwise identical properties at annual rents of £2,000 or £15,000 would offer the same sum for either." He went on to support the approach taken by the LVT which had been to deduct £130,000 from the leasehold value of £2.79m on account of the rent of £15,000 pa. The deduction was arrived at by assessing the excess rent at £13,000 pa and capitalising that excess at 10 YP.

Following *Carl*, the level of non onerous rent has been taken at about 0.1% of the leasehold value. Both the tipping point for onerousness, and the rate at which the onerous element is capitalised, have yet to be tested in any depth.

A deduction on this count impacts on the price payable by increasing the amount of the marriage value, of which 50% is payable.

---

3   In the case *Carl* v *Grosvenor* [2000] LRA/33/1999.

4.13 Income other than rent and reversions

The valuation should include all the factors which would influence a purchaser in the open market. In practice this would be not only the value of the rents during the term and the value of the reversion, but also:

- the possibility of selling a longer lease to each lessee (hope value — see section 4.17)
- any value attaching to potential development on the site and
- the value of any other income, for example, the value of staff accommodation and parking rights.

Income other than rent has been taken into account where appropriate. The LVT for Wales held (*Lynch v Castlebeg* (1987)) that, while the ground rent of £14 pa (for the remainder of the 85 years unexpired) should be capitalised at 7%, an annual insurance commission of £16.80 should be capitalised at 6.5 years purchase adding nearly 20% to the enfranchisement price.

Following that decision, profit from management fees and insurance commissions can be capitalised but only if the vendor can enforce a covenant providing that they are to be placed through an agency from which he can derive them. Care must be taken only to value the saleable assets at rates applicable to the type of business and to distinguish between fees or commissions and profit. There is some cost in administering an insurance agency and in running the services of a block of flats and in collecting service charges.

In recent times, the benefit of this sort of income has been included in the capitalisation rate.

4.14 Yields — generally

When it comes to consideration of the appropriate capitalisation and deferment rates (collectively known from 1967 to 2005 as yields), investment in property must be compared with other investments available to the investor before he decides to enter the property market. No rational investor will place his capital in property for a yield lower than current running and/or redemption yields unless that property has growth potential. However, the prospect of rental growth and/or capital growth often induces an investor to accept initial yields for property below the running or redemption yields on conventional gilts.

There are various types of property investment which may have to be acquired upon enfranchisement:

(a) a ground rent with no rent reviews or fixed increases

(b) a ground rent with provision for review to a rent linked to rental or capital value but widely spaced intervals

(c) the vacant possession value which will change over time.

The growth potential of those investments are respectively nil, moderate and good. A logical investor would require returns reflecting these differing potentials; he would not accept a return of less than say, gilts plus 2–4% for a fixed property income and, say, gilts plus 1–2% for one with moderate growth, such as with value-based reviews. It is only the final reversionary value which has real growth potential but, until recently, this was rarely taken in to account in the yields assessed.

In the early years of the 1967 Act, yields of 6–8% were almost exclusively applied, except in cases where the original ground rent had over 60 years to run, where 10% was used. This consistency persisted through the period in which gilt yields rose as high as 15% and fell no lower then 7%. The

Land's Tribunal's dogmatic approach appeared to be favoured largely for political expediency as the enfranchisement prices produced by high capitalisation rates would be lower than what are already thought to be unfairly low valuations. In 1975 WA Leach commented:

> the fact is that 6% and 8% owe more to the decisions of the tribunal than to anything else. They are applied and accepted in settlements because it is believed that these contrived valuations are what the tribunal would uphold. The contrived valuations are 'comfortable' valuations because the tenant still does very well out of them and the landlord is less harshly treated than he might be. So long as no attempt is made to justify them in any other way that may be a good thing.

However as interest rates fell below the conventional yields of 6% in central London and 7% elsewhere, landlords who were adopting differential rates as between the various elements of the landlord's interest, started to look for lower deferments rates. This culminated in two hearings in 2005 and 2006 where the Lands Tribunal considered several cases together and received evidence from a broad sweep of valuers and financial experts. The decisions in the cases of *Arbib*[4] and *Sportelli*[5] are fundamental to the consideration of deferment rates in particular, but also to the likely approach to capitalisation rates unless and until there is a decision dealing with that issue.

4.15 Yields — deferment rates

In *Arbib*, the LT determined that, where the reversion was in 19-35 years' time and was a freehold reversion on the Cadogan Estate, the deferment rate should be 4.5% for houses and 4.75% for flats. They built up their rates from a basic risk-free return of 2%, to which they added a basic risk premium of 2.5% and a specific additional management risk element of 0.25% for flats. They did not consider anything other than a freeholder's interest, so did not consider how the rate might vary for a leasehold interest.

In *Sportelli*, the Lands Tribunal received evidence from financial experts as well as valuers, which enabled them to refine and extend the *Arbib* decision. In the absence of evidence from sales, they concluded that:

$$\text{deferment rate} = \text{risk-free rate} - \text{real growth premium} + \text{risk premium}.$$

4.15.1 Risk-free rate

The Tribunal[6] defined the risk-free rate as "the return demanded by investors for holding an asset with no risk, often proxied by the return on government security to redemption". Paras 18 and 70 deal with how the Tribunal concluded that the risk-free rate should be 2.25%, assessed against index-linked gilts.

In the light of the evidence from financial and other experts, the Tribunal decided that, for terms over 20 years, the risk-free rate is constant. (The Tribunal's conclusion that the deferment rate is constant is derived from the constancy of the risk-free rate.) On the evidence, they adjusted the risk-free rate to 2.25%.

---

4   *Cadogan v Arbib.*
5   *Cadogan v Sportelli.*
6   In *Sportelli.*

4.15.2 *Growth factor*

They determined the real growth premium for property at 2%.

4.15.3 *Risk premium*

The Tribunal in *Sportelli* defined the risk premium as "The additional return required by investors to compensate for the risk of not receiving a guaranteed return". Paras 74–79 deal with this matter in detail. They were comparing the reversionary value with such comparators as long gilts, equities, individual property companies and the Act-world market in reversions. They concluded that the risk premium should be 4.5%.

Now they had:

deferment rate   =   2.25% (risk-free rate) – 2% (real growth premium) + 4.5% (risk premium)
              =   4.75% for houses.

In *Sportelli*, we are reminded that in *Arbib* the Tribunal "concluded that there should be 0.25% differential between houses and flats ... having regard to the lesser management problems of a single house as opposed to the flat market." [para 92]. Later they say "... the adjustment ... was intended to reflect both the greater management problems associated with flats and the possibility that there might be a better prospect of growth in the house as opposed to the flat market" [para 95].

They concluded "... we accept ... that any disparity between growth rates for houses and flats is likely to even out over the long term. We think however that an adjustment needs to be made to reflect the management problems ..." [para 95] and they went on to confirm the generalised addition of 0.25% for flats first made in *Arbib* on account of the potential for problems to arise in respect of the management of flats, whether the block is held on a headlease or not.

4.15.4 *Risk premium — specific factors*

In *Sportelli* several specific factors were considered.

On the length of the term, the rate was held to be constant in the range 20–75 years which was the range under consideration by the Tribunal. Of shorter terms they said "Below 20 years ... the rate would need to have regard to the property cycle at the time of valuation." Of longer terms they said "Beyond 75 years we see no reason ... to conclude that the rate would be either higher or lower".

On location they said they saw "no justification for making any adjustment to reflect regional or local considerations either generally or in relation to the particular cases before us. ... locational differences of a local nature are, in the absence of clear evidence suggesting otherwise, to be assumed to be properly reflected in the freehold vacant possession values."

4.16 Capitalisation rates

Neither *Arbib* nor *Sportelli* dealt with capitalisation rates as the capitalisation rates had not been appealed. However, it is fairly clear how the Tribunal would regard the sort of market evidence that valuers would be likely to produce; they would say it is too tainted by the Acts to be of any assistance. In *Arbib*, the Lands Tribunal said (in relation to deferment rates and in summary of their views set out in paras 96–99) "Although market evidence is usually the best evidence of value, the extent of the right to enfranchise or to a lease extension is now so wide that there is unlikely to be dependable market evidence in any particular case". In *Sportelli* the Tribunal again considered the possibility of deriving the deferment rate from market evidence. Their conclusions include the following:

The market in the real world is substantially different from the one to be envisaged in the hypothetical no-Act world.

... the market is inevitably influenced by expectations of what an LVT might determine as the enfranchisement price under the statutory assumptions. ... It may be that other considerations are more influential than the expectation of what an LVT would decide, but the possibility, indeed, as we think, the likelihood, of there being such an influence constitutes a further reason for rejecting an approach based on the market evidence.

In a subsequent case[7] they said

The factors that are relevant to the determination of the capitalisation rate (which, we accept, are correctly identified above[8]) are so manifestly different from those that are relevant to the deferment rate that there can be no valuation rationale to justify adopting a rate for capitalisation simply because that rate is being taken for deferment. Moreover the application of the factors affecting the capitalisation rate, unlike the application of the factors affecting the deferment rate, is likely to vary in every case. It is, of course, the case, that if the ground rent is small and the unexpired term is not long there will be no significant difference in adopting one particular rate rather than another. ... It would clearly be disproportionate for valuers to dispute capitalisation rate in such circumstances...

The same general approach could be followed to derive a capitalisation rate, ie to find a financial instrument that is equivalent to the interest being valued.

Where the rent is subject to increases which are a proportion of capital or rental value, it is inflation-proofed at review in much the same way as the reversion. In those cases it seems reasonable to adopt the same sort of approach as that taken to deferment rates (so capitalisation rate = risk-free rate – growth premium + risk premium), except perhaps in relation to the final step of the rent which is in effect a fixed rent.

However, if the rent is fixed or subject to fixed increases, there is no real growth to consider. Such rents are more akin to a conventional gilt or an annuity; low in amount and well secured. However, as compared with gilts, property interests are illiquid, costly to transact and require day-to-day (or at least quarter-to-quarter) management. In relation to deferment rates the Lands Tribunal said in *Arbib* "There is therefore a need to make allowance for the comparative illiquidity of an investment in a freehold reversion. In the no-Act world they could, no doubt, be sold but only at a cost and with delay. .... If the investment was regarded as long-term it would still justify an allowance of say 1% in yield as against gilts". No doubt similar considerations would apply to capitalisation rates.

---

7   *Appeal by the Trustees of the Calthorpe Estate* [2007] LRA/29/2006.
8   They were identified as "the length of the lease, the security of recovery, the size of the ground rent, whether there was provision for review of the ground rent and, if there was such provision, the nature of it".

4.17 Hope value

In *Sportelli*, the Lands Tribunal decided that hope value was not permissible as a separate item in calculations under schedules 6 and 13 of 1993 Act, but that it was permissible in valuations under section 9(1A) of the 1967 Act. This decision has been referred to the Court of Appeal.

The Lands Tribunal said that there were two alternative approaches to the quantification of hope value in valuations under section 9(1A) and that they preferred the addition of a lump sum to the adjustment of the deferment rate. The only evidence they received on how a lump sum should be added in the case of a section 9(1A) valuation, came in an uncontested appeal. The appellant's valuer simply quantified the hope value as 20% of the available marriage value. This approach was justified by knowledge and experience referring specifically to the sales of a central London estate in 1996, where advice was given to the purchasers to pay 50% of the 50% marriage value payable. He softened the resulting 25% to 20%.

The problem with this, as yet, untested approach is that the available marriage value is derived from a conventional valuation under the 1967. In other words an Act calculation has been adopted in a no Act world valuation. A preferable approach might be to follow the practice adopted in the market prior to the introduction of the 1993 Act. At that time leases were extended by agreement between the parties. The price was assessed by taking the value of the lessee's interests before and after the grant of the new lease, and paying a proportion of the increase.

In fact, of course, the price payable for a new lease or a freehold can always be expressed as a proportion of the increase in value, even when it has been calculated under the Act. The proportion can be expected to be dependent on the unexpired term of the lease and the ground rent payable.

Once the increase in value has been established, the likely premium or price can be assessed as a percentage of the increase. Part of the premium would cover the pure investment value of the rent and reversion calculated by applying the hope value free rates. Once that has been deducted, the balance would be the anticipated capital gain. An investor would be willing to pay a proportion, perhaps 50% for that potential gain.

Example:	freehold value:	£250,000
	value of present lease:	£150,000
	increase in value	£100,000
	anticipated premium @ 60%:	£60,000
	less investment value included in premium:	£10,000
	potential gain, say:	£50,000
	Hope value @ 50%	£ 25,000

Marriage value varies according to lease length, and one of the benefits of this approach is that it allows that variation to be reflected. Hope value could be reflected in deferment rates, but it is difficult to know how to adjust that rate at varying lease lengths.

4.17.1 Section 42 claims in section 13 claims

There is another, rather different class of hope value to that considered in *Sportelli*. It is the hope value which arises when a section 42 claim has been made. This was not considered in *Sportelli* because there were no such claims made by the non-participants in the cases being considered.

The no-Act assumption in schedule 6 is modified by para 3(1)(b) so that it does not "...preclude the taking into account of a notice given under section 42 with respect to a flat contained in the specified premises where it is given by a person other than a participating tenant".

Para 3(1)(d) of schedule 6 also requires an assumption that the property is being sold subject to the various encumbrances to which the conveyance will be made. A section 42 notice operates as a statutory contract for the grant of a new lease and is registerable. The conveyance of the freehold would have to be made subject to the existence of that notice whenever it is served and, accordingly, it must be taken into account in the statutory valuation. Section 25(6) enables the court to refer to a LVT modification of terms of acquisition where a change in circumstance has arisen since the time when the terms were agreed. Service of a section 42 notice would certainly amount to a relevant "change in circumstance", and accordingly regard could be had to it even if the notice is given after the terms of acquisition have been agreed but before a contract is entered into. Therefore, the section 42 notices to be taken into account are not only those given by a non-participating tenant before the date of service of the section 13 initial notice, but those given at any time during the life of the claim.

In *Sportelli* the Tribunal said that schedule 6 does not allow hope value, which they defined as the option that the freeholder has in the no-Act world to sell a lease extension to the tenant, and thereby realise part of the freeholder's share of the marriage value. It pre-supposes that, at some point in the future, the tenant will make an approach to the landlord (or vice versa) for a new lease, and the landlord will thereby have his "option to deal". However, the freeholder is not obliged to sell and, in consequence, is able to control both the timing and the terms of any deal.

Once a section 42 claim is made, the hope value is not simply speculative but, short of a deemed or actual withdrawal, an actuality. There is a very good, although not copper-bottomed, prospect of 50% of the marriage value being forthcoming in the forseeable future. This is something for which the third party investor envisaged by para 3(1) would pay an additional amount, over and above the price he would pay if there were no such claims extant. It is not general hope value in the Sportelli sense, but a proportion of the specific marriage value payable where the claims have been made, perhaps also deferred over the period until it is expected to fall in.

4.18 Marriage value

Marriage value was not mentioned in the 1967 Act. In the first cases under section 9(1) it was said that, as he is not specifically excluded, the tenant was to be held to be in the market, and therefore his overbid should be taken in to account. The valuer was therefore required to assess the marriage value which will be released if the tenant acquires the superior interests, and to apportion that marriage value between the parties. As that had not been the intention of Parliament, the 1967 Act was quickly amended to exclude the tenant from the market.

When the 1967 Act was amended again in 1974 to increase the value of the houses to which it applied and to change the basis of valuation, the tenant was, again, not excluded from the market and this time it was held[9] that Parliament must have meant it and that marriage value should be paid. In the next major case,[10] the Tribunal determined that 50% was a more appropriate percentage to be paid than 25%.

The 1993 Act specifically provides for marriage value to be assessed and shared between the parties. Originally the proportions were left open, to be agreed between the parties with the landlords

---

9   In *Norfolk* v *Trinity College, Cambridge* [1976] LR/106/1975.
10   *Lloyd-Jones* v *Church Commissioners* [1981] LR/29/1980.

receiving not less than 50%. However, since 2002, the amount payable has been fixed at 50% and that is shared among the landlords in proportion to the values (or diminutions in value) of their interests.

Marriage value (see Baum and Mackmin 1989, p144) is simply the difference between the aggregate of the values of the interests in the property before and after the transaction.

Example 4.1

Value of lessor's interest excluding prospects of 'marriage' of interests. say			£84,257
Lessor's share of 'marriage' value			
Estimated value of unencumbered freehold interest		£187,500	
Less:			
Value of lessor's interest excluding prospects of 'marriage'	£ 84,257		
Value of lessee's interest excluding prospects of 'marriage'	£ 45,000	£129,257	
Gain on 'marriage'		£ 58,243	
50% to the lessor			£29,122
			£113,379
Enfranchisement Price		say	£113,000

In all cases, there is only one calculation of marriage value. That is as between the purchaser/claimant and the reversioner/competent landlord on behalf of all the vendors. If there are intermediate interests, the reversioner/competent landlord shares the marriage value between the vendors in proportion to the values/diminutions in value of their interests.

In an enfranchisement of a block of flats, marriage value arises only in respect of the participants' flats. This is because they are the only lessees in a position to grant themselves very long terms without payment of a fine or premium.

The disregards to be taken into account in calculating the values to be adopted in the marriage value calculation are discussed in detail in the section on vacant possession values.

4.19 Compensation

The third element which must be paid in all cases, is compensation for any loss or damage suffered by a vendor in relation to any interest in other property resulting from the enfranchisement or lease extension. It follows that the vendor must own property other than the claim property which is being sold, and that it will suffer a diminution in value because of the sale of the claim property. Hence, loss of development potential in the block being enfranchised is not a matter to be dealt with under this head unless that potential is diminished by severance, but loss of development value in a neighbouring block might be.

It was successfully argued twice that a landlord should be compensated for the loss of control over other remaining blocks. Compensation at £100 per flat being enfranchised has been awarded by the Southern LVT. However this was not pursued in later cases.

4.20 Intermediate interests

In many cases, there is a head lessee, or a series of superior tenancies between the claimant(s), and the freeholder or competent landlord. Each interest is compulsorily acquired in an enfranchisement or diminished in value in a lease extension. Each lessor is entitled to be represented individually and to have the value/diminution in value of his interest, and any compensation due to him for severance, assessed separately.

However, when it comes to the share of marriage value, there is only one assessment made, and the amount payable is apportioned amongst the vendors in proportion to the value of their interests. In the 1967 Act there are no specific provisions relating to the sharing of marriage value among the various vendors. However, schedule 1 provides that the reversioner, usually the party with the biggest superior interest, acts on behalf of all the vendors, although each is entitled to be separately represented. Under the 1993 Act, the vendors share 50% of the marriage value between them, in proportion to the value, or diminution in value, of their interests.

Under lease extension the intermediate lessor's loss is likely to be greater than it is under enfranchisement. This is because in enfranchisement, the intermediate landlord loses just the profit rent but in a lease extension he loses all his income. Therefore, intermediate lessors seem likely to encourage lessees to extend their leases rather than to enfranchise, so as to obtain the maximum marriage value.

Intermediate leaseholds other than Minor Intermediate Leaseholds (MILIs — see below) must be valued in accordance with the rules for valuing the freehold.

4.20.1 Minor intermediate leaseholds

Where an intermediate interest has an expectation of possession of not more than one month, and a profit rent of not more than £5 pa, this is termed a minor superior tenancy (in the 1967 Act) or a minor intermediate leasehold (in the 1993 Act), and is commonly referred to as a MILI.

The value of a MILI is calculated in accordance with the following formula:

$$P = £ \frac{R}{Y} - \frac{R}{Y(1 + Y)^n}$$

where:

P = the price payable
R = the profit rent
Y = the yield (as a decimal fraction) from 2.5% Consolidated Stock
n = the number of years which the minor superior tenancy has to run.

The formula looks complex but is simply a single rate capitalisation of the profit rent of £R for n years, at a rate derived from the market price for 2.5% Consols. The rate is taken from the sale price on the Friday before the valuation date. It is obtainable from the Government's Debt Management Office web-site (*www.dmo.gov.uk*).

The formula accords with the recognised method of valuing a fixed income receivable for a limited period of time. The reason for the deployment of such a formula was to prevent expensive disputes over small sums of money.

4.20.2 MILIs and increasing rents

The formula only provides for a single rent to be capitalised. This may have been satisfactory in relation to low value houses prior to 1980, but it has now been extended to high value houses and flats and there are many such with stepped rents or with increases to figures linked to capital or rental value. If R in the formula is treated as the "initial (or current) profit rent" and the formula is applied to that rent only, the intermediate landlord will be seriously under compensated.

Throughout the passage of the 1993 Act through Parliament, it was made clear that the new rights were not intended to be confiscatory, and that landlords were to be properly compensated for their losses. It would be bizarre if Parliament had intended the intermediate landlord not to be properly compensated for his loss of income, which must included loss of future income as well as current income. If the current rent only is valued the intermediate landlord will not have sufficient compensation to meet future rents.

The issue has not yet been tested at the Lands Tribunal so it is with some caution that it is suggested that the simplest way of approaching the valuation of the increase would be to apply the principle of the formula, ie to capitalise at the yield derived from 2.5% Consols but to allow for a the rent on review to be deferred and for the formula to be applied more than once.

Alternatively, as the definition of R refers to "rent payable under the lease" R may fairly be said to cover all the rent payable under the lease. So, if the formula is only to be applied once, R must be the single figure that is equivalent to the actual increasing negative profit rent (the equivalent profit rent).

The equivalent profit rent is not a hard figure to assess. It is the profit rent which, when capitalised over the whole term is equal to the actual profit rents, capitalised and discounted over the periods for which they are payable.

Example 4.2

Assuming a current rent of £200, a rent on review in 2.56 years for 38.5 years of £1,055 pa and a capitalisation rate of 5.5%, the actual profit rent payable would be capitalised as follows:

Current profit rent		£200
YP 2.56 years @ 5.5%	16.7288	£482
Increase to	£1,055	
YP 38.5 years @ 5.5%	15.868	
PV 2.656 years @ 5.5%	0.8674	£14,534
Capital value of rent		£15,016

To produce the same capital value, the equivalent profit rent ("r") would be capitalised as follows:

Equivalent rent	£r	
YP 41.156 years @ 5.5%	16.1743	£15,016

In formulaic terms this is:

$$16.1743 \, r = £15,016$$

Thus R = 15016/16.1743 = £928.39. So the profit rent equivalent to the actual increasing profit rent is £928.39 pa which can be applied to the formula in 8(6).

4.20.3 MILIs in lease extensions

In an enfranchisement (of a house or block of flats), the intermediate landlord's interest is totally extinguished and his loss is simply his profit rent. The MILI formula which appears in the Leasehold Reform Act 1967 was included, without amendment, in schedule 13 of the 1993 Act in relation to lease extensions, which seems to have been an error. In a lease extension, following the deemed surrender and re-grant of the intermediate lease, the intermediate landlord continues to have an interest after the grant of the new lease (unless it covers only the flat in question and the whole reversion has been used up). So, in fact, the intermediate landlord loses his entire income (as opposed to profit) and any reversion, while his outgoings remain unaltered. He has to go on paying the rent to his immediate landlord under the terms of the intermediate lease.

Thus, while the intermediate landlord's profit rent might be £5 pa or less at the date of valuation, his immediate loss of income could be well in excess of £5 pa with a very considerable additional loss of income in the future on reviews. In these circumstances it seems inappropriate to value the intermediate landlord's interest by the MILI formula.

4.20.4 Intermediate leaseholds in lease extensions

The valuation of intermediate interests present some interesting conundrums, first, in conceptual terms. The way in which the intermediate landlord's interest after the grant of the new lease is to be valued is the subject of controversy and has not yet been considered by the Lands Tribunal.

In the case of an enfranchisement, all intermediate interests are extinguished on completion. However, under the lease extension regime, the intermediate interests do not disappear. Therefore, even though the occupational lease has been extended beyond the term of the intermediate lease and even though the rent reserved by the occupational lease has been taken away, the intermediate leaseholder's obligation to pay the full rent and, perhaps, to repair and maintain services continues through the term of the higher lease. Furthermore, this can become very burdensome as lease extensions are taken up and the adverse gap between the rent payable and the aggregate rents receivable widens. Bear in mind that there is no right to sell or transfer the interest either to the occupational lessees or to the freeholder.

The interest to be valued is not one that exist in isolation, or which can usually be created, because most headleases prohibit the assignment of part. However, valuations are themselves hypotheses and hypothetical interests often have to be valued.

If the interest is not a MILI and the valuation is to be carried out under para 8(1) of schedule 13, para 3(4) would enable an assumption to be made that the interest is sold only with the whole headlease. It has been suggested by many LVTs that such an assumption should be made, following the Lands Tribunal's decision in *Squarepoint*.[12] However, the definition in para 1 of schedule 13 seem clear that the intermediate leasehold interest to be valued is the interest in the tenant's flat.

After the grant of the new underlease, the intermediate landlord's ground rent income from the flat in question will fall to a peppercorn but, following the deemed surrender of the headlease and regrant on identical terms, the full rent reserved by the headlease will remain payable to the freeholder.

By the time the headlease is regranted, the intermediate landlord will have no expectation of possession at all, and there is a view that no expectation of possession is not the same as an expectation of possession of less than one month, so the intermediate landlord's future interest in the flat is not a

---

12   *Visible Information Packaged Systems Ltd* v *Squarepoint (London) Ltd* [2006] 30 EG 121.

MILI, and so the value of the liability must be assessed under para 8(1) to open market value as set out in para 3. A negative profit rent with no reversion can be anticipated to have a negative value.

After the grant of the new lease, the intermediate landlord's interest comprises the obligation to comply with the terms of the headlease and the new lease as regards payment of rent, management and provision of services and insurance and so on, but without the benefit of any rental income. This is not really a property interest at all, as valuers know it, having no parallels in the present property world.[13] There are no benefits of occupation in the present or in the future and no regular income. The potential for spasmodic income from licences or consents still exists, but there is no guarantee of any such income nor any predictability about its timing, if any.

The liabilities, on the other hand, are certain. The owner will have at least marginal operational costs in meeting those obligations. As much of the cost as possible will be recovered through the service charge and that the administrative costs of those elements will be similarly recoverable. However non-recoverable costs can be foreseen for such matters as:

- payment of rent and bank charges
- dealing with any rent reviews
- administrative costs
- accounting and taxation returns.

In addition to the foreseeable burdens, there may be other non-recoverable expenditure and uncertainty. For example:

- there might be service charge issues raised at an LVT and an order made under section 20(c) of the Landlord and Tenant Act 1985. Then, the costs of appearing at the LVT will not be recoverable. Indeed, to minimise this risk, the assignee must closely monitor the day to day performance of the managing agent which has cost implications
- the level of rent payable on review is uncertain (and almost bound to be much greater than the figure determined by the LVT)
- there may be correspondence on matters such as breaches of covenant or licences which are not covered by the service charge, by licence fees or which even turn out to have no basis or foundation.[14] Nevertheless, they have to be dealt with at the attendant cost. Most landlords have a capital asset or an income stream to cover these costs, but this landlord does not.
- there will be a liability to the freeholder for the adequacy of the insurance. He will expect the managing agent to be responsible for the adequacy of the sums insured etc, but this is not sufficient. He must ensure that the managing agent is doing his job properly.

Para 3(2) requires that the value of the interest is to be "... the amount which at the valuation date that interest might be expected to realise if sold on the open market ...". So a sale of the interest must be

---

13   Although these intermediate leasehold interests as a whole being pure liabilities will start to arise as claims are made and can be anticipated to come to the market in due course.

14   Most saliently in LVT cases, the intermediate landlord has costs of attending the LVT which are not covered by either the statutory recoveries or a receipt of marriage value. In fact in this case, the intermediate landlord will be out of pocket by defending his right to be adequately compensated for his on going liability and no more. In the no-Act world, the assignee has just such pleasures in store.

hypothesised. As the value is less than zero, the vendor would have to pay the purchaser to persuade him to take the valueless interest and potential liability off his hands. The vendor would have to find a new owner he could rely on:

- to pay the rents
- to manage the interests diligently
- not to abscond with the fund invested
- to remain in existence for the remainder of the headlease
- to be equally scrupulous about any onward sale.

The vendor would seek bids from that class of persons and/or companies as to the level of reverse premium required. The successful new owner will be that party having the necessary covenant strength, which requires the least subsidy or reverse premium. A cautious person or body is not normally the hypothetical purchaser because they cannot be foreseen as bidding up to the level of a more bullish purchaser. However, in this case, a cautious person or body is almost the only conceivable new owner.

It is accepted in other areas of law that reversions with negative values have to valued as on a sale to the least unlikely purchaser[15] (LUP). This is not the purchaser likely to accept the least reverse premium. In a situation where there is a real risk that no new owner could be found (ie where there is no likely purchaser at all), the LUP is the next best thing.

The old owner will want to know that the LUP will be a diligent manager, will remain in existence for the remainder of the headlease, or will be scrupulous in its choice of an onward new owner. Given the long term of the commitment, the LUP can be hypothesised as:

- a body corporate
- a substantial corporation or blue chip company
- probably in the financial services sector and possibly of the nature of a life fund.

It seems likely that there are three elements that will make up the inducement, subsidy or reverse premium that will be required by the new owner. They are:

- a net sum sufficient to induce the new owner to take on the burden of ownership
- the costs payable on the transfer
- the VAT payable on the reverse premium.

The core element of the reverse premium is a sum of money which the new owner can invest to produce an income sufficient to cover the outgoings and to provide a return from the ownership. The outgoings will include the rent payable, costs of management and tax payable on the interest received. Another approach to the valuation of the deficit income might be to capitalise it at a very secure rate, possibly well below current rates of return or of interest.

---

15   *Per* Neuberger J as he then was in *Craven (Builders) Ltd* v *Secretary of State* [2000] 1 EGLR 128.

4.21 Rentcharges

The 1977 Rentcharges Act provided for:

- the eventual expungement of rentcharges
- the compulsory redemption of most rentcharges on a set formula and
- varied section 11 of the 1967 Act so that, if the freehold in the house is subject to a rentcharge and the rentcharge exceeds the rent payable under the tenancy, when the tenant serves his notice, the landlord is responsible for procuring the release of it, at least to the extent of the excess. This has the effect of ensuring that after enfranchisement, the tenant is subject to no greater financial burden than before.

In considering whether the rentcharge exceeds the rent payable, the valuer must disregard any part of the rentcharge (a) that relates to other land, and (b) in respect of which the landlord is entitled to be exonerated or indemnified, provided these rights were passed to the tenant.

The rent with which the rentcharge is to be compared is that payable when the tenant serves his notice, but the valuer must disregard any reduction because of damage to the property, and any penal addition because of any breach of the terms of the tenancy or any collateral agreement.

The landlord can apply for the compulsory apportionment or redemption of the charge under sections 4 and 8 of the Rentcharges Act. If he so redeems the charge, it will increase the price that the tenant has to pay.

4.22 Anxiety to settle

In an enfranchisement case the tenant must pay the landlord's costs of valuation, conveyancing, and in some cases, litigation. Arguably there is less inclination on the part of tenants to take a dispute to a tribunal than that shown by freeholders and, as a result, the tenant's anxiety to settle out of court has been used as a somewhat unprincipled lever to force up the enfranchisement price. The Lands Tribunal have recognised (*Delaforce* v *Evans* (1970)) that it is necessary in some cases to adjust prices downwards to reflect tenants' anxiety to settle which has become known as "the *Delaforce* effect". The appropriate adjustment will depend on the facts of the individual case and in one case (*Wilkes* v *Larkcroft Properties Ltd* (1983)) the Court of Appeal approved no adjustment.

The matter was referred to again more recently in *Cadogan Estate* v *Hows* (1981). The Lands Tribunal preferred the landlord's evidence that, in prime central London locations, where claimants are usually represented, the *Delaforce* effect has minimal impact. They agreed with the argument originally put forward in the *Lloyd-Jones* case, that freeholders of large estates were equally desirous of avoiding references to tribunals.

4.23 Negative values

Where the value of an interest to be acquired is negative, the price to be paid for that interest is nil. However, the amount of the negativity is deducted from the value of the next superior interest until all the negativity has been exhausted.

Theoretically that rule applies to marriage value as well, but if a calculation throws up negative marriage value, there is probably something wrong. Theoretically negative marriage value simply tells you that the transaction will not take place because there will be no benefit from it.

4.24 Modern ground rents for houses

Section 15 of the 1967 Act governs the basis on which the modern ground rent (MGR) from the expiry of the current term is to be assessed. It is "the letting value of the site (without including anything for the value of the buildings on the site) for the uses to which the house and premises have been put since the commencement of the existing tenancy ...".

Essentially the MGR is usually derived from the value of the site, decapitalised at the same rate as the current ground rent is capitalised. In the absence of suitable comparables (either of letting values of sites or of site values), Tribunals have been happy to adopt what has become known as the 'standing house approach' valuation. In *Farr* v *Millersons Investments Ltd* (1971) this was described as:

> a practice of deriving the section 15 site value mainly by reference to the value of the whole premises as they stand, the site value being taken as a proportion of the entirety value. The proportion normally adopted (outside London) varies, on the evidence, between one quarter and a third the actual proportion depending on such factors as the cheapness or dearness of local values and the attributes of the site looked at in the context of the entirety. In central or near central London when other areas of highly priced residential land, the proportion adopted is commonly higher, of the order of + or –40%.

The standing house approach should, however, be regarded with some scepticism even though it is adopted in virtually all cases now. In *Miller* v *St John the Baptist College Oxford* (1977) the Lands Tribunal said:

> I would regard the standing house approach as being in the present case quite inappropriate and I would reject it but for the fact there would then be no evidence upon which I could perform my statutory duty of fixing the price.

On the use of evidence drawn from other cases decided by agreement or otherwise, Sir Eric Sachs said (*Walker* v *Gallagher Estates* (1974))

> Generally, in regard to the matter of using a multiplier I would enter a caveat against the common line of approach which introduces a fallacy. Only too often.... if one party offers a lump sum, the experts at once dissect it and refer to it as being x years' purchase of the ground rent. As a matter of arithmetic that calculation is naturally correct. The tendency, however, is for both sides thereafter to refer to the offer as if 'x years' purchase of the ground rent has been the sole or true basis of the particular offer, while the truth usually is that the sum offered simply represented in the eyes of the offeror the value to him of the reversion...

The valuer who presents evidence of comparable transactions in his cleared site approach is likely to be in receipt of considerable respect.

Example 4.3

Assess the MGR of a terraced house constructed around 1890 on a site of 500 m². A house identical to the subject property sold last month for £250,000. The site value is likely to be 35% of capital value. A capitalisation rate of 6% would be suitable for a rent of this size with a review to a market linked rent.

Standing house value	£250,000
Site value @ 35%	£87,500
Decapitalise @	6%
MGR	£5,250 pa

Valued in this way, the MGR is likely to be well above generally accepted levels of ground rent, possibly reaching what would otherwise be regarded as an open market rent for the property and certainly above fair rent levels.

MGRs can only apply to houses that were below the rateable value limits for regulated tenancies under the 1977 Rent Acts. If the long lease, the tenancy, was created before January 1989, the tenancy is a regulated one, and a fair rent can be registered for the house. The fair rent will reflect the obligations of the parties which, unusually, will probably include that the tenant is responsible for all repairs and insurance. Once a fair rent is registered, that is the maximum rent the landlord can charge, even though the MGR may be higher. The fair rent can be reviewed regularly and will probably eventually surpass the MGR. Then the maximum rent the landlord can charge will be the MGR.

However, while the MGR cannot be greater than the registered rent for the house, the landlord can serve notice on the tenant that (although he does not need to seek the cancellation of the registered rent) the registered rent no longer applies (section 16(1A)).

4.25 Valuations under section 9(1) (the original basis)

These are the valuations which apply to the original group of houses which were enfranchisable ie those which had a rateable value in 1965 of under £400 in Greater London and £200 elsewhere (£1000 and £500 by 1990). The enfranchisement price under section 9(1) is the amount which the house and premises, if sold in the open market by a willing seller, might be expected to realise on the following additional assumptions:

(a) the vendor's interest is subject to the tenancy, but the tenancy is to be extended for 50 years in accordance with the 1967 Act
(b) the extension may (notionally) be subject to the landlord's rights under section 17 (redevelopment) and the parallel right of the tenant compensation
(c) The right to enfranchise under the Act is disregarded in the valuation of the interest in order to prevent the valuation process from following ever-decreasing circles.

The valuation approach followed by the Tribunals in the vast majority of cases is to split the freeholder's interest in to three components:

(a) the right to receive the current ground rent until the original lease terminates
(b) the right to receive a modern ground rent for 50 years thereafter (with a review after 25 years)
(c) the right thereafter to vacant possession.

Section 16 of the 1967 Act excludes any further rights to the tenant after the expiry of the extended lease and therefore, the only interest to be considered beyond the MGR would be a reversion to capital value. At one time this was regarded as so far distant as to be negligible when quantified and it was often therefore, ignored or rather the MGR was taken into perpetuity.

However in the case of *Haresign* v *St John the Baptist College, Oxford* (1980), where the original lease had only three years to run, the tribunal considered it correct to value the receipt of the reversion.

Example 4.4

96 Lordswood Road, Harborne (*Brown* v *Try Best Properties Ltd* (1996)). The West Midlands LVT's valuation was as follows:

Term			
Annual rent	£4.50 pa		
YP 10.5 years @ 7%	7.26		£32.67
Reversion			
Standing House Value	£124,000		
Site Value @ 33%	£40,920		
Modern Ground Rent @ 7%	£2,864.40		
YP 50 years @ 7%		13.8007	
PV 10.5 years @ 7%	0.4917	6.7858	£19,437.25
Haresign Reversion			
Standing House Value		£124,000	
PV £1 in 60.5 years @ 7%		0.0166928	£2,069.90
Say			£21,540

Where the development potential would be of sufficient scale to have enabled the freeholder to refuse the statutory 50 year lease extension (under section 17 of the 1967 Act), that must be reflected in the valuation. In such a case the reversion is to a site value reflecting the development potential, less the compensation payable to the tenant, which is the profit rent for 50 years, ie the market rent less the MGR.

4.26 House enfranchisements after lease extension

Since 2002, the lessee of a house who has taken up the right to a section 14 lease extension also has the right to enfranchise. However section 9(1AA) provides that, if the valuation is carried out under section 9(1A), there is to be an assumption that the section 14 lease has not been granted. So, where the original term has not expired, the valuation follows section 9(1A) as usual, but as if the right to the new lease had not been taken up. However where the original term has expired, if the new lease does not exist, the price payable will be the full VP value of the unimproved house.

This will put such lessees in a quandry. A lease at a MGR for the balance of 50 years, even with a review of that MGR 25 years before the end of the lease, will still have a value because the MGR will be less than open market rental value of the house. However, if the lessee wants to take up the right to the freehold, they will have to forgo the value of the remaining lease.

Example 4.5 Valuation under section 9(1A) or 9(1C) of the 1967 Act

What price is payable to enfranchise a house held on a house held on a lease with 83 years unexpired at a rent of £50 pa, fixed throughout the term? The freehold vacant possession value of the house is £400,000.

Value of freeholder's interest

(i)	Ground rent		50	
	YP 83 years @ 7%		14.2337	712
(ii)	Reversion to freehold value		400,000	
	PV £1 deferred 83 years @ 4.75%		0.0212	8,497
	Price payable			£9,209

Example 4.6 Valuation under section 9(1A) or 9(1C) of the 1967 Act

What price is payable to enfranchise the above house if it is held on a lease with 60 years unexpired? The vacant possession value of the house held on the 60 year lease is £360,000.

Value of freeholder's interest

(i)	Ground rent		50	
	YP 60 years @ 7%		14.04	702
(ii)	Reversion to freehold value		400,000	
	PV £1 deferred 60 years @ 4.75%		0.0618	24,707

Freeholder's share of marriage value

Interests after marriage

Freehold in possession		400,000	

Interests before marriage

Freeholder's current interest	24,707		
Lessee's current interest	360,000	384,707	
Marriage value		25,293	
50% payable			12,647
Price payable			£37,354

Example 4.7 Valuation under Schedule 13 to the 1993 Act

What price is payable to extend the lease of a flat held on a house held on a lease with 83 years unexpired at a rent of £50 pa, fixed throughout the term? The value of the flat in hand to the freeholder is £400,000.

Diminution in the value of the freeholder's interest

Freeholder's interest before lease extension

(i)	Ground rent		50	
	YP 83 years @ 7%		14.2337	712
(ii)	Reversion to freehold value		400,000	
	PV £1 deferred 83 years @ 5%		0.0174	6,972
				7,684

=

Freeholder's interest after lease extension

(i)	No ground rent receivable		
ii)	Reversion to freehold value	400,000	
	PV £1 deferred 173 years @ 5%	0.0002	86
Diminution in value (price payable)			**7,598**

Example 4.8 Valuation under Schedule 13 to the 1993 Act

What price is payable to extend the lease of the above flat when it is held on a lease with 60 years unexpired. The value of the flat held on the current lease is £360,000 and held on the extended lease is £395,000.

Diminution in the value of the freeholder's interest

Freeholder's interest before lease extension

(i)	Ground rent	50	
	YP 60 years @ 7%	14.04	702
(ii)	Reversion to freehold value	400,000	
	PV £1 deferred 60 years @ 5%	0.0535	21,414
			22,116

Freeholder's interest after lease extension

(i)	No ground rent receivable		
(ii)	Reversion to freehold value	400,000	
	PV £1 deferred 150 years @ 5%	0.0007	265
Diminution in value			21,851

Freeholder's share of marriage value
Interests after marriage

Freeholder's interest	265		
Lessee's interest	395,000	395,265	

Interests before marriage

Freeholder's interest	22,116		
Lessee's current interest	360,000	382,116	
Marriage value		13,149	
50% payable			6,575
Price payable			**£28,426**

Example 4.9 Valuation under Schedule 6 to the 1993 Act

What price is payable to enfranchise a pair of maisonettes (which together form a block of flats). Both flats are held on leases with 60 years unexpired at ground rents of £50 pa fixed. Each maisonette is valued at £400,000 freehold and at £360,000 held on the current lease. Both lessees are participating.

Value of the freeholder's interest

(i)	Ground rent (2 @ £50)	100		
	YP 60 years @ 7%	14.04	1,404	
(ii)	Reversion to freehold value	800,000		
	PV £1 deferred 60 years @ 5%	0.0535	42,828	
				44,232

Value

Freeholder's share of marriage value

Value of flats held "freehold"	800,000	
Value of flats on current leases	720,000	
Value of ability to extend leases	80,000	
less value of freeholder's interest	44,232	
Gain on marriage of interests	35,768	
50% payable		17,884
Price payable		**£62,116**

Part 2

National and
Local Taxation

Taxation

In this chapter the procedure and practice relating to valuations for national taxation is examined. A valuer may be asked by a client to prepare a valuation for tax purposes. To do this the valuer needs to be aware of the purpose and taxation implications of the valuation. As well as valuers advising clients, the UK Government's tax department, Her Majesty's Revenue and Customs, also has need of valuation advice. This is provided by the District Valuers who work for the Valuation Office Agency, an executive agency of HM Revenue and Customs. The District Valuers are assisted by a number of chartered valuation surveyors who prepare and negotiate tax valuations, as well as providing valuation and estate services to government departments and local authorities.

The main national taxes are:

- Income Tax
- Corporation Tax
- Capital Gains Tax
- Inheritance Tax
- Stamp Duty Land Tax
- VAT.

All have property implications.

In addition there are the two local property taxes, Non-domestic Rates (sometimes called business rates) and Council Tax (Domestic Rates in Northern Ireland). Both do very much require statutory valuations, and are considered in the rating chapter of this book. For further consideration of rates and council tax, the reader is referred to *Rating Valuation: Principles and Practice*, Bond and Brown, Estates Gazette 2006.

The government has also proposed a new tax on development gain called the Planning Gain Supplement. While not expected to be implemented until 2008, details of the proposed workings of the tax are set out at the end of this chapter.

5.1 General principles of national taxation valuations

The earliest requirement for taxation valuations, if rating is ignored, came with death duties when Legacy Duty was extended to Real Estate in 1853. This became Estate Duty, now Inheritance Tax, in 1894.

The methodology and principles used in valuations have evolved with these taxes, and the early requirement for the valuation of unusual situations and properties for taxation purposes has been a major driver in their development. The methods and approach to valuation for taxation purposes are, therefore, essentially no different from any other valuation task. The RICS, in the Red Book, provides specific guidance for tax valuations for Capital Gains Tax, Inheritance Tax and Stamp Duty Land Tax. This is in UKGN3. It analyses the assumptions behind market value for tax purposes. Further guidance may be obtained from HM Revenue and Customs' internal guidance manuals and the Valuation Office Agency's instructions to district valuers. These are available on the respective organisations' websites (*www.hmrc.gov.uk and www.voa.gov.uk*).

There are a number of special principles, such as the requirement for a willing seller, though these may well be met when undertaking valuations for other purposes. They are set out below.

5.1.1 *Prudent lotting*

With death duties and other taxes very large estates of properties needed to be valued. One concern was whether the value to be adopted was the value of the estate sold as a single lot, or whether its division into sensible lots for sale should be envisaged, which could give a higher value.

In *Ellesmere* v *Inland Revenue Commissioners* (1918) the executors had sold a miscellaneous and scattered collection of property, comprising the whole of the estate of the Earl of Ellesmere, as a single lot. They had been advised they would probably obtain a lower total price as a single lot, but decided to sell the estate as a single lot because it avoided the risk of them being left with some unsold lots. The purchaser divided and resold the estate in lots for a much higher total consideration. The court decided that the taxation value should be based on the separate value of the various parts, estimated on the basis that the properties were sold in whatever lots would realise the best price.

In *Duke of Buccleuch* v *Inland Revenue Commissioners* (1967) the House of Lords considered the lotting should form easily identifiable natural units, and not require elaborate subdivisions involving difficulty and expense.

The breaking up of an estate could cause a flooding of the market, but the definition of market value for death duties requires any such flooding of a market caused by the notional sale of a whole estate to be disregarded.

In *Inland Revenue Commissioners* v *Gray* (1994) the Court of Appeal considered that where an estate included various assets, the hypothetical vendor must be assumed to have "taken the course which would get the largest price for the combined holding". As in *Ellesmere*, this might involve selling in lots, but equally it might involve viewing several assets as a single lot. In the case, the relevant assets were a 92.5% interest in a farming partnership, and the freehold ownership of the farmland. The court held they should be valued as one unit of property not as separate units.

5.1.2 *A sale must be assumed*

Some assets suffer from encumbrances that prevent a sale. This does not mean they have a nil value. A sale must notionally be assumed, and the price is the amount a purchaser would pay to "stand in the hypothetical vendor's shoes" and enjoy the asset subject to the same restrictions.

In *Re Aschrott, Clifton* v *Strauss* (1927) securities could not be sold because the deceased was German and his assets were frozen during the First World War. The value of the securities for estate duty was held to be the ordinary market price of similar securities.

In *Inland Revenue Commissioners* v *Crossman* (1937) the value of unquoted shares subject to stringent restrictions on transfer precluding an actual sale were considered. The House of Lords confirmed that the value for estate duty was the price that would be realised in the open market, on the assumption that a purchaser would be registered as holder of the shares subject to the same stringent restrictions.

In *Alexander* v *Inland Revenue Commissioners* (1991) the Court of Appeal considered the value of an ex-council flat, where there was a potential liability to repay a right to buy discount on a sale. The court decided the appropriate basis was the value of the flat's lease, subject to the obligation to make a repayment to the landlord in the event of a disposal, but the hypothetical sale was not to be treated as such a disposal. The hypothetical purchaser was to be, in effect, treated as stepping into the deceased's shoes.

5.1.3 Special purchaser

In *Inland Revenue Commissioners* v *Clay* (1914) the adjoining owners were in the market for the subject property in order to extend their nursing home. The question was whether the value for taxation should be the value as a private residence (£750), or its enhanced value to the nursing home (£1,000). The courts decided the fact there was a special purchaser should be taken into account. Further it should not be assumed the special purchaser would be able to secure the property for only one bid over the bids of ordinary purchasers.

The argument that a special purchaser would only need to offer a £1 more than the next best bidder is an old argument. In *IRC* v *Clay* the courts postulated speculators entering the market to compete against the special purchaser knowing they would be able to resell to the special purchaser at an enhanced price. In *Raj Vyricheria Narayana Gajapatiraju* v *Revenue Divisional Officer, Vyricheria* (the "Indian case") (1939) the Privy Council, commenting on the one bid over argument, pointed out that "the vendor is a willing vendor and not compelled by circumstances to sell his potentiality for anything he can get". The vendor will be aware of the special purchaser's interest in the property and will not accept merely £1 over any other bid, but will be able to secure a significant premium.

Special purchasers have, however, to be real. In *Walton* v *Inland Revenue Commissioners* (1996) the courts rejected a hypothetical special purchaser.

5.2 Income tax

Income tax is levied on a person's annual income. This includes income from property such as the rent from let houses or commercial property.

The taxation year runs from 6 April in any calendar year. This surprising date was adopted because it was the start of the calendar New Year until 1751, when the Gregorian calendar replaced the Julian calendar. The start of the New Year had been Lady Day, 25 March. The Calendar Act 1751 changed the start of the New Year to 1 January from 1752. Because the Julian calendar was 11 days behind the Gregorian calendar the equivalent day in the Gregorian calendar to Lady Day was 6 April, and this date stayed in use as the start of the government and tax year.

Income tax was first introduced by Pitt the Younger as a temporary tax to finance the Napoleonic Wars. Despite its importance as a source of government income it remains a temporary tax renewable each year by a Finance Act.

Income for the tax has been divided into a number of schedules. The most common was schedule E, income from employment; the familiar Pay as you Earn (PAYE), but this has been replaced by the

Income Tax (Earnings and Pensions) Act 2003. This new legislation is the work of the Tax Law Rewrite Project, which has the intention of re-writing tax legislation in a more easily understood format. Schedule E income is now referred to as Employment Income.

Schedule A income is income accruing to a person from ownership of an estate or interest in land, eg rent, Schedule C is income from government stock, schedule D is income from self employment, income from most types of investment, overseas income and miscellaneous income, and schedule F is income from company dividends. Schedule B, dealing with income from commercial woodlands, was abolished in 1988.

5.3 Corporation tax

Corporation tax is levied on the chargeable profits of all companies. A single rate is used, except for companies whose profits do not exceed a maximum level. The normal rate is 30%, with the small companies rate at 20% where chargeable profits are less than £300,000, and a marginal rate on profits between £300,000 and £1.5m.

Corporation Tax does not follow the tax year starting on 6 April; instead each company sets its own tax year.

5.4 Capital allowances

Most businesses need to use at least some capital items to operate, even if this is only a desk and a computer. The purchase of capital items cannot be set off against income for the purposes of calculating tax liability, because the expense is capital not an annual expense. Nor can a taxpayer choose to allow for the depreciation, as such, of capital items against chargeable profits. However, because capital items do depreciate, tax legislation allows for relief to be given for specified items of capital expenditure in a standardised way. These are called capital allowances and are calculated as a percentage of the capital expenditure. They are set off against income or profits for a set number of years. They are designed to ensure a fair way of calculating depreciation figures on a yearly basis, and avoiding taxpayers choosing unrealistic depreciation figures. Because the capital allowance figures are set by government, they have also been used as a tool of government policy to encourage investment in particular types of assets, by allowing accelerated depreciation.

Valuers may be asked to undertake apportionments of purchase prices where the expenditure qualifies for an allowance.

There are a number of different allowances available:

- industrial building allowances (including qualifying hotels)
- plant and machinery allowances.
- scientific research allowances on plant and machinery and buildings used for such research
- mineral extraction allowances
- agricultural and forestry buildings allowances.

Two of these allowances are examined below.

5.4.1 Industrial building allowances

A person or company can claim an allowance on capital expenditure in constructing an industrial building or structure. Allowances are confined to the expenses of construction, including site works, but not the land. Expenditure on extension or refurbishment may also be eligible for an industrial building allowance.

The definition of industrial building or structure is quite wide and can include storage where the storage is in the nature of a trade.

An annual writing down allowance at 4% of construction cost is permitted, providing the building continues to be used as an industrial building calculated on a straight line basis on the original expenditure. If the building is sold, demolished or ceases to be used for a qualifying activity within 21 years, then a balancing allowance or charge is assessed. The residue of expenditure not so far covered by allowances is compared with the consideration received by way of sale or scrap value. If the residue exceeds the amount received, then a balancing allowance of the difference can be set against tax. If the residue of expenditure is less than the amount received, then a balancing charge arises. Similar balancing allowances and charges apply to the other capital allowances.

5.4.2 Plant and machinery allowances

Allowances may be available on the provision of machinery or plant, where the machinery or plant belongs to the taxpayer and is used, or is deemed to be used, for the taxpayer's trade. This may include the purchase of a building containing plant and machinery.

Machinery or plant is not defined in the Taxes Acts, but the meaning of plant has been considered by the courts. It is a question of fact and degree as to whether any particular asset is machinery or plant. One of the earliest cases in which the meaning of plant was considered, was *Yarmouth* v *France* (1887), where the court defined plant as including:

> Whatever apparatus is used by a businessman for carrying on his business, not his stock in trade which he buys or makes the sale; but all goods and chattels fixed or moveable, live or dead which he keeps for permanent employment in his business.

The intriguing mention of plant being live or dead was because the case concerned employer's liability, and whether a horse, known to be unreliable, was or was not plant. The court decided not only was it plant, but it was also defective plant!

From the case it is clear that plant and machinery must

- not be stock in trade
- not function as the premises in which or on which the trade is carried on and
- be used for the purposes of the trade.

Expenditure on buildings and structures is generally not expenditure on machinery and plant.

The legislation includes a schedule setting out items that are specifically excluded from being machinery or plant, together with items that may or may not be plant, depending upon the facts of the case.

Taxpayers entitled to plant and machinery allowances may claim a writing down allowance on the qualifying expenditure at the rate of 25% pa, calculated on a reducing balance basis.

5.5 Capital gains tax

Capital Gains Tax was originally introduced in 1965 as a charge on actual gain accruing between the date the tax was introduced, or the date of acquisition if later, and the date of disposal. Because of inflation, the taxable gain increasingly became not just a matter of real capital gain, but also the product of inflation in money values. The tax therefore came to be seen as a tax on inflation, rather than true gain. This led to a re-basing of the tax from the date of its original introduction, 6 April 1965, to 31 March 1982, and the introduction of an indexation allowance to allow for inflation in money values.

Valuers are now rarely called upon to prepare 6 April 1965 valuations, though for disposals where the actual acquisition was prior to 31 March 1982, taxpayers may need a valuation at that date.

The principle Act is the Taxation of Chargeable Gains Act 1992.

The tax is levied on all chargeable gains made in the year of assessment. There is an annual exemption allowance. CGT is paid by individuals at their Income Tax rate, and by companies at their Corporation Tax rate.

5.5.1 Acquisition cost

The acquisition cost will normally be the price paid for the asset. If the asset was acquired before 31 March 1982, then the value at that date is used as the deemed acquisition cost.

Incidental costs of acquisition such as surveyors and legal fees, Stamp Duty Land Tax, advertising and reasonable costs of making a valuation for CGT computation, may also be deducted from the disposal proceeds. This does not extend to the cost of resolving any disagreement over the tax computation, whether by negotiation or litigation. During the period of ownership, the owner may have extended or improved the property. In effect, this is part of the acquisition cost of the asset as it now is, and can therefore be deducted from the disposal proceeds. Normal repair and maintenance and other running costs that are allowable for tax against income cannot be deducted.

5.5.2 Indexation

Indexation was introduced in 1982 to remove the inflation element from the chargeable gain. The indexation factor is calculated using the Retail Price Index (RPI) as:

$$\frac{RD - RI}{RI}$$

RD = RPI for the month in which the disposal occurs
RI = RPI for March 1982 or, if later, the month of acquisition or expenditure

In 1998 the method of allowing for inflation was simplified for non-corporate taxpayers, ie individuals, trustees and partnerships, by the introduction of a taper relief. Rather than using actual RPI inflation figures, arbitrary relief is given depending on the number of whole years the asset has been held since 5 April 1998. The relief tapers to a maximum of 10 years. Different rates of taper are applicable to business and non-business assets. Business taper applies at 50% of the chargeable gain for assets held for more than one but less than two complete years, and 25% for any period greater than two years. For non-business assets the taper applies over ten years. Indexation, however, continues for companies.

5.5.3 *Business assets taper relief*

Number of complete years after 5 April 1998	Business assets % of gain chargeable	Equivalent tax rates	
		Higher rate taxpayer	Basic rate taxpayer
0	100	40	20
1	50	20	10
2 or more	25	10	5

5.5.4 *Non-business assets taper relief*

Number of complete years after 5 April 1998	Non business assets % of gain chargeable	Equivalent tax rates	
		Higher rate taxpayer	Basic rate taxpayer
0	100	40	20
1	100	40	20
2	100	40	20
3	95	38	19
4	90	36	18
5	85	34	17
6	80	32	16
7	75	30	15
8	70	28	14
9	65	26	13
10 or more	60	24	12

5.5.5 *Part disposals*

Disposing of part of an asset can give rise to a chargeable gain or allowable loss. A formula is provided for the calculation:

$$\frac{A}{A + B}$$

A = consideration for the part disposal
B = market value of the remainder of the asset

Where the disposal is of a wasting asset, such as a lease, the expenditure incurred in acquiring, creating or improving the asset is written down due to the effluxion of time.

5.5.6 *Exemptions and reliefs*

There are a number of exemptions from CGT, including gains from disposals held by approved superannuation funds, the proceeds from the sale of trees and underwoods grown in commercial woodlands. Compensation paid to a tenant displaced under the Agricultural Holdings Act 1948 or Landlord and Tenant Act 1954 may be exempt if it arises purely from the statutory provisions.

The reliefs from CGT are more significant.

(a) *Private residence relief*

Relief is given on the disposal of a taxpayer's private residence and its grounds up to the "permitted area".

The permitted area is 0.5 ha or such larger area as is "required" for the reasonable enjoyment of the dwelling-house as a residence, having regard to its size and character.

The substantial gains often made by private individuals when they sell their homes are therefore normally exempt from CGT and free of tax. The complication for taxpayers is when there is land with the home.

The relief is only available if the dwelling house is the taxpayer's only or main residence. If a taxpayer owns a house with a large garden, but not larger than the permitted area, and obtains planning permission for a second house in the garden and sells off the plot, then the relief will apply. If, however, the taxpayer sells the house first but retains the plot, it will no longer be part of the taxpayer's residence and on subsequent disposal the gain will not be eligible for relief.

Whether the grounds exceed 0.5 ha is a matter of simple measurement: deciding whether the grounds are nonetheless within the permitted area because they are required for the reasonable enjoyment of the dwelling house as a residence, having regard to its size and character, causes much more difficulty. There is the added difficulty in determining, where the grounds exceed 0.5 ha and not all is required for the reasonable enjoyment, what parts of the grounds should be treated as the part required, and what parts not? Where a plot, or plots, are being sold off this can make a considerable difference to the initial tax liability.

The residence can include staff accommodation and may include a number of different buildings, but all must be within the same curtilage. In *Lewis (HMIT)* v *Lady Rook* (1992) a cottage 175m from the main house was held not "within the curtilage or appurtenant" to the main house.

The statute uses the word required for the reasonable enjoyment of the dwelling house as a residence. This word could simply mean that it would be pleasant or desirable to have the extra land, but in a High Court case on appeal from the Special Commissioners, *Geoffrey Longson* v *Victor Baker (HMIT)* (2001), the court equated the word "required" with "necessary". While it might be pleasant and an amenity to have plenty of land around a house, what is envisaged is that the land has to be necessary, or would be needed by anyone, in order to reasonably enjoy the particular dwelling.

(b) *Roll over relief*

Persons carrying on a trade may claim deferment of CGT on gains arising from the disposal of specified classes of business assets, providing the proceeds are exclusively spent on acquiring new assets within these classes for trade purposes.

The chargeable gain is deducted from the acquisition cost of the new assets, thereby rolling the charge on to the date of disposal of that new asset. The roll over may then be repeated if replacement assets are acquired.

(c) *Hold over relief*

Relief similar to roll over relief is available on the gift of certain qualifying assets. These assets include business assets, heritage property, works of art, gifts to political parties and gifts for national purposes.

5.6 Inheritance tax

In 1975 Capital Transfer Tax replaced Estate Duty. As its name implied this replacement taxed not just the transfer of assets on death, but those made during life as well. In 1986 the tax was altered and renamed Inheritance Tax, though it still retained an element of the idea of taxing transfers during life. In particular this includes gifts made within the last seven years of life.

5.6.1 Market value

Section 4(1) Inheritance Tax Act 1984 requires:

> On the death of any person tax shall be charged as if, immediately before his death, he had made a transfer of value and the value transferred by it had been equal to the value of his estate immediately before his death.

Section 160 Inheritance Tax Act 1984 defines market value for IHT as:

> ... the value at any time of any property shall for the purposes of this Act be the price which the property might reasonably be expected to fetch if sold in the open market at that time; but that price shall not be assumed to be reduced on the ground that the whole property is to be placed on the market at one and the same time.

The deceased's estate is defined as the "aggregate of the property to which he is beneficially entitled". This includes interests where the deceased had an interest in possession in settled property; typically a life interest.

As for any statutory valuation, it is important to know the date of valuation. For IHT the date is specified precisely as the moment before death. This was designed to ensure interests that terminate on death are treated as part of the estate. However the effect is somewhat modified by section 171 IHTA 1984. This provides that increases and decreases in value of the deceased's estate that have occurred by reason of the death shall be taken into account as if they had occurred before death. This means that a life assurance policy payment as a result of the death is treated as part of the estate. None the less, section 171 specifically states that it does not apply to changes in the value of the estate due to the termination on death of a life interest in settled property so, as intended by the moment before death valuation date, life interests remain included within the value of the deceased's estate.

While any mortgage debt will be treated as a debt owed by the estate, in valuing property for IHT the valuation should ignore the mortgage. Valuations for IHT, traditionally called probate valuations, follow the usual requirements for valuations to market value.

5.6.2 Liability

From 6 April 2006, IHT has been payable at 40% on the value of estates above £300,000. Below this figure the tax payable is nil.

Where it cannot be known which of two or more people, who died together, died first and who survived longer, the IHTA provides that they shall be assumed to have died at the same instant (commorientes). This prevents a multiple charge to tax if, say, a father and his son died together in an accident. There will be a single charge to IHT and not one on the father and then, assuming the estate had been willed to him, on the son.

5.6.3 Undivided shares

Sometimes one person owns the whole of a property; in other circumstances a property may be owned by a number of people. This could be by each person owning a defined part of a property, but there is also the situation where no one owner has a defined portion; instead each has a share in the whole. This is known as an undivided share. The legal basis is slightly complicated, but in essence there are two ways of owning an undivided share:

- a joint tenancy
- a tenancy in common.

They are quite different. With a joint tenancy, on the death of one of the tenants his or her interest automatically passes, by the doctrine of survivorship, to the remaining tenants with the ultimately surviving tenant gaining the whole interest. With a tenancy in common, each tenant has a distinct undivided share that does not automatically pass to the other tenants on death. Tenancies in common are held in trust for the tenants.

The value of an undivided share is a difficult issue. It may give the right to receive a share of the income from a property, but in the case of residential property it may also give the right to occupy with the other co-owners. Ultimately it gives a right to a share of the proceeds when the trust is ended. Generally it has not been considered that the right will be worth the share's arithmetical fraction of the whole. This is because there will usually be a degree of uncertainty with a share as to if and when the property will be sold. This will make the share less attractive to a potential purchaser, and a purchaser would expect some discount.

Sales of undivided shares are rare, and therefore evidence of actual sales of shares will not normally be available. A practice of making a 10% deduction grew out of the facts in the case of *Cust* v *Commissioners of Inland Revenue* (1917). This case involved the valuation of an estate of properties yielding an income. In a later case, *Wight and Moss* v *Commissioners of Inland Revenue* (1982), the property was a house owned jointly by Miss Wight and Miss Moss, and the Lands Tribunal had to consider the value of the late Miss Wight's share. It considered that as 10% had been appropriate for an income in Cust, it was likely a higher percentage would be appropriate where a purchaser would have the right to jointly reside in the property with the other co-owner. The Lands Tribunal adopted 15%.

Example 5.1

You have received an instruction to value a half share for probate in a 1930s built detached house. It is owned freehold by two persons as tenants in common. Both lived in the property but one has died; the other remains in occupation:

Freehold vacant possession value based on comparable sales	£700,000	
Arithmetical half		£350,000	
Allow for other half share owner being in occupation — less 15%		−£52,500
Valuation		£297,500

A factor to be considered is, while the vendor and purchaser of the share are to be regarded as hypothetical, the owner of the other share, or shares, is real. This may have an effect on the discount which is appropriate.

In a Lands Tribunal case, *Charkham* v *Commissioners of Inland Revenue* (1997) the Tribunal considered the value of minority shareholdings rather than a simple half share. The Tribunal accepted that a prospective purchaser might be concerned at the lack of control the minority status would offer, and this could increase the discount from the arithmetical proportion of the entirety value. The discounts ranged from 15% to 22.5%.

5.6.4 Lifetime transfers

As mentioned IHT was originally Capital Transfer Tax, payable on both lifetime and death transfers of value. This did not mean that transfers by sale were taxable. The idea was to tax gifts made during a person's life in the same way as bequests and legacies were taxed as part of the deceased's cumulative estate. There remains within IHT this liability or potential liability for lifetime transfers, though this is less than under Capital Transfer Tax.

The key is in the definition of transfers of value in section 3 IHTA 1984

> ... a transfer of value is a disposition made by a person (the transferor) as a result of which the value of his estate immediately after the disposition is less than it would be but for the disposition; and the amount by which it is less is the value transferred by the transfer.

It is not the value of the asset transferred which is taxable, but the loss in value to the estate. The legislation uses the word estate to mean the aggregate of all a person's property. If a man sells a property on the open market for £1m, he has made a transfer by way of sale but in return has received £1m. There is, therefore, no loss in the value of his estate. The man still has £1m, though now in currency rather than as a property.

Transfers of value include gifts, eg from aunt to niece, but will also include sales intentionally at an undervalue. In the latter circumstance there is a diminution in value of the estate because the consideration received does not balance the loss of the asset.

The loss to the estate may not be the same as the market value of the asset. This is particularly so where lifetime tax is payable, because the tax to be paid by the transferor will also be a loss to the estate. The tax payable will therefore not just be on the gift, or other transfer, but also on the tax payable on the gift. The figure transferred has to be grossed up to reflect the tax payable. If the transferee agrees to pay the tax, then the net loss to the estate is the gift only.

Example 5.2

Calculate the loss to the estate on the gift of a half share interest out of the freehold whole interest in a 1930s built detached house, from a mother to her daughter who lives with her.

Freehold vacant possession value based on comparable sales		£700,000
Value		
Arithmetical half	£350,000	
Allow for other half share owner being in occupation — less 15%	–£52,500	
Valuation of retained interest	£297,500	
Loss to the estate		£402,500

Lifetime transfers are:

- immediately chargeable
- potentially exempt transfers (PETs) or
- exempt.

Few transfers are now immediately chargeable. Before 22 March 2006 gifts into discretionary trusts or to companies fell into this category. For gifts made on or after 22 March 2006, an immediately chargeable transfer is one made to the trustees of a relevant property trust or to a company.

Transfers made during life are usually exempt providing they are made at least seven years before death. A transfer made during life is therefore potentially exempt. If the person lives for a further seven years the transfer becomes fully exempt. If the person dies the exemption is lost. The transfer becomes taxable.

If the transfer is made more than three years before death, then a reduced rate of tax is payable. A gift made up to three years before death attracts the full rate of tax, though the value of the gift is as at the date it was made.

Taper relief is at 80% of the full rate if the person died between three to four years of the transfer; at 60% of the full rate if between four to five years; 40% of the full rate if between five to six years; and 20% of the full rate if between six to seven years. Taper Relief is only available on that part of the gift, or gifts, exceeding the date of death threshold.

To avoid someone transferring an interest as a PET but retaining all the advantages of use, eg transferring the freehold of a house to a daughter but continuing to live in it, such situations are gifts with reservations and treated as if the gift formed part of the deceased's estate at death. The circumstances treated as gifts with reservations are:

- where bona fide possession and enjoyment of the property was not assumed by the donee at or before the beginning of the period between gift and death, unless the period is longer than seven years or
- at any time in this period the property was not enjoyed to the entire or virtual exclusion of the donor and of any benefit to the donor by contract or otherwise.

5.6.5 Exemptions

To deal with avoidance and the multitude of assets and interests persons may possess, the provisions for what is and what are not taxable are complex. Greater detail can be found in books dealing specifically with the intricacies of IHT.

- Foreign property owned by a person living abroad, are excluded from being treated as part of the person's estate.
- Various transfers are made specifically exempt. The main ones are set out below.
- Transfers between husband and wife or between civil partners are exempt. This applies to both lifetime and death transfers. If a spouse dies and leaves most of the estate to the other spouse, tax is only liable on bequests to others such as the spouse's children.
- There is an annual exemption of £3,000 for lifetime transfers. In addition the first £250 of any number of gifts is also exempt. This allows small gifts to be made in any one tax year without them being treated as PETs.

- Lifetime transfers made out of income are exempt if they form part of a person's normal and habitual expenditure, eg sums given to provide for a person's well being.
- There are exemptions for lifetime transfers as wedding gifts.
- Transfers to charities, political parties, housing associations, defined national purposes such as The National Trust, museums and universities are wholly exempt.
- Transfers of National Heritage property to non-profit making bodies or to private owners providing undertakings are made allowing for public access and for maintenance of the property.

5.6.6 Reliefs

(a) Quick succession relief

To prevent estates being decimated by successive beneficiaries dying within a short time of each other, relief is given on death if there was a chargeable transfer within the previous five years. This quick succession relief is given by reducing the tax payable on the death, by a proportion of the tax payable on the earlier transfer on a tapering scale of 100% if the transfer occurred in the year before death, down to 20% if it took place between four to five years before death.

(b) Agricultural and woodland relief

Relief is available for the agricultural value of agricultural land or pasture, woodlands and any building used in connection with intensive livestock, providing the occupation is ancillary to that of land.

Agricultural property is defined as:

agricultural land or pasture and includes woodland and any building used in connection with the intensive rearing of livestock or fish if the woodland or building is occupied with agricultural land or pasture and the occupation is ancillary to that of the agricultural land or pasture; and also includes such cottages, farm buildings and farmhouses, together with the land occupied with them, as are of a character appropriate to the property.

Agricultural value is the value on the basis of a perpetual covenant prohibiting any other use, ie it is a current use value excluding any non-agricultural value. Business relief may be available on the additional non-agricultural value.

The relief is at 100% for land held for at least two years with vacant possession and farmed alone, or in partnership. Where the owner does not have, and will not have, vacant possession within 12 months, but nonetheless has owned it for seven years, the relief is at 50%.

In *Lloyds TSB v Commissioners of Inland Revenue* (2001) the Lands Tribunal had to decide whether the bids of lifestyle buyers should be taken into account in determining what was the agricultural value of a large Grade II* listed, five bedroom house owned with the surrounding 50 ha of agricultural land. The taxpayer's representatives argued lifestyle buyers dominated the market for this sort of well located property. They would use the land for farming, and therefore their bids represented agricultural value. The Tribunal considered "a farmhouse is the chief dwelling house attached to a farm, the house in which the farmer lives", and the farmer is "the person who lives in the farmhouse in order to farm the land comprised in the farm and who farms the land on a day to day basis". The Tribunal therefore decided the bids from lifestyle buyers, who would not primarily be farmers but would have other careers elsewhere, should be excluded in arriving at the agricultural value. The agricultural value figure determined was 30% less than the agreed open market value.

Farmhouses, cottages and farm buildings are included in the definition of agricultural property, and gain relief where they are of "character appropriate" to the property. The phrase "character appropriate" means that not all dwelling houses and other buildings occupied with agricultural land gain exemption. The test is in relation to the property comprising the transferor's estate. What is, or is not, appropriate is a matter of judgment. Is the farmhouse or buildings proportionate in size and nature to the requirements of the likely farming activities to be conducted on the agricultural property by a typical fit farmer occupying the property for the purposes of agriculture as at the date of transfer/death?

In *Executors of David and Lady Cecilia McKenna v Commissioners for HM Revenue and Customs* (2006) the Special Commissioners decided a seven bedroom Cornish house with 45 ha of farmland and other land, was not a farmhouse and, even if it was, it was not of character appropriate to the land. It was at the very top end of the size of farmhouses in Cornwall, and farms with a house that size had more agricultural land. When the executors sold the house it was sold as a large country house with farmland, and not as a farm with a house; its value was well out of proportion with the profitability of the farm.

Where a large farmhouse is occupied with a smallholding then the reasonable view may be that it is not a house occupied with land, but rather it is primarily a dwelling which has land with it. The size of the dwelling house is not, therefore, appropriate in character to the land.

(c) Business property relief

Relief is provided for transfers of value if the whole or part of the value transferred relates to relevant business property. The conditions needing to be satisfied for an asset to be relevant business property are complex and include a condition that it has been owned for at least two years, and is of a specified description such as being an interest in a business, or shares in a company controlled by the deceased. The relief is at 100% for some property and 50% for others.

(c) Falls in value relief

Relief is provided where the transfer occurs at a high point in the market and after this date a sale occurs at a lower value. Essentially the relief applies for sales within three years of death, though sales in the fourth year at less than the value at death are treated as having occurred in the third year. The relief is given by substituting the actual sale value for the value adopted at death. Adjustments may be needed to the sale price where other property is included in the sale, or it is not at best value, or for certain other specified matters,, to ensure the sale price is on a comparable basis to the original value on death.

5.7 Stamp Duty Land Tax

Since 1694 a tax has been imposed on documents known as Stamp Duty. Certain legal documents were simply not effective until they had been stamped. From 1 December 2003 Stamp Duty has been replaced by Stamp Duty Land Tax (SDLT). The new tax retains many features of the earlier tax. It raises some £6 billion annually, which is more than IHT and CGT together, so it is a major tax.

The reason for the change brought about by the Finance Act 2003 was to modernise stamp duty ready for electronic conveyancing, to raise more revenue and to remove many opportunities for tax avoidance. The new tax has been criticised for its complexity and the burden it places on purchasers or new lessees.

The liability to SDLT rests on the purchaser who must deliver a land transaction form (form SDLT1) to HM Revenue and Customs within 30 days after the effective date of every notifiable land transaction. The form has to include a self-assessment of tax liability and be signed by the purchaser.

It is now not a payment for a stamp on a document, but a tax on land transactions. These are defined as the acquisition of a chargeable interest. These are:

- the acquisition of a freehold or leasehold interest in land
- the grant of a lease for a term of seven years or more
- any other transaction that attracts SDLT.

Mortgages, licences, and tenancies at will are exempt. The provisions are careful to ensure that leases do not masquerade as licences. The tax is interested in substance not name.

Liability arises on the legal completion or substantial performance of the contract. Substantial performance is when the purchaser takes possession or receives rent for a substantial part of the land or a substantial proportion of the consideration is paid. With Stamp Duty it was possible to exchange contracts but not complete. As Stamp Duty was only payable for stamping the conveyance and not the contract it was possible to avoid the tax by occupying and using the property under the contract only. SDLT avoids this position by providing that the transaction becomes notifiable on the substantial performance of the contract for the transaction.

The tax rate varies between residential and non-residential.

Residential

Consideration	%
Not more than £125,000	0
£125,001–250,000	1
£250,001–500,000	3
More than £500,000	4

Note for transactions in disadvantaged areas the £125,000 limit is replaced by £150,000

Non-residential or mixed

Consideration	%
Not more than £150,000	0
£150,001–250,000	1
£250,001–500,000	3
More than £500,000	4

Liability for leases arises on the earlier of:

- the grant of the lease
- taking possession
- payment of the whole or a substantial part of a premium
- first payment of rent.

As with freehold transactions the time a lease becomes notifiable is when it is effectively in operation, not just when it is finally granted.

Where premiums are payable, the tax rates are as for freehold transactions. SDLT is also payable on the level of rent at a flat rate of 1% on the net present value of the total rent payment, discounted at 3.5% pa. As with capital transactions no duty is paid if the NPV is less than £125,000 (residential) or £150,000 (non-residential) but, unlike capital transactions, duty is only payable on the amount in excess of these nil rate bands. The Finance Act 2003 provides a formula for calculating NPV. The idea is that the NPV can be calculated by a solicitor or other non-valuer. There is no requirement for the market value of the rents to be assessed. The calculation is a formula only.

$$V = \sum_{i=1}^{n} \frac{r_i}{(1 + T)^i}$$

V = net present value of rent
r_i = rent payable in year 1
i   = the first, second, third etc year of the term
n = term of lease
T is the discount rate

Rather surprisingly the Act creates uncertainty with leases, as there is a risk the tax may have to be re-calculated during the life of the lease, and the then lessee may be liable for an unexpected tax bill. Provision is made for extra tax if the rents at review show an increase of more than RPI + 5% over the highest rent payable during the previous review period.

SDLT represents a significant increase in the tax payable on the grant of leases compared to stamp duty Stamp Duty: up to eight times greater on a 25 year lease.

Relief is given for inter-company transactions and for reconstructions of companies. Purchases by charities for charitable purposes are exempt.

5.8 Value Added Tax

5.8.1 *History*

VAT replaced the earlier Purchase Tax as the UK's expenditure tax so as to bring the UK into line with the European Community. It was introduced on 1 January 1973. The main legislation is contained in the Value Added Tax Act 1994, but there are also many statutory instruments.

Up until a ruling in favour of the EC in the European Court of Justice in 1988, the UK resisted EC moves to harmonise VAT taxation across member states. The court ruled that zero rating on new housing constructed by builders was lawful on the basis that it was a social provision, but not for the construction of buildings for industrial or commercial use. Following the ruling the government changed the rules for VAT treatment of land and commercial construction from 1 April 1989.

5.8.2 *Introduction*

To the surveyor versed in the clear logic and straightforwardness of rating and other property taxes, VAT seems a strange tax whose basic concept, let alone workings, can be difficult to grasp.

For VAT to be charged there has to be:

(a) a taxable supply of goods or services
(b) made by a taxable person
(c) in the course or furtherance of a business.

All of these three elements have to be present for there to be a VAT charge.

VAT is a transaction based tax, designed to be chargeable on the end consumer. Intermediary consumers, ie traders, are able to reclaim any VAT they pay, with the final consumer, having no-one to pass the tax on to, paying the full rate of tax when making his or her purchase. The various intermediary traders pay VAT on the raw materials, parts and other goods purchased and have VAT paid to them on the goods or services they sell. These payments are called input tax and output tax respectively. The sale of goods or services, or the purchase of goods and services, are known as supplies. Traders make VAT returns, showing their quarterly input and output taxes to HM Revenue and Customs. Broadly, if their output tax exceeds their input tax they pay the surplus to HM Revenue and Customs: if input exceeds output tax they obtain a refund. Supplies of goods and services are either:

- standard rated — taxed at 17.5% (reduced rate of 5% on fuel and some other goods)
- zero rated — taxed at 0% but intermediary consumers can reclaim their input tax because zero rating is notionally output tax
- exempt — not taxable and therefore intermediary consumers cannot reclaim input tax.

For taxable traders standard or zero rating is preferable to exemption from VAT.

Businesses that are not registered for VAT, or make exempt supplies, are unable to reclaim input tax and therefore are in a similar position to an individual making a purchase — in both cases they pay VAT on the purchase of goods and services, but cannot pass the burden on. Such suppliers are described as VAT averse. They include banks, building societies, insurance companies, bookmakers, private hospitals, universities, charities, welfare services and schools — all in all quite a force in the market place.

As a general rule, input tax can be deducted only if it can be attributed, either directly or indirectly, to a particular taxable supply, eg the purchase of timber, glue and varnish (input tax) is directly attributable to the supply of a table (output tax). Where businesses make both standard and exempt supplies, then they will need to attribute the supplies they receive separately as they will only be able to reclaim the input tax on the goods and services supplied to them that are attributable to the taxable supplies they make.

5.8.3 Taxable persons

A taxable person is someone who makes taxable supplies of goods or services while he is, or is required to be, registered for VAT. Registration is compulsory if a person makes, or anticipates making, taxable supplies in excess of the prescribed figure (£64,000 as from 1 April 2007) in a year. It is possible to register, subject to the commissioners approval, when annual turnover is less than the level of compulsory registration so that input tax can be reclaimed. HM Revenue and Customs have wide powers of enforcement and can require returns where the commissioners have reasonable grounds for believing a business is making taxable supplies in excess of the current limit. There are substantial penalties for failure to register when turnover is in excess of the limit.

5.8.4 *Supplies of goods and services*

Any transaction involving the transfer of the ownership of goods, or where services are supplied for a consideration, is a supply. It is basic concept for VAT legislation. Wherever there is a supply, there may be a liability.

- Supply of goods includes outright sale, hire purchase, grant of a major interest in land, or the supply of power.
- Supply of services includes professional services.

The transfer of a business as a going concern is not deemed to be a supply providing:

- the assets are to be used by the transferee in the same type of business
- if the transferor is a taxable person, then the transferee must also be (or become) a taxable person and
- the assets transferred must be capable of separate occupation.

The sale of land and buildings in the course of development has been held to be the sale of a going concern.

The tax point is normally the earliest of the date the services are performed, or the goods delivered. Payment is made on the date the tax invoice is delivered. A tax invoice is necessary to establish the recipient's right to reclaim the tax paid, and must contain certain particulars. There is an obligation to supply a tax invoice when a supply is made by a registered person.

5.8.5 *Tax Rates*

Standard rate (17.5%) or the reduced rate (5%) applies to all supplies of goods and services by a taxable person that are not zero rated or exempt. The following are the main types of zero rated and exempt supplies.

(a) *Zero rating*

- Food, (excluding alcoholic drinks (subject to excise duty), sweets, chocolate and catering supplies), books, newspapers, water, public transport, medicines, children's clothes and shoes and supplies by charities.
- The grant of a major interest (ie the sale of the freehold or the grant of a lease for more than 21 years) in a building (or any part) providing:
  - the building is designed as one or more dwellings or is used for communal residential purposes or charitable purposes and
  - the supply is made for the person constructing the building; services (except professional services) or materials supplied in the course of construction of such a building are also zero rated. NB construction services supplied by a surveyor, architect or supervisor are standard, not zero rated.
- The grant of a major interest in a substantially reconstructed listed building, providing the building will be a dwelling or dwellings, or will be used for relevant charitable purposes.

(b) Exempt

- Insurance, betting, normal banking operations, education in a school or university, health or fund raising by charities.
- The grant or assignment of any interest in or right over land, or of any licence to occupy land, including the surrender of an interest in land, but excluding the supply of the following (which are standard rated):
  - a freehold interest in a new or partly completed building, provided that it is neither designed or intended to be used for residential or charitable purposes. New buildings are those completed after 1 April 1989 and are completed less than three years before the date of sale
  - a freehold interest in a new or partly completed civil engineering work. A civil engineering work is new if it was completed less than three years before the grant
  - accommodation in a hotel, inn etc
  - holiday accommodation in a house, flat, caravan or house-boat.

Example 5.3

A simple example of the VAT position in the making and sale of a table shows how progressively VAT is charged, and how input tax is reclaimed. The example begins with an oak tree and ends with a table:

Forester
A forester fells an oak tree and sells it to a saw mill for £400 plus VAT @ 17.5%

Output VAT on sale of tree	£70
Less input VAT incurred on growing of trees	£NIL
Balance paid to HMRC	**£70**

Saw mill
The saw mill converts the tree into waney edge boards, and puts them in stick to dry for two years, before selling them to a carpenter for £1200 plus VAT @ 17.5%

Output VAT on sale of boards	£210
Less input VAT incurred on purchase of trees	£70
Balance paid to HMRC	**£140**

Carpenter
The carpenter builds tables from the oak, and sells them to a shop for a total of £10,000 plus VAT @ 17.5%

Output VAT on wholesale of tables	£1,750
Less input VAT incurred on purchase of boards	£210
Balance paid to HMRC	**£1,540**

Furniture shop
The shop sells the tables to customers for a total of £22,000 plus VAT @ 17.5%

Output VAT on retail sale of tables	£3,850
Less input VAT incurred on wholesale purchase of tables	£1,750
Balance paid to HMRC	**£2,100**

The customers pay the final VAT bill as ultimate consumers and have nobody to pass it on to and their payments effectively includes all the sums paid by previous suppliers.

5.8.6 Residential property

New housing is zero rated both when sold freehold, or when a lease for more than 21 years is granted. Building material supplies and building services attributable to the new housing are also zero rated.

"Second hand" sales of residential property are normally outside the VAT net because they are usually made by private individuals who are not selling in furtherance of a business. If the supplier is, for example, a developer, then the supply is classed as exempt and the supplier will be unable to recover input tax.

5.8.7 Commercial property

Freehold sales of new buildings are standard rated. New buildings are those completed less than three years before the date of sale.

Second hand, ie old buildings, are an exempt supply (but it is possible to waive the exemption as explained above with the option to tax).

Sales of short or long leases in new or second hand buildings are exempt (but there is the option to tax).

The table below sets out the position for various building works, sales and grants of leases.

Residential or Charitable Buildings
Building works

Construction of a new building	zero rated
Conversion of a non-residential building to residential, providing supplied to a registered social landlord	zero rated
Alterations to listed buildings	zero rated
Renovating single household dwellings that have been empty for three years or more	reduced rate (5%)
Conversion of houses to flats or non-residential property to residential	reduced rate (5%)
Conversion of property to a care home	reduced rate (5%)
Other building works	standard rated

Sales and grants of interests in buildings:

Sale or grant of a lease in excess of 21 years by a person constructing a building, converting a non-residential building into residential, substantially reconstructing a listed building or renovating a residential building that has been empty for ten years	zero rated
Other grants	exempt

Land and Commercial Buildings
Building works

All building and civil engineering works	standard rated

Grants of interests in land/buildings:

Freehold sale of a new building or land underneath new civil engineering work	standard rated
Grant of land to an individual to build an owner occupied dwelling for the individual	exempt
Grant of land to a Registered Social Landlord who certifies it will be used for construction of homes	exempt
All other grants and assignments	exempt with option to tax
Demolition work	standard rated

5.8.8 *The option to tax*

Prior to 1 April 1989, virtually all transactions in land and buildings did not attract a VAT liability and therefore VAT was of little interest to valuers.

Now many transactions are liable, though some, such as the grant of a lease, are exempt. *Prima facie* this sounds attractive because exemption means VAT does not have to be added at 17.5% to rents, but it also means that landlords cannot recover any input tax that they paid on acquisition, construction or refurbishment.

However it is possible to waive exemption. A registered supplier can elect to waive exemption. This option to tax, permits a business to convert an exempt supply to one that is taxable at the standard rate, thereby allowing input tax to be recovered. There is, accordingly, a strong incentive for landlords to elect. Most tenants will themselves be registered for VAT, and the imposition of VAT will not significantly affect them because they will be able to recover it as input tax.

Indeed tenants may welcome such election because, as service charges are normally expressed in leases as additional rent, the service charge will cease to be an exempt supply and become taxable, allowing recovery of input tax on goods and services included in the charge. In theory this should reduce the net of VAT cost of the service charge. However payment of VAT, even when input tax can be recovered, imposes a burden on the payer and may cause cash flow difficulties. The cash flow implications of VAT should not be underestimated.

Approximately 90% of all VAT registered businesses are able to reclaim input tax paid on rent, and therefore an election should not have any effect on the rent payable or the analysis of the rent for valuation purposes. The effect of election can, however, adversely affect the remaining 10% because they are unable to reclaim the additional input tax. The remaining 10%, includes banks, building societies, insurance companies, bookmakers, private hospitals, universities, charities, welfare services and schools, so comprises bodies with substantial monetary resources and bargaining strength, which is disproportionate to their number, and can strongly influence the market.

Such tenants might when considering taking a letting either:

* reduce their bid to offset the extra tax or
* offer a higher rent in exchange for a binding undertaking by the landlord not to opt to tax.

However the rental value of property in a particular locality will only be affected if the market for a type of property in that locality is dominated by VAT averse tenants. Elsewhere VAT averse tenants will be outbid if they offer less rent than VAT registered tenants. The balance of demand between VAT averse and VAT registered tenants will determine whether there is any discernable effect on the rents agreed, and premiums paid, on assignment in the locality.

Where a market is dominated by VAT averse tenants, landlords are only likely to elect where they have incurred significant costs, eg for refurbishment, and judge the benefit of recouping input VAT outweighs the depressive effect on rental values.

Once exercised the option cannot be revoked for 20 years by the person who exercised it. If the interest is acquired by a new owner, then the option becomes exercisable afresh. Similarly if the building is demolished and a new building erected, then the option lapses and the owner can make a fresh decision.

A single landlord cannot exercise the option for part of a building, but only for the whole building. This may make the decision whether or not to exercise the option difficult where the building is, or will be, occupied by both VAT registered and VAT averse tenants.

5.8.9 *Capital Goods Scheme*

The Capital Goods Scheme was introduced in 1990 to ensure traders cannot gain the advantage of being able to recover input tax on a capital purchase but then, due to a change in their taxable status, not charge VAT or charge a disproportionately low amount of VAT on their outputs. The scheme requires an adjustment to the original first year treatment of VAT incurred on the acquisition of certain capital goods, where the degree of their use in making taxable supplies changes in subsequent years. The capital goods concerned are:

- computers and items of computer equipment where the value of the supply is £50,000 or more and
- land and buildings (or parts of buildings) where the value of the supply is £250,000 or more eg freeholds, leaseholds and extensions increasing the floor area by 10% or more.

In both cases the scheme only applies where the goods were acquired on or after 1 April 1990 and are used to some extent for non-taxable purposes.

Self supplies are also included where within ten years of the completion of construction either:

- a zero rated building (other than a dwelling) is first used for a non-residential or non-charitable purpose or
- a commercial building is first subject to an exempt supply, or is first used by a person who is not fully taxable.

If the scheme applies then the initial VAT treatment is reviewed annually and an adjustment made where the business either:

- started off wholly taxable and becomes wholly or partly exempt during the review period or
- started off wholly or partly exempt, if the extent of its input tax recovery varies during the review period.

The adjustment is made using the formula:

$$\frac{\text{Total input tax on capital item}}{\text{Number of years of review period}} \times \text{the adjustment percentage}$$

The review period is five years for computers, or ten for land and buildings.

The adjustment percentage each year is the percentage by which the use of the capital item in making taxable supplies increases or decreases in the later year, by comparison of the first year. This results in an additional VAT charge or rebate.

Example 5.4

Assume a trader purchases an asset for £300,000 plus £52,500 VAT for use in the business. The trader is at present partially exempt to the extent that 60% of non-attributable input tax can be reclaimed, but each year the trader's recovery percentage changes. The following table shows the initial amount of input tax the trader can reclaim and the further reclaims and repayments the trader will make over the ten year period as the recovery percentage changes.

Year	Recovery	VAT Reclaimed/ Repaid	Calculation
1	60%	+31500	£52,500 × 60%
2	50%	−525	£52,500 × 10% ÷10
3	50%	−525	£52,500 × 10% ÷ 10
4	60%	0	
5	80%	+1050	£52,500 × 20% ÷ 10
6	60%	0	
7	50%	−525	£52,500 × 10% ÷ 10
8	30%	−1575	£52,500 × 30% ÷ 10
9	60%	0	
10	80%	+1050	£52,500 × 20% ÷ 10

5.9 Planning gain supplement

The Barker report, *Review of Housing Supply: Delivering Stability — Securing our Future Housing Needs* 2004, recommended financing the infrastructure needed to achieve a substantial increase in the supply of housing by capturing a portion of the increase in land value occurring when planning permission is granted.

The idea of taxing the value over and above the existing use of land has a long history. The rationale is that the extra value, the windfall gain, has been created by the community as a whole in the provision of infrastructure, and indeed the demand for land, rather than by the owner. Therefore it is right for the community, or the state, to take a large share of this gain. The facts will, of course, vary and the owner may sometimes have had to do a great deal to devise or originate the development scheme and secure permission. Usually the idea of taxing development gain can be seen as a matter of political viewpoint, though it should not be forgotten that it was a Conservative government that introduced Development Gains Tax.

The original scheme to tax development gain, or betterment, was the Development Charge in the Town and Country Planning Act 1947. This act nationalised development rights and provided for a 100% charge on development value; the intention being that the vendor received the existing use value, and the state the development value. It was abolished in 1954.

Betterment Levy, introduced in 1967, was seen as a temporary measure while the new Land Commission purchased all land needed for development. The tax on transactions and actual development was set at 40% with the intention of this rising in the future. It was repealed in 1971.

Substantial gains made on the sale of land for development led to the introduction of Development Gains Tax from December 1973. This was levied whenever there was a disposal or notional disposal of land or buildings with development value. DGT was superseded in 1976 by Development Land Tax.

Development Land Tax was brought in as a joint measure together with the Community Land Act that had the aim, at some future point in time, to vest all development rights in the state. DLT was assessed on the net disposal proceeds less one of three bases calculated from acquisition cost or current use value. DLT was abolished in 1985.

In considering whether to look again at betterment, the present government has been mindful of the failure of the previous attempts to tax this land value uplift. One contributing factor was undoubtedly the high rates at which it was levied; 100% for the Development Charge, 110% for Betterment Levy, between 52% and 82% for DGT and 80% for DLT.

It should not be forgotten that CGT does, to an extent, tax betterment. Indeed DGT was very much built on CGT. However, with the re-basing of the tax and the introduction of indexing, the impact of CGT has been reduced. None the less a taxpayer who is able to secure a valuable planning permission that could not have been obtained at the base date in 1982, will on disposal have a tax liability not far off 30%, even allowing for indexing and annual exemption.

At the time of writing the government is consulting and developing ideas for a planning gain supplement. The consultation paper suggests that "PGS would be set at a modest rate to capture a portion of the land value uplift created by the planning process". The likely level of the tax appears to be around 20%. At first sight this is very much lower than the earlier taxes though, in some cases, this will be additional to Capital Gains Tax at 30% on the chargeable gain.

Since the abolition of DLT local authorities have been increasingly using section 106 of the Town and Country Planning Act 1990 to achieve some contribution from developers to the cost of local infrastructure. In granting planning permission local planning authorities are permitted to make agreements, section 106 agreements, with developers imposing specific requirements, eg for the developer to provide a proportion of the development as affordable housing, open spaces, transport, education or community facilities. Planning guidance states that obligations secured under section106 should be relevant, necessary, reasonable, fairly and reasonably related in scale to the proposed development and directly related to the proposed development. In practice different authorities use these powers in different ways, some even adopting tariff based systems and arguably going beyond being directly related to the planned development. While section 106 agreements can be seen as flexible and capable of accommodating local needs, a developer will find different treatment and cost depending on where the proposed development is to be undertaken. This variation between different local authorities can be seen as inconsistent and unfair.

The government's current proposals do not abolish section 106 agreements, but retain them in a reduced form to mitigate the impact of development and provide affordable housing. They are seen as part of a set of measures, together with the PGS, to generate extra resources to invest in infrastructure.

PGS is not intended to arise on the sale of land, but at the time full planning permission is granted. So if X sells land to a developer, X may be liable for Capital Gains Tax but not PGS. When the developer gains full planning permission the level of PGS is assessed. To take account of cash flow it is intended that the developer's payments of PGS will be phased as the development progresses.

The proposed way of calculating PGS liability is to subtract the current use value (CUV) of the site from its value (its planning value or PV) at the moment after full planning permission has been granted, and then by multiplying the difference , the land value uplift, by whatever rate, possibly 20%, is set for PGS:

$$PV - CUV \times PGS \text{ rate } (\%) = PGS \text{ liability}$$

The consultation paper defines CUV as:

> the market value of the land the moment before full planning permission is granted assuming that it is and will continue to be unlawful to carry out any development other than development permitted under the General Development Order 1995 or development in accordance with planning permission granted before an appointed day.

The intention is to make the tax self assessment, with developers submitting the two valuations, CUV and PV, and with the district valuer sampling and monitoring the valuations supplied.

Clearly, if the tax is introduced, valuers will need to assist developers in preparing valuations on both the CUV and PV bases.

Rating and Council Tax

6.1 It's all cloud-cuckoo land

Don't let anyone tell you that valuation is a science; it is more of an art, and, according to Godfrey JA in *China Light and Power Company Ltd* v *Commissioner of Rating and Valuation* (1996), encompasses fiction and imagination:

> The world of rating appears, to one unfamiliar with the arcana, to be cloud cuckoo land, a world of virtual unreality from which real cuckoos are excluded (although it seems that permission to land will be granted to a cuckoo flying in from the real world if it can demonstrate that its presence in cloud cuckoo land is essential, not merely accidental) ... A valuation for rating must be based on hypothetical, not real, facts. Nevertheless, the authorities establish that such a valuation is itself to be treated as a matter of fact, impeachable only if the person responsible for the valuation has fallen into some error law in arriving at it.

6.2 Overview of local taxation

Many countries around the world separate their central government functions (generally strategic issues, for example legislative and monetary policy) from local government functions (often social-type issues, for example providing education, health care, housing and street cleaning). Ignoring external revenues, the funds to perform the central and local government functions, and the infra-structure that is necessary, largely come from the people and businesses that are resident in that country.

When we speak about local government in the UK, we are including such bodies as district councils, borough councils, metropolitan borough councils, the States of Jersey, and so on. In the UK, examples of local government provided services include:

- schools and nursery facilities; community education and youth services
- libraries, parks and open spaces; leisure facilities including swimming pools and recreation centres
- social care for the elderly, children and other vulnerable members of the community
- galleries and museums
- support to the voluntary sector
- planning and building control

- refuse collection; street cleaning
- maintenance of roads and bridges; traffic management and road safety
- elections, registrars of births marriages and deaths; cemeteries, crematoria and mortuary services
- consumer protection
- economic development and regeneration; community development services
- housing, including the provision of social housing, housing strategy and advice and services for the homeless.

Note that this chapter is mainly written about the English system of local taxation, with attention drawn to differences from systems in other parts of the UK.

6.2.1 Sources of income

Local government bodies receive their finance from various sources.

(1) Monies from central government, a charge from national taxation, makes up about 75% of the revenue on average. This is known as the revenue support grant, and is split into two types: grants towards the cost of specific services, and a revenue grant which in effect diminishes the amount to be raised by local taxation. The amount of revenue support grant will reflect how much money central government perceives is needed for local authority services, less the income which it is considered should come from council tax (paid by residents) and rates on non-domestic property (paid by commercial/business occupiers and/or owners).

(2) The local authority raises some income by charging for services or facilities it provides, such as markets, abattoirs, cemeteries and crematoria, car parks, on-street parking, council house rents, property lettings and sales, some forms of refuse collection, licensing dues etc. These sources are sometimes
  - discretionary — a local decision is made to levy the charge or not or
  - mandatory — the local authority is required to make charges
  but in general the yield is small in relation to total finance required. One new form of local authority charge is the congestion charge in central London, where vehicular traffic within a defined area is charged a flat rate per day, (£8 at the time of writing).
  A problem with income from these sources is that they are inflexible; the local authority cannot always raise its charges in line with inflation, yet it cannot stop providing the service, and often direct costs are not even covered — the service is a loss-maker.

(3) Local taxation makes up the remaining income. In the UK, this local taxation is in two parts, known as business rates (also called the national non-domestic rate); and council tax. Together business rates and council tax provide just under 24% of local government revenue on average. The non-domestic rate is collected locally but pooled by the Exchequer and redistributed to local authorities based on a per-capita formula.

Local authorities act as the collecting agency for various Business Improvement District (BID) schemes across England and Wales, which have begun operating since 2004. These do not contribute to local authority income, but are specifically allocated to the BID management body which is elected locally for defined local projects. BIDs are discussed in more detail in section 6.9 below.

The question arises: when rates are essentially re-allocated by central government, and a relatively small proportion of income is raised and retained locally, why all local authority expenditure is not directly funded from central government? This is one of the questions currently being considered by the Lyons review, ie whether the retention of a link between the spending body (the local council) and the electorate (who vote for the local councilors) is an important element in the UK structure.

6.2.2 Business consultation with local authorities

Local authorities are under an obligation to consult with local businesses, in order to make the link with local businesses more substantial. Otherwise it could be argued that there is little point in a business not paying its non-domestic rate directly to central government.

6.2.3 Lyons review of local government

The Lyons review of local government is on-going at the time of writing. One of the issues it is considering is the "gearing effect". Due to the fact that locally-raised and locally-retained income makes up a relatively small proportion of local government expenditure, if a local authority wishes to raise more revenue, it has to raise its council taxes by a ratio of approximately four times. This is known as the gearing effect, and is politically a sensitive issue. It links to the concept of control of local authorities via the ballot box.

6.3 Who's who in local taxation

6.3.1 Department of Communities and Local Government

At the time of writing, the Department of Communities and Local Government (DCLG) has legislative responsibility for local government, including the rating and council tax systems; and the pooling and redistribution of the locally collected non-domestic rates. DCLG was a new government department created in 2006; reorganised from the Office of the Deputy Prime Minister.

The Welsh Assembly Government (WAG) carries out the same role in respect of Wales.

6.3.2 Demarcation between assessor and collector of local taxes

Since the LGFA 1948, the bodies responsible for setting the local tax (then the general rate) and collecting the tax, have been separate. The Valuation Office Agency sets the assessment for local taxation, and the various local councils collect the charges.

6.3.3 Valuation Office Agency

The Valuation Office Agency (VOA) was part of Inland Revenue, and is now an executive agency of Her Majesty's Revenue and Customs (the merger of the Inland Revenue and Customs & Excise).

The VOA operates through various group offices, the senior officer in each being the district valuer and valuation officer (DV/VO). Although operating under different titles, the DV/VO is the same person responsible for assessing both domestic and non-domestic properties in the area, for council tax and rating respectively. Dealing with council tax, the DV/VO is known as a listing officer, (LO), and draws up a valuation list. Dealing with rating, the DV/VO is known as a valuation officer (VO) and draws up a rating list. The VO is responsible for assessing rateable values, preparing local rating lists, and maintaining the lists until the next revaluation (non-domestic rating); and the LO is responsible for assessing council tax bands, preparing valuation lists, and maintaining the lists (council tax).

In Scotland, the assessors carry out this function.

6.3.4 Scottish assessors and local authorities

Each of the 32 local councils in Scotland is a valuation authority, responsible for appointing an assessor. There are, however, only 14 assessors in Scotland. Four are appointed directly by a single council and the remaining ten are appointed by valuation joint boards comprising elected members appointed by two or more councils. The system of assessors was established by the Lands Valuation (Scotland) Act 1854.

The assessor is responsible for the valuation of domestic and non-domestic properties within one or more council areas. The assessor provides a valuation roll listing non-domestic properties, and a council tax valuation list for domestic properties. The assessor then has the responsibility for maintaining the roll and valuation list.

6.3.5 Billing authority

The Billing Authority (BA) is the body which collects the rates and council tax payments in respect of the properties in its area. It is generally the borough council/metropolitan council/district council. The BA is also responsible for calculating any reliefs from rating or council tax. They are also known as local authorities.

6.3.6 Precepting authority

Although the BA collects the charges, some of money raised, a precept, is collected on behalf of other bodies, which are known as precepting authorities. These include county councils, town and parish councils, fire authorities and police authorities.

6.3.7 Ratepayer

Put simply, the ratepayer is the person responsible for payment of rates (ie on non-domestic or composite properties). The ratepayer is the occupier of the premises, or where there is no occupier, the owner. Certain properties are exempt from payment of rates, but are still included in the rating list, for example vacant industrial properties. In these cases there is no ratepayer.

6.3.8 Council taxpayer

The person responsible for the payment of council tax (ie on domestic or the domestic element of a composite property), is generally the resident. However, there is a hierarchy as follows:

(a) freeholder in residence
(b) leaseholder in residence

(c) statutory or secure tenant in residence
(d) licensee in residence
(e) the resident (who has no formal interest in the dwelling) or
(f) the owner where the dwelling is vacant (or deemed to be vacant).

There is joint and several liability for council tax, so if there is more than occupier with the same status, the BA can seek payment from any of them.

6.3.9 *Valuation Tribunal*

The valuation tribunal is the appeals body; an independent judicial body which deals with disputes between ratepayers/council taxpayers (or their representatives) and the VO/LO, where discussions between the two sides have not produced an agreed settlement. The valuation tribunal is quasi-judicial in role, and at the time of writing there is, broadly, one valuation tribunal per county, although there are plans for the valuation tribunal to become a single national body. The members of the valuation tribunal are laypersons, but assisted by a clerk and other staff, who are employed by the valuation tribunal service.
    There is a separate but similarly structured valuation tribunal service in Wales.
    In Scotland, the valuation appeal committee carries out the appeals function.

6.3.10 *Valuation Appeal Committee in Scotland*

Appeals against the valuation roll and council tax valuation list are initially to the relevant assessor. If a settlement is not reached, the appeal is heard by an independent valuation appeal committee. Members are drawn from an appeal panel appointed by the sheriff principal and are independent of the assessor, local council or valuation joint board.
    Committees are assisted by a secretary, usually a qualified solicitor (also appointed by the sheriff) who provides advice on eg the law or procedure, but who takes no part in the decision making process. Members of the panel are usually unqualified in valuation or legal matters but are appointed because they are local ratepayers or council taxpayers.

6.3.11 *Lands Tribunal*

The Lands Tribunal (LT) is the next tier appeal body after the valuation tribunal, and the highest court on disputes relating to facts. The Tribunal is an independent judicial body set up to resolve disputes concerning land (not only rating). The Tribunal consists of a President, who is the judicial head, and legal and surveyor members.
    The offices and permanent courtrooms are in London. When necessary hearings can be arranged anywhere in England and Wales subject to the availability of courtrooms. There are separate Lands Tribunals for Scotland and for Northern Ireland.
    The Lands Tribunal does not act as the appeal body for council tax. An appeal from a valuation tribunal decision on council tax, can be made to the High Court. Otherwise an appeal against the decision of the Lands Tribunal can be made to the Court of Appeal, but only on a point of law.

6.4. Technical difference between domestic and non-domestic

6.4.1 Introduction

Until 1990, when the community charge (also known as the poll tax) was introduced, general rates applied to all properties. The 1988 LGFA separated local taxation into domestic and non-domestic elements. These can broadly be thought of as residential for domestic properties or commercial for non-domestic properties. However, there are some subtle permutations, particularly where residential properties are run as businesses, for example short-term accommodation.

6.4.2 Definition of domestic

Domestic property is not rateable under the 1988 LGFA. Domestic property is defined as:

- wholly used as living accommodation
- a yard, garden, outhouse or similar, used along with the living accommodation
- a private garage of 25m² or less, or which is used for a private motor vehicle or
- private storage premises used for domestic articles.

This means, for example, that ordinary houses and flats and their gardens are domestic if they are not used for any business or other purpose. Second homes, including holiday houses and flats, are also domestic, as is time-share accommodation.

Where someone works from home, and uses a bedroom or a lounge as an office, it will remain a domestic property as long as the office use is relatively minor. If the office use is significant, so that the character of the room and the nature of its use are changed, the property will be a composite hereditament, with entries in both the rating list (non-domestic office element) and the valuation list (the dwelling).

Section 66 of LGFA 1988 deals with the boundary between domestic and non-domestic property.

6.4.3 Short-term accommodation

Short-term accommodation, such as hotels and guest houses, holiday caravan parks, chalet parks, holiday camps and self-catering accommodation, are not domestic property. These are instead liable to non-domestic rating if they are:

- available for letting for at least 140 days per year
- not entirely a person's sole or main residence, (although part of the property may be, in which case it is a composite hereditament) and
- not subject to a lease or licence to one person for more than a short period letting (perhaps 31 days).

6.4.4 Bed and breakfast accommodation

The property will be domestic and therefore subject to council tax as long as:

- the short-stay accommodation is not let to more than six persons at any one time and
- the property is the proprietor's sole or main residence and
- the bed and breakfast use is subsidiary to the private use.

Subsidiary is not defined, but the VO is likely to consider factors such as whether the property has been adapted so that it no longer has the character of a private house, (eg the installation of additional wash basins, bathrooms or en-suite facilities, and fire precautions such as fire doors, alarms and extinguishers), and/or whether the property is open all year round, serves evening meals or has a licence for serving alcohol.

6.4.5 Caravan sites and moorings

Caravan parks are non-domestic and liable to rating, except that a domestic caravan used as someone's sole or main residence is excluded from the value of the caravan park, and is instead a dwelling subject to council tax.

Although a caravan park can comprise many separate caravans, the whole site is deemed to be a single hereditament, being occupied for rating purposes by the site operator.

Moorings are domestic if they are occupied by a boat which is someone's sole or main residence.

6.4.6 Halls of residence

Students' halls of residence are domestic property, since they are not within the definition of short-stay accommodation for business purposes. Other short-stay hostels may also be domestic property and so are excluded from the non-domestic rating list.

6.4.7 Non-domestic premises

The definition of non-domestic is much simpler than the domestic definition. A hereditament is non-domestic if it either:

- consists entirely of property which is not domestic or
- is a composite hereditament.

It is worth noting that all entries in valuation list (for council tax) are entirely domestic, whereas entries in rating list can be either entirely non-domestic, or composite. A composite hereditament includes a mixture of domestic and non-domestic uses. Typical examples include:

- a shop with the owner's flat above
- an office with a director's flat
- a house with a garage converted to form an office or
- a hotel with staff living accommodation.

In such cases, the whole hereditament is valued for rating and the value apportioned to the non-domestic section.

6.5 National non-domestic rating

6.5.1 Introduction

Non-domestic rates are a mechanism for businesses and other non-domestic occupiers (or the owner of the property, if it is vacant) to contribute financially to the local authority for the area in which their property is located.

Rates are payable on the formula: rateable value multiplied by the uniform business rate, (RV × UBR), subject to adjustments for exemptions, various reliefs and a transition scheme.

6.5.2 Rateable value

This is essentially an estimate of the net rental value of the non-domestic property, at a fixed point in time. The statutory definition of rateable value is contained in schedule 6 (2(1)) of the LGFA 1988, amended by section 1 of the Rating (Valuation) Act 1999:

1. The rateable value of a non-domestic hereditament none of which consists of domestic property and none of which is exempt from local non-domestic rating shall be taken to be an amount equal to the rent at which it is estimated the hereditament might reasonably be expected to let from year to year on these three assumptions:
   (a) the first assumption is that the tenancy begins on the day by reference to which the determination is to be made;
   (b) the second assumption is that immediately before the tenancy begins the hereditament is in a state of reasonable repair, but excluding from this assumption any repairs which a reasonable landlord would consider uneconomic;
   (c) the third assumption is that the tenant undertakes to pay all usual tenant's rates and taxes and to bear the cost of the repairs and insurance and the other expenses (if any) necessary to maintain the hereditament in a state to command the rent mentioned above.

(Similar statements follow for composite properties, ie partly domestic and partly non-domestic, and for properties which are partially exempt from rating, so that the valuation is only carried out for the appropriate part of the property.)

Hence, when valuing a non-domestic property for rating, the valuer must try to find the rent it would command in the open market, assuming that it was in reasonable repair, (reasonable meaning reasonable for its age, character and location), under a full repairing and insuring tenancy.

The "day by reference to which the determination is to be made" refers to the date which is fixed as a baseline for all rating valuations. Rating valuations are carried out five-yearly, the current rating list having come into effect on 1 April 2005. The reference date for these valuations was 1 April 2003, called the antecedent valuation date. Since 1990, all rating list valuation dates have been two years before the date the list came into effect, essentially to allow the valuation office agency time to collect, analyse and rental data. (In the past, valuation dates coincided with the date the list came into effect, meaning that the VO had to project values forward when preparing the list.)

6.5.3 Exactly what is rated?

(a) Hereditament

Business rates are charged on each hereditament. There is a statutory definition of hereditament, contained in section 115(1) of General Rate Act 1967, applied by the LGFA 1988, section 64(1):

> A hereditament is "property which is, or may become, liable to a rate, being a unit of such property which is, or would fall to be, shown as a separate item in the valuation (now rating) list".

So, the rating list identifies the hereditaments, but a hereditament is defined as something that is included in the rating list. This circular definition is not particularly helpful, therefore case law is used to identify exactly what constitutes a hereditament.

This chapter is not intended to deal with the historical development of the definition of a hereditament, but over the years, case law has developed to the point that for a property to be a hereditament, it must be capable of being occupied in a manner which is:

- actual (some use is possible, even if that use is for a limited purpose)
- beneficial (some benefit is derived from occupation, though not necessarily monetary benefit: it could be for the fulfilment of a duty)
- exclusive (the ability to exclude others from using the same property in the same way)
- not too transient (generally minimum nine–12 months, as an annual rent is being considered as the basis of rateable value).

These four points are known as the tenets of rateability, and relate to the capability of the property, not to the nature of any one particular occupier. Rates are chargeable on a daily basis, so occupiers can come and go but the property will remain a hereditament and entered into the rating list, providing the property still fulfils the above criteria.

A hereditament can consist of land, buildings, mines and advertising rights.

(b) Plant and machinery

It is worth discussing the situation where a property contains plant and machinery. A large proportion of property in many sectors, for example industrial property, is occupied by tenants paying a rent, rather than occupied by the owner. In these cases, the occupier's own equipment is excluded from the rent charged, but could be included in the rateable hereditament (one of many reasons why rateable value is not necessarily the rent passing on any one property).

The Wood Committee periodically reviews the criteria of plant and machinery which are included or excluded from the rateable hereditament. The Wood Committee's recommendations then tend to be adopted into regulations. The most recent regulations are the the Valuation for Rating (Plant and Machinery) (England) Regulations 2000, (amended in 2001), which continue to apply for the 2005 rating list. Separate regulations have been introduced for England, Wales and Scotland, but all reflect the recommendations contained in the Wood Report.

Items of plant and/or machinery which are specified in the regulations are rateable; all others are disregarded for rating valuation. A hereditament cannot consist of plant and machinery in isolation: there must be occupation of something else, eg land or buildings.

Examples of rateable plant and machinery are: power generation within the hereditament; heating, cooling and ventilation systems; lifts and hoists; furnaces; gantries; silos; and wind tunnels. Full details are contained in the regulations (referred to above).

(c) How many hereditaments?

It is generally the case that each individual non-domestic property is a separate hereditament, but it is not always clear, for example where one person or business occupies more than one adjacent factory unit. Should the factory units each be separately assessed and entered as separate items in the rating list, or should they form a single assessment?

The leading case in identifying how many hereditaments are present is *Gilbert (VO)* v *Hickinbottom & Sons Ltd* (1956). Denning LJ held:

(1) Where two or more properties are within the same curtilage or contiguous to one another, and are in the same occupation, they are as a general rule to be treated for rating purposes as if they formed parts of a single hereditament...

(2) Where the two properties are in the same occupation but are not within the same curtilage nor contiguous to one another, each of them must as a general rule be treated as a separate hereditament...

This has become known as the "ring fence" test: if an imaginary fence could be erected around the extent of the same occupation, without having to cross any other occupation, or a barrier such as a road, then the premises will be a single hereditament. The concept can be continued into three dimensions, for example where one business occupier uses several floors of the same office block. Where the floors are adjacent, there is a single hereditament; but if other floors separate the occupations, they must be separate hereditaments.

An exception is where separate premises, which would appear to require treating as separate hereditaments, are functionally essential to one another. These are treated as a single hereditament. An example would be a golf course, where the land might cross roads or other barriers, but still effectively forms a whole.

(d) Contents of central rating list and local rating list

The LGFA 1988 places the responsibility for drawing up and maintaining the rating list onto the VO. A rating list (properly known as a local rating list) is drawn up for each BA. The rating list can be seen at the VO's office, or viewed online (*voa.gov.uk*).

Each rating list lasts for five years, with all non-domestic property being re-valued five yearly. The most recent rating list took effect on 1 April 2005.

The rating list holds the following information for each non-domestic hereditament, (as prescribed by the LGFA 1988 section 42, and the Non-Domestic Rating (Miscellaneous Provisions) Regulations 1989):

- the address
- whether it is a composite hereditament, ie partly domestic
- whether any part is exempt from rating
- the rateable value
- the effective date of the entry in the list, where there has been an alteration from the original entry
- a description of the hereditament

- the VO's reference number and
- if either the valuation tribunal or Lands Tribunal has given a decision about the hereditament, which body made the alteration.

As well as the local rating list, the VO also has responsibility for drawing up and maintaining the central rating list. This is also re-valued on a five-yearly basis. The central rating list deals with various national premises, including the canal network, electricity supply network, gas, water, telecommunications, railway network and long-distance pipelines. The type of properties which are included in the central rating list, are those which by their nature cross many billing authority boundaries and are regional/national concerns.

The central rating list differs from the local rating list in two main ways: first, central list hereditaments are rated en-bloc rather than individually; and, second, the central list identifies and designates a person (a corporate body) as the liable person for rate payments in respect of that hereditament. (The local list does not identify the person who is liable for the rate payments.) The entry in the central rating list for each hereditament also includes its (aggregated) rateable value and its prescribed description.

Note that certain parts of these properties are specifically excluded from the central list, and instead included in the local rating list. These parts are units which could be separately occupied and let. These fall under the description of relevant hereditament for the purposes of the LGFA 1988. So, for example, offices and workshops owned by the railway authority concerned (but not an intrinsic part of the utility's undertaking) would be included in the local rating list, whereas the railway network itself would be in the central rating list. In contrast, gas and electricity meters are deemed to be part of the network supplying the gas or electricity, and therefore included in the central list, not as part of the hereditament in which they are situated, which is likely to be included in the local rating list.

Properties within the central list have historically been valued by formula (a mechanism for avoiding particularly difficult valuation issues), but since rating has been nationalised by the NNDR (national non-domestic rate), formula rating has effectively been just a means of passing funds from one government department to another. The issue under the NNDR system is that the funds raised have to be re-distributed to the various billing authorities, and therefore each central list hereditament must have a value ascribed to the parts in each billing authority area. Therefore, formula rating is being phased out, and conventional valuation (which can include the contractors test for many central list hereditaments) is being re-instated. Classes already returned to conventional valuation include the property of British Waterways and British Railways Boards, The Tyne & Wear Metro, Docklands Light Railway and the national telecommunications networks of BT and Cable & Wireless Communications (Mercury) Ltd. Further classes of central list hereditaments are expected to be conventionally rated in the future.

This chapter is too brief to discuss the central rating list in more detail; the remainder of this section on non-domestic rating therefore relates only to local rating lists.

In Scotland, the equivalent of the local rating list is the valuation roll. The valuation roll is a public document which contains an entry for all non domestic properties in the assessor's area except those specifically excluded by law. Each entry in the roll includes the names of the proprietor, tenant and occupier as appropriate, the net annual value which has been set by the assessor and the rateable value. The rateable value is derived from the net annual value and as legislation currently stands, for the majority of properties, rateable value and net annual value are the same.

Values are established every five years at revaluation which (since 1990) takes place at the same time throughout Scotland, England and Wales. Revaluation results in the production of a new valuation roll which contains revised values for all non domestic properties in the assessor's area. Following a revaluation new values will generally remain unchanged until the next revaluation, unless the property is altered or other changes take place. New properties are added to the roll as they become occupied and entries for demolished buildings are deleted.

The assessor is required to notify proprietors, tenants and occupiers of all changes which he makes to the valuation roll by issuing a valuation notice.

(e) Exemptions from local rating list

The LGFA 1988 schedule 5 lists various exemptions from non-domestic rating. Wholly exempt properties are not included in the rating list.

(1) Crown property
It used to be the position that, unless an Act specifically stated that the Crown was bound, the Crown was not subject to the Act. The LGFA 1988 does not make Crown property liable to non-domestic rating, but the Local Government and Rating Act 1997 altered the situation. Crown property is now subject to non-domestic rating in the same way as any other hereditament.

(2) Sovereign and diplomatic immunity
The LGFA 1988 has not changed the general position under common law, that foreign sovereigns and sovereign states are immune from rates.

Diplomats have specific immunity under the Diplomatic Privileges Act 1964, as long as the property concerned is being used for the purposes of the diplomatic mission.

Consular premises, property occupied by visiting armed forces, and property of certain international organisations, are also exempt from rating.

(3) Agricultural land and buildings, and fish farms
The LGFA 1988 provides that agricultural land, agricultural buildings and fish farms are exempt from rating.

The definition of agricultural land is quite precise:

2.–(1) Agricultural land is–
(a) land used as arable, meadow or pasture ground only,
(b) land used for a plantation or a wood or for the growth of saleable underwood,
(c) land exceeding 0.10 hectare and used for the purposes of poultry farming,
(d) anything which consists of a market garden, nursery ground, orchard or allotment
(e) land occupied with, and used solely in connection with the use of, a building which ... is an agricultural building.
(2) But agricultural land does not include–
(a) land occupied together with a house as a park,
(b) gardens (other than market gardens),
(c) pleasure grounds,
(d) land used mainly or exclusively for purposes of sport or recreation, or
(e) land used as a racecourse.
3. A building is an agricultural building if it is not a dwelling and—
(a) it is occupied together with agricultural land and is used solely in connection with agricultural operations on the land, or

(b) it is or forms part of a market garden and is used solely in connection with agricultural operations at the market garden.

4. (1) A building is an agricultural building if it is used solely in connection with agricultural operations carried out on agricultural land

5. (1) A building is an agricultural building if–

(a) it is used for the keeping or breeding of livestock, or

(b) it is not a dwelling, it is occupied together with a building or buildings falling within paragraph (a) above, and it is used in connection with the operations carried on in that building or those buildings

6. (1) A building is an agricultural building if it is not a dwelling, is occupied by a person keeping bees, and is used solely in connection with the keeping of those bees.

Livestock includes any mammal or bird kept for the production of food or wool or for the purpose of its use in the farming of land.

Fish farming means the breeding or rearing of fish, or the cultivation of shellfish, for the purpose of ... transferring them to other waters or producing food for human consumption.

Shellfish includes crustaceans and molluscs of any description.

(4) Churches and other places of worship

The Local Government Act 1988 schedule 5 paragraph 11 makes the following exempt from rating:

• churches, chapels and other places of public religious worship
• church halls, chapel halls and similar buildings
• ancillary administrative premises.

The property must be for public religious worship, of either the Church of England or Church in Wales, or has been certified as a place of religious worship for another faith. Halls used in connection with the place of religious worship will be exempt if they are used for the purposes of that organisation, or for carrying out administrative activities to support the organisation of public worship.

"Worship" in this context can be defined as:

have some at least of the following characteristics: submission to the object worshipped, veneration of the object, praise, thanksgiving, prayer and intercession.

R v Register General, ex parte Segerdal and Church of Scientology of California (1970)

The worship must also be public. In *Church of Jesus Christ of Latter Day Saints v Henning (VO)* (1963), where a Mormon Temple had been certified as a place of religious worship, the House of Lords held that because the public at large and members of the Mormon Church not recommended by their bishop could not enter the temple where certain sacred ceremonies were performed, it was not a place of public religious worship and did not qualify for exemption.

(5) Sewers

Schedule 5, paragraph 13 of the LGFA 1988 exempts sewers, manholes, ventilating shafts, pumping stations and other ancillary elements from rating.

(6) Property of drainage authorities

Schedule 5, paragraph 14 of the LGFA 1988 exempts land used by drainage authorities, which is part of a river or watercourse, along with structures controlling the flow of water.

(7) Highways and bridges

Land which is dedicated to the public, for example the highways, cannot be a rateable hereditament, as the public at large cannot be a rateable occupier. Schedule 5, paragraph 18A of the LGFA 1988 specifically exempts bridges, viaducts and tunnels.

(8) Public parks

As for roads and bridges, as the public at large cannot be a rateable occupier, a public park cannot be a rateable hereditament. They are therefore exempt from rating. Statutory exemption is given under schedule 5, paragraph 15 of the LGFA 1988, but previously case law gave the exemption, (notably *Lambeth Overseers* v *London County Council* (1897) — often called the *Brockwell Park* case, after the location concerned).

(9) Certain property used for people with disabilities

The LGFA 1988, paragraph 16(1) of schedule 5 exempts hereditaments which are used wholly for any of the following:

(a) the provision of facilities for training or keeping suitably occupied, persons who are disabled or who are or have been suffering from illness;
(b) the provision of welfare services for disabled persons;
(c) the provision of facilities under Section 15 Disabled Persons (Employment) Act 1944;
(d) the provision of a workshop or other facilities under Section 3(1) Disabled Persons (Employment) Act 1958.

A person is considered to be disabled if they are

(i) blind, deaf or dumb,
(ii) suffering from mental disorder of any description or
(iii) substantially and permanently handicapped by illness, injury, congenital deformity or any other disability prescribed for the purposes of section 29(1) National Assistance Act 1948.

This exemption from rating could, for example, include holiday accommodation adapted for disabled persons and let to disabled persons even when accompanied by an able bodied person, *Chilcott (VO)* v *Day* [1995] RA 285.

(10) Swinging moorings for boats

Schedule 5, paragraph 18 of LGFA 1988 makes these exempt, if they are only equipped with a buoy anchored to the sea bed, and which is designed to be raised from the sea bed from time to time.

(11) Property of Trinity House

Trinity House is the body that operates lighthouses, buoys and beacons. Schedule 5, paragraph 12 makes these exempt from rating.

(12) Air raid protection works

Schedule 5, paragraph 15 of LGFA 1988 makes these exempt if they are not occupied for any other purpose.

(13) Property in enterprise zones

Schedule 5, paragraph 19 of LGFA 1988 exempts property that is in an enterprise zone. These exemptions have mostly elapsed in time, unless the government designates further enterprise zones.

(f) Reliefs from rating

Although reliefs from rating are a billing issue, rather than a valuation issue, it is still worth briefly covering them, for completeness.

(1) Charities and registered sports clubs

At one time charitable organisations were often held not rateable by the courts on the grounds of absence of beneficial occupation. But in *London County Council v Erith Parish (Churchwardens) and Dartford Union Assessment Committee* (1893), the House of Lords determined that the absence of profit was no ground for the exemption from rates. Thereafter, charitable institutions were held liable. In practice, however, charities continued to enjoy reduced rates or to have their liability remitted in full.

Charities have received relief against rates payable through the Rating and Valuation Act 1961, the General Rate Act 1967 and the LGFA 1988. Currently, mandatory relief for registered charities reduces their liability to 20% of the full charge. From 1 April 2008, charities are exempt from empty property rates, as a result of the Rating (Empty Properties) Act 2007.

The LGFA 1988 provides for two forms of relief: the first is given as of right (mandatory) and the second lies entirely at the discretion of the billing authority.

Mandatory relief (LGFA 1988 section 43) 80% of rates is given in respect of the following hereditaments:

(a) a hereditament occupied by a charity, or wholly or mainly used for charitable purposes; this can include public schools, friendly societies etc;
(b) a hereditament occupied by a non-profit organisation whose aims are charitable, philanthropic or religious or concerned with education, social welfare, science, literature or the fine arts;
(c) a property used wholly or mainly for recreation, occupied by a non-profit club or organisation; this includes amateur sports clubs registered under the Finance Act 2002 as a community amateur sports club.

Charitable owners of unoccupied property pay at the same 20%, but of the empty property rate, providing the hereditament will be used for charitable purposes when next in use.

Discretionary relief (LGFA 1988 section 47) may extend a charity's relief above the mandatory 80% to 100%. It may be awarded by the billing authority for hereditaments:

- already in receipt of mandatory relief
- occupied by institutions or organisations not established or conducted for profit, whose main objects are charitable or otherwise philanthropic or religious, or concerned with education, social welfare, science, literature or the fine arts
- occupied for the purposes of a club, society or other organisation not established or conducted for profit and wholly or mainly used for recreation (eg Scout huts).

For unoccupied hereditaments, the anticipated future use must be considered: if it is intended to use the property for charitable purposes, the relief applies.

Several major charitable organisations run shops with the objective of fundraising (this is one of Oxfam's major sources of income, for example). These shops are specifically included in the relief for charities.

The relief available to charities, both occupied and unoccupied liablities, is also available to community amateur sports clubs, which are registered for the purpose under schedule 18 of the Finance Act 2002.

(2) Non-agricultural business on agricultural land or former agricultural land
This applies to business premises with rateable values of less than £7,000, whether they are run by the farmer or not. The land or buildings in which the business is located must have been in agricultural use for at least 183 days (half a year) during the year ending 14 August 2001, and the property must have been included in the rating list before 1 April 2005. This relief offers a possible 50% reduction which can be increased to up to 100% by the local authority if the business is of benefit to the local community. This particular scheme was intended to run for a five year period and ended on 14 August 2006.

(3) Rural shops, public houses and petrol filling stations
If the village has a population under 3000, these businesses have a mandatory 50% reduction in the rates bill, which can be increased to 100% at the discretion of the local authority, if it is:

- the only village general store or post office, as long as it has a rateable value of less than £7,000
- a food shop with a rateable value of less than £7,000
- the only village pub or the only petrol station (rateable value of less than £10,500).

If the business is in a qualifying rural village, and has a rateable value of less than £14,000, the local authority has discretion to award up to 100% relief from rates if the business is of benefit to the local community.

(4) Small business rate relief
This relief came into effect on 1 April 2005.

Eligible businesses with rateable values of below £5,000 will get 50% rate relief on their liability. This relief will decrease on a sliding scale by 1% for every £100 of rateable value over £5,000, up to £10,000.

The relief is available to ratepayers with either:

- one property or
- one main property and other additional properties, providing the additional properties do not have individual rateable values of more than £2,200, and the combined rateable value of all the properties is under £15,000 (or £21,500 in London). The relief then applies to the main property.

In addition to this relief from rates, eligible businesses with rateable values of between £10,000 and £14,999 (or between £10,000 and £21,499 in London) will have their liability calculated using the small business multiplier, (which is slightly lower than the standard UBR multiplier).

The small business rate relief scheme is funded by a supplement on the rate bill of those businesses not eligible for the relief. This supplement is built into the standard multiplier.

Eligible ratepayers initially had to apply for the relief each year, but the latest regulations allow businesses to apply once for a five-year period, with the ratepayer being under an obligation to advise the local authority if eligibility changes.

Assuming a business meets the eligibility criteria, the relief can only be granted if the property the business occupies is in the rating list from 1 April. The date of occupation of the property is irrelevant, the key date is the effective date given to the property in the rating list. If the property has an effective date after 1 April, then the relief can only be applied for from 1 April of the following year.

(5) Relief from hardship

A billing authority has power to reduce or remit any rates on account of the poverty of the person liable to pay them. This provision stems from the earliest days of rating, when rating was intended to relieve poverty (now section 49 of the LGFA 1988).

Similarly, a billing authority has further discretion to reduce or remit a ratepayer's liability where they are satisfied that the ratepayer would suffer hardship, and it is reasonable having had regard to the burden placed on local council taxpayers (section 49 LGFA 1988). Local authorities point out that their ability to exercise this discretion is restricted, as the relief is partly funded by council taxpayers in the area. The government only reimburses 75% of hardship or poverty relief granted in these circumstances, so a quarter of the cost falls directly on council taxpayers. For this reason, billing authorities usually only grant this relief where it is evident that it is in the interests of the community that the ratepayer remains in business because either:

- he provides a unique, regularly-required amenity or
- the loss of employment provided by a ratepayer would be severely damaging to the local community.

(6) Unoccupied properties

Until 2008 unoccupied non-domestic properties are liable to empty property rates at 50% of the full rate bill (or of the transitional bill where the transitional arrangements apply), under section 45 of the LGFA 1988. However, the NDR (Unoccupied Property) Regulations 1989 provides 12 exempt classes from empty rates where:

a. the whole hereditament has been unoccupied for less than three months
b. the owner is prohibited by law from occupying it, or from allowing it to be occupied, (eg under Health and Safety legislation)
c. the hereditament is kept vacant by Crown or local or public authority action, which has been taken with a view to prohibiting its occupation or to acquiring it (eg a compulsory purchase order has been served)
d. it is subject of a building preservation notice under section 58 of the Town and Country Planning Act 1971 or is a listed building
e. it is a monument under section 1 of the Ancient Monuments and Archaeological Areas Act 1979
f. it is a qualifying industrial hereditament — discussed further below
g. its rateable value is less than £2,200

h.   the owner is entitled to possession only in his capacity as the personal representative of a deceased person

i.   the owner is bankrupt under Parts VIII to XI of the Insolvency Act 1986

j.   the owner is entitled to possession of the hereditament in his capacity as trustee under a deed of arrangements to which the Deed of Arrangement Act 1914 applies

k.   the owner is a company which is subject to a winding-up order made under the Insolvency Act, 1986 or which is being wound up voluntarily under the Act

l.   the owner is entitled to possession of the hereditament in his capacity as liquidator by virtue of an order made under section 112 or section 145 of the Insolvency Act 1986.

The major exempt category comprises unoccupied qualifying industrial hereditaments. This term is wide in its scope. Certain industrial hereditaments are exempt from empty rates, where the property is constructed for one of the purposes specified in the regulations:

From 1 April 2008, the Rating (Empty Properties) Act 2007 provides that empty property rates relief will be increased from 50% to 100% (although with a provision for the government to vary this between 50% and 100% if there is a change in property market conditions). The exemption for qualifying industrial hereditaments is reduced to six months, and charities and community amateur sports clubs become exempt from empty property rates.

(7)   Partly occupied properties. If a property is only partly occupied on a temporary basis it is not always possible to amend the entry in the rating list to split the property into two or more hereditaments, which would allow separate rating bills for each hereditament and hence empty relief on the unused element.

In this situation, the local authority can ask the VO to determine the rateable values of the occupied and unoccupied parts. The rates payable on each part can then be calculated so that the ratepayer only pays the full rates for the occupied part plus 50% of the rates for the unoccupied part (unless an exemption is applicable).

This relief is given as laid down in section 44A of the LGFA 1988 and the government reimburses the council in full.

This relief is granted either to the end of the financial year or to the date on which the property ceases to be only partly occupied.

6.5.4 *Who is the ratepayer?*

Business rates are charged on a daily rate, albeit the local authority will issue an annual bill and then amend it if the ratepayer changes within the year. Therefore, the ratepayer is the person who is the occupier, or the person entitled to occupy the property, on a daily basis. This has the intriguing effect that trespassers can be liable.

If the property is empty, there is no occupier in the ordinary use of the word, but instead the owner is deemed to be the rateable occupier and is the ratepayer. The LGFA 1988 section 65 defines the owner as "the person entitled to possession". Where the property is empty, a 50% charge is levied after the first three months, except where the property is exempt from empty rates.

6.5.5 Basis of valuation

(a) Statutory definition of rateable value

Although this has been outlined at section 6.5.2 above, it is worth elaborating. The statutory definition of rateable value is contained in schedule 6, paragraph 2(1) of the LGFA 1988, as amended by the Rating (Valuation) Act 1999:

> The rateable value of a non-domestic hereditament none of which consists of domestic property and none of which is exempt from local non-domestic rating shall be taken to be an amount equal to the rent at which it is estimated the hereditament might reasonably be expected to let from year to year on these three assumptions —
>
> (i)   the first assumption is that the tenancy begins on the day by reference to which the determination is to be made;
> (ii) the second assumption is that immediately before the tenancy begins the hereditament is in a state of reasonable repair, but excluding from this assumption any repairs which a reasonable landlord would consider uneconomic;
> (iii) the third assumption is that the tenant undertakes to pay all usual tenant's rates and taxes and to bear the cost of the repairs and insurance and the other expenses (if any) necessary to maintain the hereditament in a state to command the rent mentioned above.
>
> The rateable value of a composite hereditament none of which is exempt from local non-domestic rating shall be taken to be an amount equal to the rent which, assuming such a letting of the hereditament as is required to be assumed for the purposes of sub-paragraph (1) above, would reasonably be attributable to the non-domestic use of the property.
>
> The rateable value of a non-domestic hereditament which is partially exempt from local non-domestic rating shall be taken to be an amount equal to the rent which, assuming such a letting of the hereditament as is required to be assumed for the purposes of sub-paragraph (1) above, would, as regards the part of the hereditament which is not exempt from local non-domestic rating, be reasonably attributable to the non-domestic use of the property.

The essential items included in the definition of rateable value are that:

- the hereditament is deemed to be vacant, and available for a new tenancy
- potential tenants would pay a rent for the hereditament
- the hypothetical tenancy is from year to year, but the hypothetical tenant can reasonably assume that it will continue after the end of the year
- the hereditament is in a reasonable state of repair immediately before the hypothetical tenancy begins and
- the hypothetical tenant pays all the costs of insuring and keeping the hereditament in that state of reasonable repair.

Put simply, the basis of valuation for rating is the open market rental value. However, in rating, as in other cases, the rental value depends on the specific terms upon which the property is deemed to be let. To compare, in a rent review situation, the lease would define the hypothetical tenancy to be assumed. In rating, the LGFA 1988 gives the terms of the tenancy. In a rent review, the lease would define the date on which the valuation is to be based. In rating, the LGFA 1988 specifies the date on which the valuation is based. To some extent the rating valuer has to conjure up a world of make-believe.

The hypothetical tenant for the purposes of rating includes the actual occupier, (*LCC* v *Erith Parish (Churchwardens)* (1893)). Although the statutory definition states that the tenancy only continues from year to year, case law has established that "a tenant from year to year is not a tenant for one, two, three or four years but he is to be considered as a tenant capable of enjoying the property for an indefinite time, having a tenancy which it is expected will continue for more than a year, but which is liable to be put to an end to by notice" (*R* v *South Staffordshire Waterworks Co* (1865)). Therefore, when considering rental evidence for rating valuation, the length of lease (or more specifically, the frequency with which the rent is reviewed) does not need to be adjusted into year to year terms, if it fits the usual market pattern for that type of property in that location.

The statutory definition requires the hereditament to be considered to be in reasonable repair at the start of the hypothetical tenancy, as provided by the hypothetical landlord, with the hypothetical tenant then responsible for maintaining it in that state. The meaning of reasonable repair can be drawn from a landlord and tenant case, *Anstruther-Gough-Calthorpe* v *McOscar* (1924). The guidance from this case was relied on in the rating case of *Brighton Marine Palace and Pier Co* v *Rees (VO)* (1961), which considered the standard of repair to be assumed in determining a rating assessment.

Four points of guidance emerged.

1. The landlord is a reasonably minded person and the premises are to be managed accordingly.
2. The age of the building is of consequence. Naturally the state of repair that an ingoing tenant would expect will vary according to the age of the property.
3. The locality will have a direct bearing on the standard of repair. Poorer localities may generally have a poorer quality of building situated in them, which will affect the expected state of repair of the buildings.
4. The type of tenant likely to occupy the property also needs consideration. This concept is closely linked to the age of the buildings, the nature of the accommodation and expectations of the ingoing tenant.

Where the property to be valued is in a reasonable state of repair, no adjustments are necessary when carrying out the valuation. However, where it is not in a reasonable state, the valuer must mentally adjust it into a reasonable state, and consider what rent it could then command. This has implications when relying on rental evidence from comparable properties: whether these and the rent passing on them reflect reasonable repair or not.

However, if the property is in such poor repair that it would be uneconomic to repair it, and the landlord would realistically prefer to let it in its poor condition, rather than repair it and let it for a higher rent, then the actual state of repair of the hereditament can be taken into account.

(b) *The date of valuation*

The statutory definition refers to the "day by reference to which the determination is to be made". This essentially allows a single point in time to be fixed, for all rating valuations to be carried out — which ensures fairness and uniformity. For the 2005 rating list, the date for valuation was 1 April 2003, ie two years before the list came into effect. This is known as the antecedent valuation date (AVD). The effect is that all rating valuations, for the 2005 rating list, must be made as if the hypothetical tenancy started on 1 April 2003. This requires the valuer to value the property with the market forces, and economic conditions, of 1 April 2003.

However, the valuer looks at the physical property at the material day, which is not the same as the AVD. For the initial entry into the 2005 rating list, all properties had a material day of 1 April 2005. Therefore, initial entries into the 2005 list were valued based on the physical property at 1 April 2005, as if it had been rented in the market of 1 April 2003.

An additional complication is that the material day can change. For example, if a property is demolished or extended, some time after 1 April 2005, its rental value and hence its rateable value is likely to change. The valuer must then take the new physical circumstances at the date of the change, (the material day), but still transport it back to the economic world of 1 April 2003, the AVD, to establish the rental value in the open market.

The LGFA 1988 schedule 6 lists the material circumstances which are to be taken at the material day rather than at the AVD.

1. Matters affecting the physical state or enjoyment of the hereditament. A property must be valued in the physical state it is in on the material day, but at the level of rents passing at the antecedent valuation date.
2. Mode and category of occupation. A property must be valued in the use to which it is put on the material day. For example, a wine bar is to be valued as a wine bar and not a shop, but on the basis of the rent as at the AVD. Or, applying *rebus sic stantibus* and *Fir Mill* v *Royton*, (see below for a discussion of these principles), a factory is valued as a factory, but not one producing any specific item.
3. Quantity of minerals or other substances in, or extracted from, the hereditament. This affects relatively few hereditaments, such as mines and quarries. The quantity of minerals extracted will usually vary from year to year: hence the RV may be altered by the VO or the ratepayer on such grounds, but again relating to values at the AVD.
4. Quantity of refuse or waste material which is brought on to and permanently deposited on the hereditament. Thus rubbish tips can be reassessed annually in the same way as mineral-producing hereditaments.
5. Matters relating to the physical state of the locality in which the hereditament is situated, or matters which, though not affecting the physical state of the locality, are nevertheless physically manifest there. In the former category, any physical factor which affects a property at the material day must be taken into account, such as a new shopping centre or bypass road. Examples of the latter category are numbers of pedestrians, traffic flow, noise and fumes.
6. Use or occupation of other properties situated in the locality of the hereditament. This refers to the use of all types of property in a locality, and also means that any competing uses which enhance or diminish AVD rental values must be taken into account.

To summarise the relationship between AVD and material day, the valuer must take matters relating to value (which will include economic, planning and legal factors) as they were at the AVD. The physical factors of the hereditament, are taken as they exist on the material day. Effectively, the valuer performs time-travel. He takes a photograph of the hereditament, and the above six factors, at the material day. He then travels back in time to the AVD, and asks himself what rental value it would have had then.

The material day for alteration of a rating list is always the day of the VO alteration to the list or the day a proposal is served on the VO except when the VO alteration or the proposal relates to the following:

Reason for the alteration	*Material day*
1. Correction of an inaccurate compiled list entry	1. Compilation date
2. Correction of inaccuracy made in previous list alteration	
3. Insertion/deletion of hereditament in local or central lists	
4. Hereditament ceased to be/became domestic/exempt property (in whole or part)	2. to 6. Day of the event
5. Hereditament changes relevant billing authority	
6. Division/merger of assessments	
7. Completion notice served	7. Day stated in notice

The following examples illustrate the application of the material day.

1. If the VO finds that the original entry in the rating list in respect of a property is wrong (eg because there was an arithmetical error in the valuation, or there is a re-consideration of rental evidence), any alteration to the list must be on the basis that the material circumstances were as they existed at the date the list was compiled, *not* as at the date when the alteration is made.
2. If a new property has to be entered into the rating list, the value must reflect the material circumstances at the date the property is completed, *not* at any subsequent date on which the list is altered to include the property.
3. If the assessment of a hereditament has to be split because it has to be rated in parts, or if several assessments have to be amalgamated because they have all come into one occupation, the material day is the day upon which the changes in occupation occurred, *not* the date of proposal or alteration of the list.
4. If a new occupier decides that the assessment of his property in the rating list is incorrect due to a material change of circumstances, he may make a proposal to alter the list. Here the material day for valuation and to which the material circumstances must be related is the date upon which the proposal is served on the VO. (Hence, proposals to alter the rating list due to a temporary change, such as road-works affecting a property, must be made while the works are still taking place, so that the material day includes the presence of the road works.)
5. Similarly, if a ratepayer demolishes part of his property and seeks a reduction in his assessment at the correct time, the material circumstances to be reflected in any revised assessment are those as at the date of service of his proposal.
6. Even a temporary disturbance such as roadworks may result in a reduction in the rating assessment. The material day of a proposal which was served while the works are underway will be the date of service on the VO. The work must be of sufficient size and duration to affect rental value under the rating hypothesis (*Fielder* v *Baker (VO)* (1970)), so that a hypothetical tenant would reduce his rental bid, on a year-to-year basis, due to the disturbance.

(c) Physical property to be rated

Section 6.5.5(a) above explains that the statutory definition of rateable value requires an assumption that the hereditament is in reasonable repair. However, in all other respects, the property must be taken to be exactly as it stands. This simple statement has evolved into the principle of *rebus sic stantibus*, which means the thing as it stands.

Rebus sic stantibus has two major aspects: the physical nature of the property, and its mode of use.

Rating is a tax borne by the occupier on the current value of his occupation. Thus, where a shop is being valued for rating purposes, it would be wrong to value the shop on the assumption that it was capable of being structurally altered and used more profitably as an office. If the shop is in fact being used as a shop, it must be valued on the assumption that that use will continue. Equally, a factory must be valued as such.

However, it is not necessary to assume continuance of the actual existing trade or business. Therefore a factory making shoes can be considered as a factory capable of making any product, provided the change could be achieved without structural alteration. This would still be the case even if the real-world occupier of the factory operated under a lease with a covenant restricting its use to a shoe-making factory.

Again, it would be legitimate to value a cotton mill which was being unprofitably used, at the higher rent it might command if it were to be used for other industrial purposes, provided always that the change could be achieved without structural alteration. All this is consistent with the assumption of vacant and to let, but taking into account the last use of the property.

Although *rebus sic stantibus* precludes structural changes, it does not prevent minor changes of a non-structural nature being assumed — such as rearranging demountable partitioning, rewiring etc. In *S &P Jackson (Manchester) Ltd v Hill (VO)* (1980), an occupier used a shop, constructed as a shop and located in a parade of other shops, as a warehouse. The rating assessment based on shop use was confirmed on the basis that, vacant and to let, it would fetch the higher rental value of a shop, and:

- no structural alteration was necessary
- demand existed for the use as a shop
- planning permission could reasonably be expected.

The theory in rating is that the hypothetical landlord will let the hereditament to the hypothetical tenant who is prepared to pay the highest rent, provided that no structural change or change of use is considered. The *S&P Jackson* case does not conflict with this theory because, although the last occupier was using the property as a warehouse rather than as a shop, the property was within a parade of other shops, and any reasonable potential tenant would look at the property vacant and to let, and see it as a shop. But the doctrine that the mode or category of occupation restricts rating valuation (*Fir Mill Ltd v Royton UDC and Jones (VO)* (1960) ie factory valued as factory, shop as a shop, not any particular kind of shop) is a basic and very important concept.

Two recent cases help to explain the rebus sic stantibus principle. In *Scottish & Newcastle Retail Ltd v Williams (VO)* (2000) and *Allied Domecq Retailing Ltd v Williams (VO)* (2000), the Lands Tribunal considered the valuation of two public houses in Milton Keynes. The City Duck and the Rose & Castle opened in 1979 within a shopping centre. Retail rental values in the shopping centre grew strongly between 1979 and 1988, whereas rents for public houses rose very little. When revaluing the properties for the 1990 rating revaluation, the VO described each as a public house, but valued them as if they were shops.

The Lands Tribunal heard the two cases together. The parties agreed alternative valuations of the two properties: as public houses and as shops. In each case the value as a public house was less than a quarter of the value as a shop.

The VO argued that the properties should be valued as shops because the cost of the work to convert them to shops was small in comparison to the increased value of the property after conversion, and that in the real world demand for shops would ensure that the conversion work would be carried out. The VO's view was that, vacant and to let on the rating hypothesis, the hereditaments would be rented at the same rent as retail shops, because the potential tenants would include retailers who were prepared to strip the units down to their shells and then fit them out and use them as shops. Change of use from a public house to a retail shop (including a fast-food restaurant) would be permitted within the current planning legislation.

The ratepayers argued that the change of use from a public house to a shop broke the *rebus sic stantibus* rule described in *Fir Mill v Royton*.

The Lands Tribunal decided in favour of the ratepayers. On the mode and category of occupation, it was decided that a pub was in a completely separate category from a shop, restaurant or fast-food takeaway, even though planning law would allow the change between uses. Different valuation methods and the existence of different values was evidence of this. The Lands Tribunal upheld *Fir Mill v Royton*.

The judgment stated:

We believe that *Fir Mill* identifies with adequate clarity the basis on which hereditaments are to be categorised for the purposes of the *rebus sic stantibus* rule. The assumption is that the premises will continue to be used for the same general purpose as that for which they are used at the material date, so that a dwelling house is assessed as a dwelling house, a shop as a shop, but not any particular kind of a shop; a factory as a factory, but not as any particular kind of a factory. For the purposes of the *rebus sic stantibus* rule it is thus the principal characteristics of the actual use that are relevant — those features of the occupation that reflect the general purpose of the use — rather than the particular operations of the individual occupier.

The Lands Tribunal decided that:

The *rebus sic stantibus* rule ... rests on the concept that what has to be determined in rating is the value to the occupier of the hereditament, measured by the rent on an assumed yearly tenancy. In carrying out a valuation under the rating hypothesis the following assumptions are to be made about the hereditament:

(a) that the hereditament was in the same physical state as on the material day. Alterations which the hypothetical tenant might make to the hereditament may be taken into account, if, taken overall, they are minor. All other prospective alterations to the hereditament are to be ignored.

(b) that the hereditament could only be occupied for a purpose within the same mode or category of purpose as that for which it was being occupied on the material day. Any prospective change of use outside that mode or category is to be ignored. In determining to what mode or category a particular use belongs, it is the principal characteristics or the use and the methods of valuation commonly applied by rating surveyors to which regard must be had; and shops, offices and factories serve as examples. Some uses may not fall within any such broad category, however, and are to be regarded as *sui generis*.

Summarising the assumptions discussed so far, which a valuer has to make when valuing a property for rating:

* the physical property, and its locality, are taken as at the material day
* the property is assumed to be in reasonable repair
* in all other respects, the property is assumed to be as it stands, and used for its current use
* the property may include rateable plant and machinery, (even if these items are not included in any rent passing on the property)
* the economic and rental market is taken as at the antecedent valuation date.

Although rateable value is defined as an assumed open market rent, there are various reasons why the rateable value of a property can be different from its actual rent (if any). Some of them have already been touched on, for example:

- the rateable value is based on rental values at the AVD, whereas the rent passing is likely to have been set at a different point in time
- the rent may be for a shell building, whereas the rateable hereditament may include plant and machinery
- the tenant may have made alterations to the property which are excluded from the rent, but that are included for the purposes of rating. To the rating valuer, there is no distinction between items supplied by the landlord, or the tenant, when carrying out the valuation
- the property may not be in reasonable repair, which might be taken into account in the rent, (for example by a reduced rent, or a long rent-free period), but which could not be taken into account in the rateable value.

Other reasons why a rent and rateable value might differ for the same property could include lease issues, such as the rent being based on turnover, or the lease being for an unusually short or long period, or there being a restrictive covenant in the lease.

The valuer is trying to reach a rateable value as "the rent at which it is estimated the hereditament might reasonably be expected to let". This estimate of reasonably expected rental value requires the valuer to bring together a basket of evidence, and from this make a judgment as to what would be realistic. A lovely phrase, the higgling of the market, arose from *Robinson Bros (Brewers) Ltd* v *Houghton and Chester-le-Street Assessment Committee* (1937). Scott LJ said that the rateable value should represent:

> the figure at which the hypothetical landlord and tenant would... come to terms as a result of bargaining for that hereditament, in the light of competition of its absence in both demand and supply, as a result of "the higgling of the market".

6.5.6 *Valuing the hereditament*

(a) *Five main methods*

There are five main methods of valuation available to the rating valuer:

- rental method
- profits test
- contractor's test
- formula
- using comparable assessments.

(b) *Choice of method*

No matter which valuation method is used, the valuer should keep in mind that what is being sought is the rateable value, ie the net rental value at the AVD, of the physical property at the material day, on the statutory basis. In theory, all the methods should give the same result, although practitioners will find this is demonstrably not the case in practice.

The valuer is valuing the actual property, as it stands *rebus sic stantibus*, (subject to the statutory assumption of reasonable repair). However, the statutory definition gives specific full repairing and insuring terms for the hypothetical lease, so if there is an actual lease on the property, valuation adjustments might need to be made if the real lease terms differ from the hypothetical terms.

There is no statutory requirement to use any particular method in order to arrive at rateable value. A court would be entitled to accept the opinion of a qualified surveyor or valuer that the rateable value of a hereditament was £x without inquiry as to how or why he arrived at that result and in the past many cases were decided at quarter sessions on "spot" valuations of that kind. However, usually a qualified person will use one or other of several established methods of valuation. In *Port of London Authority* v *Orsett Union*, Lord Dunedin said:

> What is the inquiry which ... quarter sessions are engaged on? It is to find the net annual value of the hereditament in question... . What will the hypothetical tenant give for the subject? If the subject is an ordinary one similar in character to other subjects which have stood the test of the markets, or better still, if it has stood the test of the market itself without disturbing circumstances, the inquiry is simple. But when the nature and circumstances of the hereditament in question do not admit of such a test, some other way must be found. Now there are several ways of attacking the problem. One way is to consider what profit the hypothetical tenant could make out of the hereditament, not in order to rate that profit, but in order to find out what he was likely to give in order to have the opportunity of making that profit. Another way is to see what it would cost an owner to produce the hereditament in its present form and then to see what a tenant who had not himself the money to be an owner, would give the owner yearly, it being assumed that the sum must bear some relation at ordinary rates of interest to what has been spent. No question of law is necessarily involved in either of these methods.

In *Garton* v *Hunter (VO)* (1969), the Lands Tribunal was looking at the valuation of a caravan camping site, and was presented with rental information, a contractor's test valuation and a profits basis valuation. Winn LJ said:

> we do not look upon any of these tests as being either a "right" or a "wrong" method of valuation; all three are means to the same end; all three are legitimate ways of seeking to arrive at a rental figure that would correspond with an actual market rent on the statutory hypothesis and if they are properly applied all the tests should in fact point to the same answer; but the greater the margin for error in any particular test, the less is the weight that can be attached to it.

(In this particular case, the Tribunal placed considerable weight on the profits test, not much weight on the rental test and very little weight to the contractor's test.)

Therefore, the valuer can use whichever method is most appropriate, or can use several different methods, but should place different weight on different methods in different situations.

(c) The rental method

The rental method of valuation essentially consists of applying rental information (evidence) to the hereditament, in order to reach the rateable value. The rental evidence can be either direct, (from the hereditament itself), or indirect, (from other, comparable, properties). The rental method is generally considered the best method, since rateable value is defined in terms of a net rental value. As mentioned above, valuing for rateable value can, to some extent, be treated similarly to carrying out a rent review, where the assumed terms are dictated by the statutory definition rather than the specific lease.

The valuer's ability to rely on direct rental evidence (if any) from the subject hereditament to reach rateable value has shifted over time. It used to be held that, where actual rents are available and cannot be impeached, these are the best guide to rateable value. Scott LJ said in *Robinson Bros (Brewers) Ltd* v *Houghton and Chester-le-Street Assessment Committee* (1937):

Where the particular hereditament is let at what is plainly a rack rent or where similar hereditaments in similar economic sites are so let, so that they are truly comparable, that evidence is the best evidence and for that reason is alone admissible: indirect evidence is excluded not because it is not logically relevant to the economic inquiry but because it is not the best evidence. Where such direct evidence is not available, for example, if the rents of other premises are shown to be not truly comparable, resort must necessarily be had to indirect evidence from which it is possible to estimate the probable rent which the hypothetical tenant would pay.

However, in *Garton* v *Hunter*, the Court of Appeal held that, in keeping with modern practice, all relevant evidence was to be admitted. Lord Denning MR said:

Nowadays we do not confine ourselves to the best evidence. We admit all relevant evidence. The goodness or badness of it goes only to weight and not to admissibility.

Also in *Garton* v *Hunter*, Winn LJ suggested that Scott LJ's dictum should be amended as follows:

Where the particular hereditament is let on what is plainly a rack rent and there are similar hereditaments and similar economic sites which are so let that they are truly comparable, that evidence should be classified in respect of cogency as a category of admissible evidence properly described as superior; in some, but not all, cases, that category may be exclusive. Any indirect evidence, albeit relevant, should be placed in a different category; reference to the latter category may or may not be proper, or indeed necessary, according to the degree of weight of the former kind of evidence.

In *R* v *Paddington VO, ex parte Peachey Property Corporation Ltd* (1964), Widgery J said:

It is clear on authority that the actual rent payable at the date of assessment is not conclusive of the rateable value even if the terms of the current letting are the hypothetical statutory terms.

Even where there is a rent on the hereditament being valued, it is unlikely that the lease terms precisely match the statutory terms. Thus, most actual rents must be adjusted to match the statutory terms before they can be used as rental evidence. For example, rates must be deducted if the rent is inclusive of rates. If the landlord is responsible for the whole or any part of the repairs, insurance, or other expenses necessary to maintain the hereditament, a suitable adjustment must be made to the rent.

Despite all these possible adjustments, actual rents must be the starting point when valuing to rateable value for all hereditaments which are of a kind which are normally offered and taken on lease in the open market.

Direct rental evidence is the rent of the hereditament being valued. Although this is evidence of rental value, it is by no means conclusive. It is seen as *prima facie* evidence liable to be rebutted (*Baker Britt & Co Ltd* v *Hampsher (VO)* (1976)).

The hierarchy of rental evidence is discussed in *Lotus & Delta Ltd* v *Culverwell (VO) and Leicester City Council* (1976). The following procedure was established for weighing rental evidence for use in rating valuation:

- where the subject hereditament is actually let that rent should be taken as a starting point
- the more closely the circumstances under which the rent was agreed as to time, subject matter and conditions related to the statutory requirements contained in the definition of rateable value, the more weight should be attached to it
- where rents of similar properties are available, the valuer should also look at these, in order to confirm or otherwise the level of value indicated by the actual rent of the subject hereditament
- rating assessments of other comparable properties are also relevant. When a rating list is prepared these assessments indicate the valuation officer's estimate of value. In subsequent proceedings on that list therefore, they can properly be referred to as giving some indication of that opinion
- in the light of all the evidence an opinion can then be formed of the value of the appeal hereditament, and the weight to be attributed to the different types of evidence, depending on the one hand on the nature of the actual rent and, on the other hand, the degree of comparability found in other properties.

Where there are no rents available from comparable properties, a review of other rating assessments may be helpful, but it would clearly then be more difficult to reject the evidence of the actual rent.

Some rents are not suitable for use in rating valuation. These include:

- rents which do not, or cannot, be made to conform with the rating hypothesis; for example turnover rents, or rents indexed to the Retail Price Index (RPI)
- rents agreed between related persons or connected companies
- sale and leaseback rents (*John Lewis & Co v Goodwin (VO)* (1980)). Rents agreed under such arrangements may not reflect open market rental value (OMRV) for various reasons, typically:
  - financial arrangements whereby the rent reflects the cost of borrowing rather than open market value (OMV)
  - initial rent being higher than OMV to increase the value of the freehold
  - leaseback being essentially a funding operation, with the rent being geared to profitability
- Rents which require a great deal of adjustment to make them accord with the statutory definition of rateable value. For example, adjustment may be required to take into account different repairing and insuring liabilities, rent free periods, premiums, stepped rents, improvements, overage, restrictive user clauses or start dates, ie rent set at some time distant from the antecedent valuation date (AVD).

However, it is not a perfect world and it may be that the only available rental evidence will require a lot of adjustment. The reality is that non-domestic property is rarely let on an annual tenancy; responsibility for repairs and insurance is not always as envisaged in the hypothetical tenancy; tenants make alterations and improvements which are not reflected in the rent; landlords grant incentives to secure high headline rents and few rents are likely to have been fixed on the AVD. In rating, a "clean" rent can be rare. In such circumstances, the valuer will have no choice but to use such evidence.

The quality of rental evidence, drawn from comparable properties, depends on:

- how closely the rented property resembles the hereditament to be valued
- how closely the lease terms for the rented property resemble those of the hypothetical tenancy in the definition of rateable value
- how close the time the rent was fixed is to the AVD.

As a general rule the less a rent has to be adjusted, ie the more closely it resembles the statutory definition, the more weight can be placed on it to find rateable value.

The valuer will therefore gather a collection of rental information, to be adjusted and analysed. This is often called the basket of evidence, from which the valuer will reach his or her opinion as to the rental value, and hence rateable value, of the hereditament. Although the actual rent of any hereditament is *prima facie* evidence of its value, the valuer must always:

- look at the rental evidence as a whole
- compare the results of analysis using adjusted rents
- attach weight to each piece of evidence, depending on how reliable the valuer considers it to be
- draw a conclusion from all the evidence available as to what rent a hypothetical landlord and tenant might agree under the rating terms.

Examples of adjustment of evidence

(a) Repairing and insuring liabilities
   The statutory definition assumes that the property is already in reasonable repair at the start of the lease, and that the tenant is then responsible for maintaining it, and paying the insurances. This is generally called full repairing and insuring (FRI) terms. If a lease being analysed is not on FRI terms, the tenant will be paying more in rent to compensate the landlord for the repair and insurance costs. It is very common for valuers to adjust a rent to take account of non-FRI repairing liability by subtracting 10% (5% for internal and 5% for external repairing liability). Insurances are generally taken at 2.5%. Sometimes these percentages are varied to take account of the characteristics of a particular property, as in *Dawkins (VO) v Royal Leamington Spa Corporation and Warwickshire County Council* (1961). Alternatively, the actual cost of repairs can be referred to, as in *Phillips Bros Character Shoes Ltd v Childs (VO)* (1959).

(b) Service charges
   In many shopping centres, office blocks and sometimes modern industrial estates, a landlord will charge tenants for services. These may include security, cleaning, lighting, provision of chilled water, maintenance of the building, etc. If the service charge is in addition to the rent, then no adjustment needs to be made to the rent. However, if the service charge is included in the rent, then it will need stripping out to the extent that the services go beyond the statutory definition.

(c) Premium/rent free period
   When taking a lease, if a tenant pays a premium in addition to rent, it often means that the rent passing is below the open market rent. Alternatively the premium could represent "key money".
   Premiums where the rent is below the open market rent, are analysed by being decapitalised and added to the rent passing. The period of amortisation is generally over the period to the next rent review or lease renewal, where the rent will become open market rental value.
   If, instead, the premium was key money, (the tenant paying to obtain a specific location), opinion is divided on how it should be approached. It could be argued that the interaction of supply and demand simply results in the payment of key money, in the same way that interaction determines rental values. Alternatively, key money can be seen as a payment solely to obtain occupation and, with it, the ability to trade and earn a profit. The premium therefore represents

the opportunity cost of occupation and not profit rent. A way to treat key money could be to amortise the premium over the expected occupancy of the tenant.

As an example, a tenant pays a premium of £10,000 and a net rent of £20,000 pa; the lease is for 20 years with a review to OMRV after five years.

Rent actually paid (net of repairs etc)		£20,000
Premium	£10,000	
÷ YP 20 years @ 6%	3.3255	£ 3,007
Rent in terms of RV		£23,007

Any rent which includes a significant element derived from an amortised premium must be treated with caution, as the reasons for the premium being paid are rarely clear and are often confidential.

An incoming tenant could also pay a premium to receive the fixtures and fittings left by the outgoing tenant. As this does not relate to rent, it should not be decapitalised.

(d) **Improvements not included in the rent passing**
The valuer should remember that actual rents are likely to exclude tenant's improvements, whereas the value for rating is *rebus sic stantibus* and includes all rateable improvements (including any rateable plant).

The effect, if any, on rental value of the improvement should be considered. Some alterations may be personal to a tenant, for example shop-fitting works, and will therefore not add to open market rental value. These improvements by definition will not affect the rateable value. Where the alterations will affect rental value, the adjustment could be the addition of the annual equivalent of the cost of the works.

There are no set rules for the period of amortisation, but the valuer may consider the following:

(a) where the improvements are a lease requirement — the shorter of the following from the date of the improvements:
   - the end of the useful life of the improvements or
   - either the next rent review where the improvements are not specified to be disregarded (this would be unusual), or the end of the lease.
(b) voluntary improvements — the shorter of the following from the date of the improvements:
   - the end of the useful life of the improvements or
   - the next rent review where the value of the improvements is included in the rent
   - the date of the next lease renewal which is neither the first renewal since the improvements were carried out nor within 21 years of their completion
   - exceptionally, where it is known that the lease will not be renewed, the end of the current lease.

The valuer will of course be aware that cost does not necessarily equate to value. In *Edma Jewellers Ltd v Moore (VO)* (1975) the Lands Tribunal considered this issue and concluded that "all expenditure must be looked at on its merits". At the end of the exercise, the valuer must take an objective look at whether the calculation he has reached is a realistic one as an open market rent.

(e) **Year-to-year assumption**
The statutory definition requires a lease "from year to year". Any adjustment to a real lease, which is not from year to year, should only be made if there is evidence in the rental market that the lease

length does affect rental value. Hence, In *Cresta Silks Ltd* v *Peak (VO)* (1958), the rents paid under leases of up to 21 years were accepted without adjustment. But in *Naylor's (Reading) Ltd* v *Gaylard (VO)* (1958), the rents paid under a 14 year lease and a 21 year lease had a deduction applied, in order to relate them to the statutory definition, as there was definite evidence regarding the actual circumstances of the lettings.

Generally speaking, if the lease length, and more particularly the rent review frequency, is the norm for the rental market for that type of property, then no adjustment needs to be made for the year to year assumption.

(f) Rent-free period or other concessions
In recent years varying forms of rental incentives have become commonplace, including rent-free periods, stepped rents, and reverse premiums on taking over a tenant's previous lease. These changes have made it much harder to interpret some market transactions and to determine exactly what the deal equates to, in rental terms.

A rent can be adjusted by amortising the incentives. Adjustments for some of the more common incentives are dealt with below, but remember that if a rent requires significant adjustment, this casts doubt on its worth as a useful piece of rental evidence.

Adjustment periods
Views differ as to the period over which an incentive should be amortised. One view is that an incentive is a payment to the tenant for taking the lease and should be rentalised over the whole length of the lease. An alternative view is that, as the rent is normally subject to review to OMRV, the incentive should only be amortised to the first review. A third view is one of compromise, suggesting a period between the two. The correct period over which to rentalise the incentive will ultimately be a question of judgment based upon the type of incentive and local market conditions.

Rent-free periods
Short rent-free periods are often granted to tenants to enable them to complete the fitting out of the property. No adjustment should be made for such a rent-free period, as the concession is only given to allow a tenant to complete works which, in the world of rating, are assumed to be carried out by the landlord. Where the rent-free period does not relate to fitting out or putting into repair, this should be treated as an incentive from the landlord.

It is possible to make the adjustment for a rent-free period into a long and complex operation. However, the practical and accepted method of dealing with it is to consider the length of the rent-free period in the context of possible adjustment times. A long rent-free period may appear to be an incentive to take a long lease, whilst a short period may be a reflection of a slightly over-rented property. It may be necessary to do more than one calculation before drawing a conclusion.

A warehouse is let for 10 years at a rent of £25,000, with a review after five years. A six-month rent-free period has been granted by the landlord. This is not for fitting out (the unit is already fitted out), nor is it for disrepair.

From the above information, the rent-free period appears to be an incentive and should be amortised to find the true rental level. The calculation below shows the two options, first assessing the cash flow benefit to the occupier, over the period to the first rent review, (when it is anticipated that the rent will reach full open market rental value), and the second assessing the cash flow benefit to the occupier over the length of the lease. In order to find the true cash flow benefit, years purchase (YP) and present value of £1 (PV) are used. In the first, YP of 4.5 years is used, as out of the first five years to review, the tenant will only pay 4.5 years rent; then discounted by PV 0.5 year, as the tenant will not begin paying rent for six months; and then divided back across the five years of the review period. Similarly, in the second option, the present cash value of 9.5 years rent payments (YP 9.5 years) but starting in six months' time (discounted by PV 0.5 year) is spread over the full ten years of the lease.

Over 5 years to review:		
Period to first review		
Rent		£25,000
YP 4.5yrs 11%	3.3991	
PV £1 in 0.5yrs 11%	0.9505	×3.2309
		£80,773
Divide by YP 5yrs 11%		÷3.6959
		£21,855

Over whole lease:		
Rent		£25,000
YP 9.5yrs 11%	5.7131	
PV £1 in 0.5yrs 11%	0.9505	×5.4303
		£135,758
Divide by YP 10yrs 11%		÷5.8892
		£23,052

With such a short rent-free period, it is advisable to take a conservative view of the true rental level by taking the first option.

(g) VAT

Where the rent is inclusive of VAT, this should be deducted. For example, if a rent is £100 inclusive of VAT, the £100 is 117.5% of the net rent, therefore the net rent is £100 divided by 117.5%, equating to £85.11.

(h) Domestic element of rent

Any part of a rent attributable to residential accommodation, such as where a shop is let together with a flat above, is excluded from business rates and must be stripped out. The amount to be taken is the amount that is considered to be included in the total rent for the domestic part. It is not the amount that the residential accommodation could be sublet for.

(i) Date of rent

It is unusual for rental evidence to be conveniently dated at the AVD. As discussed above, the more a rent needs adjusting, the less weight should be placed on it for rateable value evidence purposes. Ideally, then, rents from up to a year either side of the AVD are particularly worth analysing. However, the hypothetical tenant would not know about rents post-AVD, and therefore these should carry less weight.

The valuer will still find it useful to adjust and analyse all rents, no matter what date they were set on, as they are helpful in indicating trends in values.

The valuer must therefore consider whether the rent needs adjusting (upwards or downwards) to reflect rental movements over time. The amount of any adjustment should be derived from the basket of rental evidence gathered, which relates to the type of property in the locality. Alternatively, published data on rental movements can be used to adjust rents for date, although this should be used cautiously, as the indices are rarely specific to the location or the type of property.

(j) Deduction of rates

Under the statutory definition, rateable value is on the basis that the "tenant undertook to pay all usual tenant's rates and taxes". If a rent being analysed is on the basis that the landlord pays the rates, taxes, and/or water-rates, a deduction from the inclusive rent must be made, to strip out the rent which the landlord will pay in the form of these rates and taxes. The deduction should be calculated as follows: if the inclusive rent is (for example) £100 and the rates are 25p in the pound, then £100 represents 125% of the net annual value. £100 divided by 125% gives £80 as the rateable value.

(k) User restrictions

Under the *rebus sic stantibus* principle, although a property is to be considered as vacant and to let, it must be considered with the existing (general) use continuing. The *Fir Mill and Royton* case requires a factory to be valued as a factory, but not for any specific product. Therefore, user clauses in an actual lease which restrict occupation to a specific occupier, or else exclude certain uses, must be ignored. If these have any effect on rental value, this effect must be reversed by the valuer when translating the rent into the rateable value equivalent. This can only be done if there is market evidence, for that type of property, of the impact on rent of the restrictive user clause.

Having built up the basket of evidence, the valuer needs to apply this to the subject property. It is beneficial to put the rental evidence into a pattern of dates and or locality. When dealing with retail properties, for example, the rental value of unit shops can vary markedly over a short geographic area, whereas department stores can be compared across a much wider area, and will take into account catchment populations as well as the rental levels in the immediate area. Offices and industrial buildings can also be compared to similar categories of buildings across a wider area.

The valuer should always remember that when reaching a rateable value, he or she should stand back at the end, and ask: would this be a level of rent which could reasonably be expected for the subject property, in the open market, as at the antecedent valuation date?

Example rental valuations

A large, modern office premises:

Floor	Description	Size m² (NIA basis)	£/m²	£
Ground	Reception	290.60	200.00	58,120
First	Offices	711.30	200.00	142,260
Second	Offices	2155.10	200.00	431,020
Third	Offices	2155.10	200.00	431,020

Floor	Description	Size m² (NIA basis)	£/m²	£
Fourth	Offices and Café	2155.10	200.00	431,020
Fifth	Offices	1968.90	200.00	393,780
Sixth	Offices	1408.30	200.00	281,660
Seventh	Meeting Rooms and Kitchen	532.20	200.00	106,440
Eighth	Meeting Rooms	224.90	200.00	44,980
Total Area		11601.50		
Sub-total				2,320,300
Plant & Machinery				4,772
Car Parking		159 spaces	£2000/space	318,000
Total Value				2,643,072
Adopted Rateable Value			£2,640,000	

Unit shop in a town centre:

Floor	Description	Size m² (NIA basis, Zoning)	£/m²	£
Ground	Retail Zone A	64.07	300.00	19,221
Ground	Retail Zone B	66.86	150.00	10,029
Ground	Retail Zone C	3.34	75.00	251
Ground	Staffroom & Stores	27.71	30.00	831
Total Area		161.98		
Subtotal				30,332
Plant and Machinery	Air conditioning system	134.27	7.00	940
Total Value				31,272
Adopted Rateable Value			£31,250	

Factory unit:

Floor	Description	Size m² (GIA basis)	£/m²	£
Ground	Bakery	8595.30	20.48	176,032
Mezzanine	Store	478.70	4.10	1,963
Ground	Store	647.50	20.48	13,261
Ground	Warehouse	4970.20	20.48	101,790
Ground	Workshop	1000.60	20.48	20,492
Ground	Plant Room	175.40	20.48	3,592
Ground	Workshop	117.80	20.48	2,413
Ground	Office	548.90	24.57	13,486
Ground	Canteen	536.00	22.52	12,071
Ground	Locker Room	163.50	22.52	3,682
Ground	Office	711.20	24.57	17,474
Ground	Store	63.50	20.48	1,300
Ground	Plant Room	101.70	20.48	2,083
Ground	Gas Meter House	11.70	20.48	240
Ground	Warehouse	1560.10	26.51	41,358
Ground	Warehouse	648.90	26.51	17,202
Ground	Warehouse	102.10	26.51	2,707
Ground	Goods Lift	17.80	32.00	570
Ground	Production Area	112.50	32.00	3,600

Floor	Description	Size m² (GIA basis)	£/m²	£
Ground	Warehouse	1597.30	32.00	51,114
Ground	Office	160.60	40.32	6,475
Total Area		22321.30		
Subtotal				492,905
Plant & Machinery				3,818
Total before Adjustment				496,723
Adjustment	Poor layout		–5%	–24,836
Total Value				471,887
Adopted Rateable Value				£470,000

Large shops warrant some additional discussion. Shops are normally valued by reference to actual rents, as there is usually sufficient rental evidence from these to provide a sound basis. However, valuation for larger shops, such as department stores, can be more difficult, as there may be little rental evidence available, or few rents that can be appropriately adjusted to the statutory definition of rateable value. Some valuers therefore base the valuation of larger shops on smaller shops in the vicinity, but there are problems in relating the two, and an end allowance for size, often called quantity allowance or quantum has been developed.

Quantity allowance has been defined by the Lands Tribunal, in *F W Woolworth and Co Lt*d v *Peck (VO)* (1967) as:

an allowance (normally but not necessarily an end allowance) which is sometimes made where the rent or rental value of a large shop is being arrived at by reference to the rents or rental values of small shops, and which is unrelated to any allowance which may appropriately fall to be made for some other reason, such as for redundant space.

The Lands Tribunal has also held that any rental effect related to size was not produced by size but by the demand (or lack of demand) for that size. It went on:

The term "quantity allowance," although convenient and much used, is misleading to the extent that it masks this essential distinction. We do not think any rental inference at all can be drawn merely from a knowledge of the size of the property being valued; to carry some significance that fact of size must in our view be coupled with other information — which often conveniently takes the form of rental evidence in respect of similar properties of about the same size.

In *Hill & Co Ltd* v *Thompson (VO)* (1955) the Lands Tribunal considered the correct approach to quantity allowance for shops as follows:

It is this question of evidence which we regard as the kernel of the problem and as the reconciling key to past decisions of the tribunal. Each case must depend on its own facts and the onus of proof is always on the appellant. We do not accept, therefore, the submission that there is a presumption in favour of a quantity allowance for large shops when valued on the zoning method; but we do accept that as the method was originally based on factual rental evidence it requires comparatively slender evidence to establish that in any particular shopping centre quantity is still allowed for in the market. Once that is done, it must be answered by positive evidence to displace it. It is not enough in our view simply to say, "I do not find the evidence sufficient or sufficiently convincing".

Department store in a major shopping centre:

Floor	Description	Size m² (GIA basis)	£/m²	£
Lower Ground	Shop main area	3174.80	130.00	412,724
Lower Ground	Shop main area	250.90	130.00	32,617
Ground	Shop main area	3465.40	130.00	450,502
First	Shop main area	3387.30	130.00	440,349
Second	Shop main area	3414.80	130.00	443,924
Total Area		13693.20		
Total Value				1,780,116
Adopted Rateable Value				£1,780,000

In this example valuation, rather than an end allowance being deducted, the store has been valued on an overall rate; that rate taking into account the rental demand for this type of property.

(d) The profits basis

Usage of the profits basis

The "profits basis" of valuation is used to find the rental value of properties which are seldom, if ever, let in the open market. It is also called the "receipts and expenditure" basis. As mentioned previously, no method of valuation has any statutory basis, but the profits basis has developed as a means of reaching a valuation in the absence of direct evidence.

An early explanation of the theory behind the profits basis is found in *Edinburgh Street Tramways Co* v *Lord Provost of Edinburgh* (1894):

> It is a mistake to suppose that valuation by rental is a process disassociated from the idea of profit ... The questions whether a hypothetical tenant could be found, and what rent he might reasonably be expected to give if he were found, cannot easily be solved, if at all, except by estimating what amount of profit the line had yielded in the past and was likely to yield in the future. An intending lessee, whether real or hypothetical, would hesitate to pay a rent which was not based upon these data.

Principle

The profits basis essentially consists of finding a net profit from the trade at the premises, and assuming that the tenant would be prepared to pay part of this in rent, while retaining the other part as profit. First, the gross profit derived from occupation of the hereditament is calculated by deducting the cost of purchases made from gross receipts. It is usual to look at three years trading figures, leading up to the antecedent valuation date. The working expenses, including an allowance for renewal of the tenant's assets, are deducted from the gross profit to give the divisible balance. The divisible balance represents the amount to be shared between the tenant (tenant's share) and the landlord (rent, or rateable value).

- Gross receipts minus cost of sales (ie the cost of the product sold, and any stock reduction) gives the gross profit .
- Gross profit minus expenditure:
  - labour/wages

- rates payable
- directors' remuneration
- pension schemes
- insurance
- repairs/sinking fund to replace the hereditament
- repairs and renewals of tenant's non-rateable equipment
- head office expenses (if appropriate)
- other running costs, eg lighting, heating, bank charges

Gives the divisible balance.

- Divisible balance minus the tenant's share (which includes an allowance for interest on capital; profit, and risk) leaves the landlord's share, which is the rent and by definition is equal to the rateable value.

There are four basic methods of estimating the tenant's share:

(i)   percentage on tenant's capital
(ii) percentage on receipts
(iii) proportion of divisible balance
(iv) spot figure, (least likely method).

Whichever method is used, the valuer should always stand back and consider whether the sum produced is reasonable compared to each of tenant's capital, turnover and divisible balance (net profit), taking into account the motive(s) for occupation, the degree of risk involved and likely competing bidders.

As mentioned above, the profits basis is used for properties which are in a class that are not generally let in the open market, and also have an element of legal or factual monopoly, such as public houses or petrol filling stations.

The tenant is the hypothetical tenant, so if the specific tenant operating the business is either particularly adept or particularly poor, the trade figures should be compared with other similar businesses to establish a reasonable level or fair maintainable profit. For this reason, the VOA has agreed scales of values with major classes of business, such as the hotel trade and the public house trade.

Historic houses were considered in *Hoare (VO)* v *National Trust* (1997) and *National Trust* v *Spratling (VO)* (1997). The Lands Tribunal found that the National Trust was the only potential hypothetical tenant of two historic houses (Petworth in West Sussex, and Castle Drogo in Cornwall). Having regard to the profits basis of assessment, the costs of repair and administration made the occupation of the hereditament unprofitable. None the less, the Lands Tribunal held that it could have regard to the Trust's overall financial resources, and its motive to preserve historic houses. The Lands Tribunal concluded that the National Trust would be prepared to pay a positive rent for the benefit (in terms of its motives) of occupying the hereditaments.

Although the Court of Appeal went on to reverse the Lands Tribunal's decision, holding that the National Trust would not be prepared to pay a rent in addition to taking on the responsibility for repairs of the hereditament, the profits method was still applied. The Court of Appeal found that only a nominal value was appropriate under the profits method.

An example of a profits test valuation is set out below: the subject premises are a private leisure complex, including restaurant, bars, billiard room, squash court, swimming pool and tennis court.

Recent accounts for years to 2003

Income		2000–2001	2001–2002	2002–2003	Note
Gross receipts:	Bar	£78,250	£69,575	£95,234	(1)
	Restaurant	£133,240	£105,463	£123,250	
	Outside catering	£10,400	£11,675	£11,980	
	Members' subscriptions	£4,250	£3,975	£4,500	
	Rooms	£6,650	£6,155	£6,375	(2)
	Other	£3,750	£3,600	£3,875	
		£236,540	£200,443	£245,214	
Purchases		£101,515	£91,020	£120,166	(3)
Gross profit		£135,025	£109,423	£125,048	
Expenditure					
Wages and national insurance		£32,300	£35,200	£33,600	
Print, post, and stationary		£16,200	£15,500	£17,000	
Drawings		£9,635	£15,325	£13,800	(4)
Loan interest		£2,400	£2,750	£1,545	(5)
Bank charges		£1,225	£1,375	£1,475	
Accountancy fees		£1,200	£1,320	£1,470	
Telephone		£620	£770	£645	
Electricity		£8,325	£8,634	£8,743	
Business rate		£7,420	£7,720	£7,935	
Environmental charge		£1,125	£1,340	£1,423	
Repairs to land and buildings		£250	£5,675	£1,234	
Vehicle servicing		£675	£705	£746	
Petrol		£875	£970	£1,123	
Licensing		£125	–	–	
Insurance	— premises	£1,365	£1,625	£1,700	
	— contents	£573	£789	£891	
	— other liability	£150	£185	£195	
Bad debts		£310	£435	£411	
Laundry		£575	£508	£568	
Advertising		£1,250	£1,345	£1,765	
New furnishings		£5,750	–	–	
Total expenditure		£92,348	£101,971	£96,269	
Net profit		£42,677	£7,452	£28,779	

Notes

1. Average for three years: £81,020, but trends would be considered.
2. Can be checked by number of rooms × rate per night × occupancy.
3. Approximately 50% of sales of food and drink.
4. Not an expense, but part of profit.
5. Not allowed as an expense — interest is reflected in the final calculation for rent.

Adjusted accounts for valuation

Income

	Bar	£90,000
	Restaurant	120,000
	Outside catering	12,000
	Members' subscriptions	4,500
	Rooms	6,500
	Other	3,875
	Total	236,875
Less	Purchases @ 50% food and drink sales (comparing previous years)	105,000
	Estimated gross profit	131,875

Expenditure

Wages & NI (as last year, plus inflation of say 3%)	34,600
Office expenses	17,000
Drawings — no allowance/ part of profit	
Loan interest — no allowance	
Bank charges	1,475
Accountants	1,470
Telephone	645
Electricity	8,743
Business rates (+ 2%; assume no transition)	8,094
Environmental charge, say	1,500
Repairs: reasonable average	2,400
Vehicle servicing	750
Petrol (budget increase 3p/litre)	1,200
Insurance (check adequacy of cover)	1,800
Contents	1,000
Other	200
Bad debts	450
Laundry	568
Advertising (note increase in Year 2 reflected in higher income)	1,765
New furniture — no allowance:	
Depreciation on furniture and equipment @ 10% of total value, say	5,000
Total expenses	88,660
Expected future net profit	43,125
Interest on operator's capital @ 8% (stock, cash float, furniture etc) say	3,450
Divisible balance	39,675
50% tenant's share	£19,838
50% rent	£19,838
Rateable value say	£19,800

The last complete year's audited accounts before 1 April 2003 (the antecedent valuation date for the 2005 List) should provide the basis of the valuation, but the valuer may take into account:

- any changes in the hereditament or the surroundings between the date of the last accounts and the year in which the valuation entry goes into the list

- any information from previous years' accounts which indicate that a particular item in the accounts being used is not an average annual expenditure.

The divisible balance contains both the tenant's share and the rent. If tenant's share is 50%, the rent is £19,838. As rateable value is, by definition, equal to the rent, then the RV is £19,838, say £19,800.

The method used here has been used by the Lands Tribunal in a number of cases. The interest on tenant's capital varies, depending on current interest rates. 50% of the divisible balance is frequently used for tenant's remuneration and risk. In practice, market evidence should be sought of actual rents paid by hoteliers occupying leasehold properties, and those rents analysed as a proportion of profit. The valuer should always remember that, despite the arithmetical nature of the exercise, they are looking for a rent which might reasonably be expected in the open market for the subject property as at the antecedent valuation date.

Shortened methods

An example of a shortened method of the profits basis can be to simply take a certain percentage of the receipts of the business, as the amount which the hypothetical tenant would pay in rent. This is often done for the sake of simplicity, once a full profits valuation of a sample of businesses has established the appropriate rateable value, and this has been analysed as a percentage of turnover. For example:

Income:	2000/2001	£917,746
	2001/2002	£529,191
	2002/2003	£711,641
Average:		£719,526
Taking @ 7% =		£50,366
Say Rateable Value		£50,250

The valuer would of course enquire why the year 2001–2002 was so much lower than the preceding year, and make any adjustments as appropriate before relying of the figures.

Similar shortcut approaches can be adopted for different types of business, not always based on a proportion of turnover.

Petrol stations tend to be valued on a modified profits basis, using a price per unit of maintainable throughput. As there are few open market rents of petrol stations which do not have an element of tie to them, (ie where the particular petrol company holds title from the oil company), the VO has negotiated a national valuation scheme with the petrol industry. A sliding scale of £/1,000 litres of maintainable throughput is applied.

As well as the standard receipts and expenditure method, hotels can also valued on a modified profits basis, either using a percentage of gross takings, or a price applied to the number of double bed units, including the occupancy rate. Again, the VO has negotiated a national scheme of valuation with the industry.

Cinemas are occasionally rented, but generally there is little open market evidence, and so a receipts and expenditure method is used. Comparison between one cinema and another can be made by a percentage of gross receipts of the RV calculated.

The valuation of public houses has been the subject of recent case law. In *Scottish and Newcastle Retail Ltd* v *RF Williams and Allied Domecq Retailing Ltd*, (two cases heard together in 2000), the Lands Tribunal stated:

"We find that public houses are valued by reference to trading receipts ... We were referred to the RICS Red Book. Guidance Note GN7 gives guidance on trading - related values and goodwill. GN7.2.14 states:

Proposals have been put forward that rental value of public houses, particularly those let free of tie, should be assessed on a floor area basis rather than by reference to trading potential. It will be appreciated that both methods rely upon the analysis of comparable transactions. At present the assessment of pub rents on a floor area basis is not a proven or generally accepted method of valuation, although further research on the subject continues.

(The Lands Tribunal was presented with agreed alternative valuations for the public houses in this case, so did not have to carry out a valuation itself.)

Profits test valuations are also sometimes carried out for mineral producing hereditaments, refuse tips, markets, racecourses, cemeteries and crematoria, and docks and harbours (those which are not valued by formula — see below).

As mentioned above, the valuer should always bear in mind that the end figure, however arrived at, should be a net rent as at the AVD.

(e) Contractor's basis

Main principles of the contractor's basis

The main principle of the contractor's basis is that, where a property is in a class which is not let in the open market, and which is unsuitable to be valued by reference to receipts and expenditure, it can instead be valued by reference to its construction cost. However, the concept that cost automatically equates to value is patently untrue, hence the contractor's basis is often called the method of last resort. The hypothesis is that the hypothetical tenant would acquire the land, build the premises, pay interest on capital to build, and that this interest would be the same amount that he would be prepared to pay in rent, had the building been available to rent.

The classic explanation of the contractor's basis was given by Lord Denning in *Dawkins (VO)* v *Royal Leamington Spa Corporation* (1961) as follows:

The hypothetical tenant has an alternative to leasing the hereditament and paying rent for it; he can build a precisely similar building himself. He could borrow the money, on which he would have to pay interest; or use his own capital on which he would have to forego interest to put up a similar building for his own owner-occupation rather than rent it, and he will do that rather than pay what he would regard as an excessive rent — that is, a rent which is greater than the interest he foregoes by using his own capital to build the building himself. The argument is that he will therefore be unwilling to pay more as an annual rent for a hereditament than it would cost him in the way of annual interest on the capital sum necessary to build a similar hereditament.

As for any rating valuation, the valuer is trying to find the rent at which the hypothetical landlord would let the property to the hypothetical tenant. The rating hypothesis requires various assumptions about the physical state of the hereditament, for example repair, but otherwise the valuer is looking for a rent for the actual property.

Under the contractor's basis, the valuer is essentially testing the rental value of the property by considering the annualised equivalent of the estimated cost of construction.

Initially, the valuer must decide whether to cost the actual property or a substitute. Where it is unrealistic to consider that a hypothetical tenant would re-build the property as it actually stands, (for example because of age, design or type of construction), the valuer should instead estimate the cost of a modern substitute. In most cases, however, building costs will relate to the actual property.

Where the substitute approach is used, then it would be usual practice to base building costs on the actual building's floor area. However, where the reason for adopting the substitute approach is because the actual building is larger than required, (for example due to changes in technology, not for reasons that are personal to the actual occupier) then the substitute should be costed on the basis of a size to reflect modern trade and business practices.

Use of the contractor's basis

As mentioned above, the contractor's basis can be used to value properties which are not generally rented in the open market, and which are not operated for profit, (in which case a receipts and expenditure method could be used). Examples include local authority leisure centres, libraries, town halls, fire stations, sewage works and oil refineries.

An outline of the method is as follows:

1.   estimate the cost of construction of the premises, or a modern substitute
2.   adjust the cost of construction to reflect differences between the actual hereditament and the modern substitute. This gives the Effective Capital Value (ECV)
3.   add the estimated value of the land
4.   apply the relevant percentage to decapitalise the value of the buildings and the land
5.   make final adjustments to ensure that the result reflects what the hypothetical tenant would be willing to pay in the open market.

These five stages were established by the Lands Tribunal in various cases, including *Gilmore VO v Baker-Carr (No 2)* (1963).

At stage 1, the valuer must consider the construction cost starting at a bare site. To reach the ECV at stage 2, the valuer makes adjustments for the differences between a modern substitute, and the actual hereditament. For example, if the hereditament is an 1800s town hall, it may be of more ornate style, and with less efficient use of space, than would be constructed in a modern equivalent function. Hence a deduction for obsolescence would be appropriate. Plant and machinery is of course added to reach the ECV at stage 2.

The value of the land is often difficult to find, as it must be considered *rebus sic stantibus*, ie available for that particular category of use. The type of building (for example a public authority leisure centre) being considered under a contractor's basis may be of a type where other land uses would be much more valuable; for example open market land costs for either residential or commercial uses could be much higher than those restricted to the particular use.

Decapitalisation rate

The decapitalisation rate is used to convert the ECV, plus land capital value, into a rental value. The rates used to be intensely argued, but are now prescribed in regulations. The appropriate rates for the 2005 rating list are:

Educational and health hereditaments	3.33%
All other hereditaments	5.00%

Examples of contractor's basis valuations

Leeds District Magistrates' Courts, upheld by Lands Tribunal

Stage 1

Actual tender as adjusted		£11,720,087
Less Costs attributable to		
Probation Service hereditament	£376,687	
and Social Services hereditament	£163,390	
Actual apportioned tender price of the appeal hereditament		£11,180,010
Fees 10%		£ 1,118,001
Total Stage 1 cost		£12,298,011

No Stage 2 adjustments

Stage 3

Value of land (7.5% of Stage 1 cost)	£ 922,351
Total Value	£13,220,362

Stage 4

Decapitalisation Rate 6% (this was earlier than the 2005 List)	£793,222

No Stage 5 adjustments

Rateable value say	**£792,500**

In two cases heard together, *Eastbourne Borough Council and Wealden District Council* v *Allen (VO)* (2001), the Lands Tribunal considered the valuation of local authority operated leisure centres. It rejected the use of a profit's test, and confirmed the use of the contractor's basis.

Sovereign Leisure Centre, Eastbourne, (1995 List)
Contractor's basis valuation by the lands tribunal:

Stage 1 Estimated Replacement Cost (ERC)
Adopt simple substitute building

		m²
Area		5,712
Less diving pool	126	
competition and teaching pool hall	39	
dry changing area	77	
viewing areas	41	
reception area	90	
internal walls	82	455
		5,257

Total Floor Area 5,257m² @ £784 per m² =	£4,121,488
Reduction for external envelope 5,257m² @ £46 per m²	−241,822
Reduction for internal finishes 5,257m² @ £16.50 per m²	−86,740
Total building costs	£3,792,926
Add external works — agreed	504,510
	£4,297,436
Professional fees @ 10% — agreed	429,744
ERC	£4,727,180

Stage 2 Adjustment of ERC to reflect age and obsolescence

Age Allowance	8.5%	
Functional/technical obsolescence	4.0%	
Total allowances	12.5%	590,898
		4,136,282

Stage 3 Land value agreed		225,000
		4,361,282

Stage 4 Statutory Decapitalisation rate 5.5%	£239,870

Stage 5 Adjustment	NIL
	£239,870

RV as at 1/4/95 Say	**£240,000**

As referred to elsewhere, no matter the mathematical exercise of the contractor's basis, the valuer must keep in mind that the result should be a rental value that could reasonably be expected in the open market.

To this end, in *Imperial College of Science and Technology* v *J H Ebdon (VO) and Westminster City Council* (1984) the Lands Tribunal considered a sixth stage: whether the relative bargaining strength of the hypothetical landlord and tenant would affect the result of the fifth stage.

(f) Statutory formulae

Hereditaments which are included in the central rating list are currently valued by formula prescribed by statute. Non-operational premises, eg electricity and gas showrooms and offices not on operational land, are rateable in the usual way. Dwelling houses are deemed to be excluded from the value arrived at by the formula. Where a statutory formula is provided, no other method may be used.

The government has made it clear that it intends to value all such properties by conventional methods of rating valuation as soon as practicable, and has already removed the use of a statutory formula from the hereditaments of British Waterways, British Telecom, Docklands Light Railway and many others. It is the government's intention that eventually all formula-rated properties will be conventionally valued.

(g) Reference to comparable assessments

Where there is no rental evidence, a valuation may be carried out by considering how similar properties have already been assessed for rating, in the current rating list. However, until the comparable assessments have been tested at the valuation tribunal or Lands Tribunal, it should be remembered that the rating list entries are simply the VO's opinion as to rental value.

(h) Measurement for rating

When valuing for rating, valuers should have regard to the RICS/IRRV Code of Measuring Practice. Shops and offices are measured for net internal area, except large shops which are measured to gross internal area: warehouses and factories are also measured to GIA.

6.6 Domestic property: the council tax

6.6.1 Introduction

General rating was abolished in March 1990, and was replaced by two taxes: national non-domestic rating on commercial properties; and community charge on domestic properties. Due to its huge unpopularity, community charge (often called poll tax) was replaced by council tax from 1 April 1993. The main legislation governing council tax is the LGFA 1992.

To some extent council tax is similar to rating, in that there is a link between a specific property value (in this case capital value rather than rental value) and the amount of the charge levied locally, relative to that value. Allowing for a range of assumptions, the council tax values relate to open-market values at 1 April 1991, and 1 April 2003 for the Welsh revaluation effective 1 April 2005.

The basic differences between non-domestic rating and council tax can be summarised as follows.

- Non-domestic rating system: the rating list shows the actual or equivalent rental value of each property with regard to the statutory definition of rateable value. The non-domestic rating multiplier, the Uniform Business Rate, is applied to each rateable value in the rating list, to calculate the rates payable. Generally, the higher the rental potential of a property, the higher its rate demand will be.
- Council tax system: the valuation list shows each property allocated to a particular valuation band covering a range of capital values. Within each billing authority area, the charge will be the

same for all properties within the same valuation band. The amounts levied on properties in different valuation bands are in fixed proportions to each other. So, unlike in the rating system, many taxpayers will pay identical amounts of tax.

Essentially, council tax is a locally set tax that is payable on all domestic properties. The basic council tax bill is made up of two parts: 50% property and 50% personal. The 50% personal part is based on the assumption that two or more people live in the property. If only one person, aged 18 or over, lives in the property (or is deemed to live in the property), the personal element is reduced to 25%, (ie the total bill to 75%); and if there are no adult residents, (or none deemed to live there), only the property element remains, ie the bill is reduced to 50%.

In most cases, there is one council tax assessment per property, whether it is a house, bungalow, flat, maisonette, mobile home or houseboat. As for NNDR, it is irrelevant whether the property is owner-occupied or rented.

Council tax provides direct income to the local authority for local services such as schools, roads, libraries, police, fire and rubbish collection, among many others (unlike NNDR where the revenue is pooled nationally and then re-distributed by central government).

The LGFA 1992 makes chargeable dwellings subject to council tax. Basically, these are properties that would have been a hereditament prior to April 1990 (when general rating was abolished), but are not non-domestic properties for NNDR purposes.

Chargeable dwellings are entered into the Valuation List. A property that is a composite hereditament will have an entry in both the rating list and the valuation list. Each dwelling is valued by the LO (the same person as the VO for NNDR purposes, carrying out a different role), and placed in one of eight valuation bands, A to H. In Wales, there are nine bands, A to I.

6.6.2 *Properties exempt from council tax*

Under the LGFA 1992 section 4, council tax is payable "in respect of any dwelling which is not an exempt dwelling". The meaning of exempt dwelling is given in the Council Tax (Exempt Dwellings) Order 1992, which lists 12 classes of exemptions. The Council Tax (Discount Disregards and Exempt Dwellings) (Amendment) Order 1995 and other statutory instruments has since increased this total to 22. Note that the exempt property classes include some which relate to occupied dwellings, and some which relate to unoccupied dwellings.

- Class A: the property is unoccupied and unfurnished, and needs major repairs or structural alterations; or is having major repairs or structural alterations done; or major repairs or structural alterations have been finished on the property within the last 12 months.
- Class B: a charity owns the property and nobody lives there, (for up to six months).
- Class C: the property is unoccupied and unfurnished, (for up to six months).
- Class D: the property is unoccupied because the person who usually lives there is in prison or held somewhere else by the authorities, (for as long as the person is in prison or held somewhere else by the authorities or until someone else moves into the house).
- Class E: the property is unoccupied because the person who usually lives there is living in a residential care home, nursing home, hospital or hostel so that they can be cared for, (for as long as the person is being cared for or until someone else moves into the house).

- Class F: the property is unoccupied and the person who paid the council tax has died and probate has not yet been granted, (for up to six months after probate is granted as long as nobody lives in the property).
- Class G: the property is unoccupied by law (eg because of the condition of the property), or it is being bought by the authorities (compulsory purchase), (for as long as nobody can live there). The scope of this class was widened from April 2007.
- Class H: the property is unoccupied because it is being made ready for a minister of religion to live in to carry out his or her duties, (for as long as nobody lives there).
- Class I: the property is unoccupied because the person who usually lives there is living with, and being cared for by, another person, (for as long as nobody lives there).
- Class J: the property is unoccupied because the person who usually lives there is living with, and caring for, another person, (for as long as nobody lives there).
- Class K: the property is unoccupied and the last person who lived there is a student, or will be a student within six weeks of leaving, (for as long as nobody lives there.) The student must own the property.
- Class L: the property is unoccupied and the mortgage company has re-possessed it (for as long as nobody lives there and the mortgage company still has possession).
- Class M: The property is a students' hall of residence (for as long as it is a hall of residence).
- Class N: All the people who live in the property are:
  - students;
  - school or college leavers;
  - non-British husbands or wives or dependants of students; and
  - the property is where they live during term time, (for as long as the only people who live there are students)
- Class O: The Secretary of State for Defence owns the property and keeps it for UK armed forces to live in, (for as long as the Secretary of State for Defence owns the property and UK forces live there).
- Class P: A person connected to a visiting force lives in the property, (for as long as such a person lives there).
- Class Q: the property is unoccupied and the person who should pay the Council Tax is a trustee in bankruptcy, (for as long as nobody lives there and the trustees for the bankrupt person are responsible).
- Class R: The "property" is a caravan pitch without a caravan on it, or a boat mooring without a boat moored to it, (for as long as there is no caravan or boat there).
- Class S: All the people who live in the property are under the age of 18, (for as long as no-one who is 18, or older, lives there).
- Class T: the property is unoccupied, but it is part of another property which is lived in. This applies to a property which cannot be let separately from the other property, for example a "granny flat" which is not lived in, (for as long as nobody lives there).
- Class U: All residents in the property are severely mentally impaired and at least one of them pays Council Tax, (for as long as nobody else lives there). This exemption also applies if a student, who is not severely mentally impaired, also lives in the property, or owns the property but doesn't live there.
- Class V: The property is the main home of a person with diplomatic privileges or immunity, (for as long as this person lives in the property, and they would otherwise have to pay Council Tax).

- Class W: The property is part of another property and a dependent relative lives in it; for example, a granny flat which is lived in by an elderly or disabled relative, (for as long as the dependent relative lives there).

Detailed conditions apply in each case.

6.6.3 How many council tax assessments?

In most instances it is clear how a property should be banded. A typical house or flat, for example, will each be allocated a single council tax band. However, there can be instances where it is more difficult to establish the number of council tax bands to be applied, particularly when there are a number of dwellings within one building.

If a property contains more than one self-contained unit of accommodation, it will be divided into as many dwellings as there are self-contained units for council tax purposes. A self-contained unit means "a building or part of a building which has been constructed or adapted for use as separate living accommodation".

There are several key points that help to identify whether any living accommodation is a self-contained unit, and therefore is a chargeable dwelling.

- The physical character and layout: a self contained unit must be physically capable of use as separate living accommodation.
- The physical identity of the accommodation: a self-contained unit will normally not be spread over different parts of a building. For example, accommodation consisting of a living room and a kitchen, with a bedroom and a bathroom situated across a common hallway is unlikely to be a self-contained unit.
- The provision of standard facilities: a self-contained unit must have areas capable of use for living, for food preparation, for washing and a WC.

A recent case to identify whether a property consisted of one or two chargeable dwellings has heard at the High Court. In *Jorgenson (LO)* v *Gomperts* (2006), the LO argued that the subject property comprised two dwellings: a basement, ground and first floor maisonette; and a second floor flat. The High Court held that the relevant test to apply was an objective bricks and mortar test. Intention and use, actual or prospective, were not relevant. The question to be answered was whether each part of the property, in terms of its objective physical structure, was constructed or adapted as separate living accommodation.

6.6.4 Council tax bands

As mentioned above, chargeable dwellings are entered into the Valuation List with a record of the band of value into which they fall. There are eight bands in England and Scotland; and nine in Wales.

The valuation bands for England prescribed in the LGFA 1992 section 5(2) are as follows:

Valuation band	Range of values
A	Not exceeding £40,000
B	Between £40,001 and £52,000
C	Between £52,001 and £68,000
D	Between £68,001 and £88,000
E	Between £88,001 and £120,000
F	Between £120,001 and £160,000
G	Between £160,001 and £320,000
H	Exceeding £320,000

The Scottish bands are the same as the English structure.

Wales had a revaluation of its council tax system, effective 1 April 2005, which added a further band of value. The Welsh bands are:

Valuation band	Range of values from 1 April 1993	Range of Values from 1 April 2005
A	Not exceeding £30,000	Not exceeding £44,000
B	Between £30,001 and £39,000	Between £44,001 and £65,000
C	Between £39,001 and £51,000	Between £65,001 and £91,000
D	Between £51,001 and £66,000	Between £91,001 and £123,000
E	Between £66,001 and £90,000	Between £123,001 and £162,000
F	Between £90,001 and £120,000	Between £162,001 and £223,000
G	Between £120,001 and £240,000	Between £223,001 and £324,000
H	Exceeding £240,000	Between £324,001 and £424,000
I		Exceeding £424,001

6.6.5 Valuation definition

Council tax bandings are based on capital values. The basis of valuation, set out in the Council Tax (Situation and Valuation of Dwellings) Regulations 1992 (SI no 1992/550), is the amount a property might reasonably be expected to realise if it had been sold in the open market (on 1 April 1991) by a willing seller, on the following assumptions:

- the sale was with vacant possession
- the interest sold was the freehold or, in the case of a flat, a lease for 99 years at a nominal rent
- the size and layout of the property, and the physical state of its locality, were the same as at the time when the valuation of the property was made
- the property was in a state of reasonable repair
- the use of the property would be permanently restricted to use as a private dwelling, and
- the property had no development value other than value attributable to permitted development.

(The valuation date for the Welsh revaluation is 1 April 2003, rather than 1 April 1991. There was to have been an English revaluation with a valuation date of 1 April 2005, and an effective date of April 2007, but the government postponed it. This is expected to be an issue in the Lyons report, due spring 2007 — see section 6.10 below.)

Where a property is a composite hereditament, ie having both both domestic and non-domestic elements, it is given a valuation band appropriate to that part of the total capital value which is "reasonably attributable to domestic use of property". This requires an apportionment of full value, similar to the apportionment to non-domestic use of property of the rental value for purposes of non-domestic rating.

Given that the English and Scottish valuation dates are now somewhat historic, any current banding exercise which is necessary, is carried out by looking at comparable sales data for 1991. Recent sale prices are rarely a reliable guide to the correct band for a property. In some cases, the LO takes recent sales data and adjusts it backwards, using indices, but this is not actually a method of valuation. Use of indices should be treated with caution, as their scope is wide, and they are rarely specific to the type of property or to the locality.

The valuation definition refers to an open market sale. This means a market where the property is offered openly with adequate publicity being given to the sale. A willing vendor is someone who sells the property as a free agent and not someone who is forced to do so. The RICS/IRRV Red Book definitions of open market and willing vendor are relevant.

A brief discussion of state of reasonable repair is appropriate. The valuer assessing a dwelling for council tax purposes must assume that it is in reasonable repair. This does not mean that the dwelling is assumed to be in a good state of repair. Instead, the valuer must decide what state would be reasonable to expect for a dwelling, having regard to its age, character and locality. Character is particularly important, for example, where one dwelling in a street of similar properties has not been maintained, but allowed to deteriorate in contrast to the majority of others. Its basic character is likely to remain the same as that of its neighbours. The valuer would then assume a state of reasonable repair which is the same as actually exists for most of the nearby properties. The property's disrepair is therefore not reflected in its banding.

However, modernisation and improvements are not "repairs". So, if a dwelling has not been modernised, despite the majority of its neighbours having been, the property will be valued having regard to its lack of modernisation. It is still to be assumed to be in reasonable repair.

6.6.6 Calculation of council tax payable

(a) The basic charge

Each billing authority fixes the amount of council tax to be charged, under the LGFA 1992. The authority considers the total council tax it needs to raise for the year's revenue. The amount to be charged for each dwelling according it its valuation band is in the following proportion:

	6:	7:	8:	9:	11:	13:	15:	18
Bands:	A:	B:	C:	D:	E:	F:	G:	H

The council counts how many "band D equivalent dwellings" it has in its area. Then, assuming a basic council tax of £x has been fixed for band D:

- the tax for a property in band A would be £x × 6/9, while
- the amount in band H would be £x × 18/9.

So the tax on a property in band H will be three times as much as for a property in band A, and twice as much as for a property in band D.

(b) Liability to council tax

Liability to the council tax is fixed by the LGFA 1992. Section 6(1) provides that the person liable is the person who falls within the first paragraph of section 6(2) taking (a) to (f) in turn.

According to section 6(2), a person is liable to tax if he is:

a.   a resident of the dwelling and has a freehold interest in the whole or any part of it
b.   such a resident and has a leasehold interest in the whole or any part of the dwelling which is not inferior to another such interest held by another such resident
c.   both such a resident and a statutory or secure tenant of the whole or any part of the dwelling
d.   such a resident and has a contractual licence to occupy the whole or any part of the dwelling
e.   such a resident or
f.   the owner of the dwelling.

The council taxpayer is therefore the person with the strongest right to occupy the dwelling. If two or more persons within the same household fall within the same level of the hierarchy, they are jointly and severally liable. (An exception to this rule is where one or more of such persons fall to be disregarded under the discount provisions — see below.)

The owner is always liable to pay the council tax in the following cases:

- residential care homes and nursing homes
- certain hostels providing care and support
- houses of religious communities
- houses in multiple occupation
- residences of ministers of religion
- accommodation occupied by asylum seekers.

(c) Reductions in the amounts payable

Council tax charges may be reduced in three different ways: by "discounts", by benefits, and by reductions for disability adaptations.

Discounts

Council tax assumes that each dwelling is occupied by two adults as their sole or main home. If there is only one resident (or where there is only one resident when persons are disregarded — see below), the tax is reduced by 25%. If there is no resident (or if all residents fall to be disregarded), the tax is reduced by 50%. Effectively, therefore, an unoccupied domestic property is charged half the full council tax for the relevant band. The amount of the full charge is not affected if there are more than two residents.

The LGFA 1992 gave Welsh billing authorities the power to reduce the 50% discount for an unoccupied property to 25%, or to remove the discount (this tended to be applied for long-term empty properties which were only used as holiday homes).

Section 75 of the Local Government Act 2003 added a section 11A to the LGFA 1992, giving English billing authorities discretion to reduce the discount in certain cases. (A billing authority may decide to amend the discounts for all the properties of that class in the whole of its area, or only in a part of its area.) This has tended to be used in cases of long-term unoccupied dwellings, as part of an incentive for all the housing stock to be utilised. The Council Tax (Prescribed Classes of Dwellings) (England) Regulations 2003 prescribed three classes of dwellings where this discretion can be exercised. Where a property is unoccupied, the billing authority can decide to reduce the discount from 50% to a minimum of 10%, (ie 90% payable instead of 50% payable).

The classes are:

- Class A — property is unoccupied and furnished, and a planning condition restricts the right to occupy for a continuous period of at least 28 days in the year. Examples are purpose-built holiday homes or chalets, subject to a planning condition restricting year-round occupancy.
- Class B — property is unoccupied and furnished, where there is no planning restriction limiting year-round occupation. However, billing authorities cannot reduce the discount if the reason the dwelling is unoccupied is that the council taxpayer has to live elsewhere, in accommodation provided by his employer, as part of their employment contract. Examples are service personnel who live in accommodation provided by the Ministry of Defence, or publicans who are required to live at the licensed premises (public house).
- Class C — property is unoccupied and substantially unfurnished. These are long-term empty homes, since unoccupied and substantially unfurnished dwellings are exempt from council tax for the first six months. In this Class C, the billing authority can decide to remove the discount altogether, not just limited to a minimum of 10%.

Disregards

The council tax provisions are further complicated by the "disregarding" of certain residents, so that even where there actually are two or more persons resident in a dwelling, less than two are counted for the purposes of calculating the amount of council tax payable.

The persons who are disregarded for purposes of calculating the number of residents are those set out in the 1992 Act, Schedule 1, and include:

1. persons detained in prison and on remand (but not those imprisoned for non-payment of a fine or their council tax)
2. the severely mentally impaired
3. persons over 18 years of age, but in respect of whom child benefit is payable
4. persons under 20 and still at school, or who have just left school
5. persons who are foreign language assistants
6. students (including student nurses, apprentices and youth training trainees), extended by Amendment Regulations 1995 to include a spouse or dependant of a student if he or she is not a British citizen and is prevented by the terms on which he or she enters or remains in the United Kingdom from taking paid employment or claiming benefits
7. hospital patients (when treated as having their sole or main residences there)
8. persons treated as having their sole or main residences in residential care homes, nursing homes, mental nursing homes or hostels
9. care workers providing care and/or support to another person

10. residents in certain communal residences for persons of no fixed abode.

Various stringent conditions apply in each case, details of which are contained in schedule 1 of the Act or in the Council Tax (Discount Disregards) Order 1992. Conditions for additional disregards for members of international headquarters and defence organisations, members of religious communities, and school leavers, are provided in the Council Tax (Additional Provisions for Discount Disregards) Regulations 1992.

Examples of how the disregarded persons rules affect the calculation of the amount of council tax payable include:

- if two people, aged 18 or over, live at a property and one is a student, a discount of 25% is applied, as the student is disregarded
- if two people, aged 18 or over, live at a property and one is a student nurse and the other is a Youth Training Trainee, a discount of 50% is applied, as both persons are disregarded
- if three people, aged 18 or over, live at a property and one is a student, no discount applies, because there are still two people resident in the property who are not disregarded.

Benefits

Council tax benefit may be payable under the 1992 Act to persons with low or no income. Benefit is means tested, and may be paid up to 100% of the council tax chargeable.

Reductions for disabilities

Under the 1992 Act and the Council Tax (Reductions for Disabilities) Regulations 1992, provision is made for relief in respect of dwellings occupied by a person who is substantially and permanently disabled, (adult or child), subject to detailed conditions set out in the regulations. Where these conditions are satisfied, the dwelling is treated for council tax purposes as if it were in the valuation band below that in which it appears in the valuation list.

After a considerable public outcry, the same entitlement to similarly disabled band A taxpayers, who occupy adapted dwellings, was granted with effect from 1 April 2000. Council tax demands are adjusted from 6/9 to 5/9 of the local band D charge.

If a disabled person is already entitled to a discount disregard or to council tax benefit, the amount payable is calculated accordingly.

The detailed conditions are that:

(1) a disabled person lives in the property and
(2) the property has one of the following:
- a room, which is mainly used by the disabled person, and is essential for the disabled person's needs, but is not a bathroom, kitchen or lavatory
- an extra bathroom or kitchen, which is essential for the disabled person's needs
- extra space inside the property so that a disabled person can use a wheelchair indoors. The disabled person must need to use the wheelchair indoors.

The guidelines state that a person must be substantially and permanently disabled and the facilities must be of major importance to their well-being. *R (on the application of Hanson)* v *Middlesbrough Borough*

Council (2006) considered exactly what this means. Overturning a valuation tribunal decision, the High Court allowed disabled relief for an additional en-suite bathroom. Deputy Judge James Gouldie QC held that the statutory test of being essential or of major importance should not be translated as a more stringent requirement of being physically or extremely difficult to occupy without it. It is not appropriate to give the relief only if a future purchaser would be able to detect that the property had been altered to meet the needs of a disabled person. The deputy judge determined that the en-suite bathroom was of major importance to the appellant because it reduced the risk of her being injured while bathing.

In contrast, any special fixtures designed to make a dwelling suitable for a person with a disability, and which add to its value, are ignored when the banding is assessed. Effectively, the dwelling is valued as if the special fixtures were not there, so that the property is not placed in a higher band because of them.

6.7 Challenging the assessment

In order for any tax to be popularly accepted; to be, and to be perceived to be, fair and accurate; and for inaccuracies in the taxation base to be corrected, there has to be a mechanism for changes and appeals. The UK system provides limited rights to make an appeal.

Different appeal mechanisms apply for the local taxes:

- non-domestic rating: the assessment can be challenged via the valuation tribunal, (Valuation Appeals Committee in Scotland); liability for rating is determined by the magistrates' court.
- council tax: the assessment can be challenged via the valuation tribunal; liability is also dealt with by the valuation tribunal (other than issues involving non-payment and enforcement, which go to the magistrates' court). The Valuation Appeals Committee deals with both these appeal types in Scotland.

6.7.1 Non-domestic rating

Changes to the local rating list can be initiated by the VO or by an appellant.

1. If the VO considers that an entry in the rating list needs amendment, he is under a statutory duty to alter the list, and the revised entry then becomes part of the list. Notification of the alteration is sent to the billing authority and the ratepayer.
2. The billing authority and any interested person have limited rights to make proposals to alter rating lists. Making a proposal is the process by which a ratepayer can, for example, challenge his rateable value or propose a split or a merger with another hereditament. A proposal which is not settled becomes an appeal when it is sent to the valuation tribunal, which the VO does automatically three months after receiving the proposal. The proposal may be resolved through discussions during this three-month period (rarely), or through discussions after it has becomes an appeal to the valuation tribunal. The proposal may also be resolved through a valuation tribunal hearing.

The NDR appeal mechanism operates under section 55 of the LGFA 1988, with detail given in the Non-Domestic Rating (Alteration of Lists and Appeals) (England) Regulations 2005. Similar legislative provisions exist for Wales as for England, and the Scottish appeal mechanism is outlined below at section 4.

(a) Who may make a proposal?

The right to make a proposal is restricted under regulation 4(2) of the 2005 regulations to:

- the billing authority, but only in certain circumstances where:
  - it seeks to include a new hereditament or
  - delete an existing hereditament from the list or
  - where it considers there has been a material change of circumstances or
  - where it considers the RV is wrong due to a decision of a valuation tribunal, Lands Tribunal or Court of Appeal.
- an interested person.

An interested person (an individual or a company), is:

- the occupier
- the owner (where he is not the occupier) or
- a company within the same ownership as the occupying company.

(b) Grounds for proposals

Proposals challenging aspects of an entry in the rating list may only be made on specific grounds, but these are quite wide:

(a) the rateable value is wrong, from the date the rating list came into force
(b) there has been a "material change of circumstances" which affects the rateable value
(c) the regulations dealing with the valuation of rateable plant and machinery have changed, which affects the rateable value
(d) the VO has changed the entry in the rating list, and the rateable value is now wrong
(e) a valuation tribunal, the Lands Tribunal or a court has made in decision in connection with another hereditament, which suggests that the rateable value (or any part of the rating list entry) of the hereditament is wrong
(f) the VO has altered the rating list, and the effective date of the alteration is wrong
(g) a hereditament which is not already in the rating list, should be included
(h) a hereditament in the rating list should be deleted
(i) a hereditament is partly exempt or partly domestic, but the rating list entry does not show this
(j) the entry in the rating list shows that some part of a hereditament is domestic property, or is exempt from non-domestic rating, but should not
(k) property which is shown in the rating list as two or more hereditaments should be shown as one hereditament, or a different combination of hereditaments
(l) a hereditament should be split into more than one hereditament
(m) the address shown in the list for a hereditament is wrong
(n) the description shown in the list for a hereditament is wrong and
(o) the entry in the rating list for a hereditament is missing something required under section 42 of the 1988 Act.

Grounds (d) and (f) relate to disagreements with an alteration made by the VO. A proposal cannot be lodged on these grounds, if the reason for the VO's alteration is simply fulfilling the decision of a valuation tribunal, the Lands Tribunal or court.

(c) Time-limits

The time-limits for making proposals have been made less restrictive, (with the aim of reducing the number of proposals that used to be made purely on a protective basis). However, the number of proposals that any one person can make has been reduced.

Basically, an interested person can make a proposal at any time, on any of the grounds (a) to (o) listed above, but can only make one proposal for any one event. Although the same occupier cannot make a second proposal on the same grounds, a new occupier of the hereditament can make a fresh proposal, even if it duplicates the previous occupier's proposal. The only exception is where the valuation tribunal, or the Lands Tribunal, actually heard and determined the appeal arising from the previous person's proposal.

Where a new ground for proposal is found, (for example, the first proposal was on ground (a), challenging the original entry in the rating list, and now there has been a material change of circumstances, ground (b)), another proposal may be lodged, whether it is by the same or a different occupier.

In summary, the rule is "one proposal, per person, per hereditament, per event".

General revaluations of non-domestic property occur every five years. A proposal to alter an entry in the 2005 rating list can be made at any time while the list is in force, ie up to 31 March 2010. The only exceptions are proposals on grounds (d) and (f), which relate to disagreement with a VO's alteration. Here, a proposal may only be made within six months of the VO's alteration, even if this is after the end of the rating list. Similarly, proposals on ground (e), which relate to the decision of a valuation tribunal, Lands Tribunal or court can be made up to six months after date of the decision, even if this is up to 30 September 2010.

As there is a rule of "one proposal per person per event", there may be situations where a person makes a second proposal on the same grounds. Assuming that the first proposal was made in the correct form, (see section (d)), the VO will automatically deem the second proposal as "invalid", and will serve an "invalidity notice". There is a right of appeal against the invalidity notice, which is dealt with by the valuation tribunal.

(d) Form of proposals

The 2005 regulations have detailed requirements for the form and content of proposals. Proposals must:

1. be made by notice in writing served on the VO (can be served electronically)
2. state the name and address of the proposer
3. give the status of the proposer, ie interested person; billing authority etc
4. identify the property(ies) to which it relates
5. explain how the proposer thinks the rating list should be altered and
6. set out the grounds (a to o) of the proposal, and the reasons to support those grounds.

Specific requirements for certain grounds

Grounds	*Proposal requirements*
1 Material change of circumstances ground (b)	Identify change and date it occurred.
2 Querying alteration by VO, grounds (d) and (f):	Identify alteration concerned; and if the effective date is being queried, the date the proposer thinks should be given instead
3 Querying effective date, ground (f):	Statement of amended date required.
4 Following relevant tribunal or court decision, ground (e):	a Identify hereditament concerned in decision b Name tribunal or court and give date of decision c Reasons for considering decision relevant to subject property d Reasons for believing that the entry in the list for the subject property is inaccurate.
5 Grounds (a) to (g), (i) and (j)	Where the hereditament has an occupier under a lease, licence or easement, the rent/annual amount payable as at the date of proposal must be given.

(e) Effective date of proposals

For the 2005 rating list, the effective date of the proposal is the date the event giving rise to the proposal occurred.

- Where the proposal is due to an inaccuracy in the rating list, the entry in the list will be corrected from the date it became inaccurate. If the inaccuracy was there when the list was compiled, the alteration has effect from 1 April 2005.
- Where the proposal is following a material change of circumstances, the alteration to the rating list has effect from the date on which the circumstances arose.
- If the day on which the circumstances arose is not identifable with reasonable certainty, the effective date is the date of the proposal (or, where the alteration is made on the VO's own initiative, the effective date is the day the VO alters the list).

Where the VO alters an entry in the rating list, and the rateable value is increased, the increase only has effect from the date the alteration is carried out, ie increases cannot be back-dated, (the only exception is where the reason the list was wrong was due to an error or default on the part of a ratepayer, although it is difficult to imagine exactly what this means, unless it imples a fraud.)

The date of the effective change is included in the rating list entry for the hereditament.

(f) Proposal and appeal procedures

Assuming that a proposal is made in the correct form, it is acknowledged by the VO. If the VO agrees with the proposal, ie he thinks it is "well-founded", he simply alters the entry in the rating list as set out in the proposal.

Although the regulations provide for the agreement/withdrawal of proposals, from the time that the VO acknowledges their receipt, in practice it is rare for them to be discussed and settled quickly. If the proposal has not been settled within three months, it is automatically passed to the valuation tribunal, and becomes an appeal (technically, it is an appeal against the VO's refusal to alter the rating list). This happens to the majority of proposals, particularly those made against an entry in a new rating list, as the VO does not have time to consider and discuss all the proposals received within those first three months. The valuation tribunal will list the appeal for a hearing, although most appeals are settled during pre-hearing discussions and do not eventually need the valuation tribunal to determine them. Appeals can be agreed or withdrawn at any time up to a hearing.

If the appeal does proceed to a valuation tribunal hearing, the valuation tribunal gives at least four weeks' notice of the hearing date and venue. Any party to a hearing may appear in person or be represented by a chartered surveyor, lawyer or other representative. If all parties agree, an appeal can be dealt entirely through written representations, although this is rare in practice.

If the appeal seems to be unusually complex or contentious, a chairman of a tribunal (see below for a description of the tribunal structure), can order a pre-hearing review. The aim is to "clarify the issues to be dealt with". A party can apply to the tribunal to ask for a pre-hearing review to be held.

The VO is restricted in what information he can present at a valuation tribunal hearing, and has obligations regarding supplying information to the appellant. Most rental evidence that the VO relies on in his valuation is based on forms of return, which are the VO's means of gathering data. In order for the VO to be able to use this information at the hearing, he must give the appellant three weeks' notice of the details. The details include the address, rateable value, summary of the lease, etc. The appellant has the right to request rental information on other properties from the VO. He can request details of as many properties as the VO has said he will use, (if the VO is using less than four, the appellant can request up to four). The VO must allow the appellant to inspect/copy the documents at a reasonable time.

Valuation tribunals are established under the LGFA 1988, as amended by the Local Government Act 1992, (although this was essentially a re-naming, with added functions, of the appeal body that existed in various forms since 1948).

Valuation tribunals are appointed to cover local areas (usually several billing authorities' areas, and often a whole county). A tribunal, the collective body, comprises a number of lay members. The tribunal members elect a president from among themselves, and arrange for a number of members to act as chairmen for hearings. The Valuation Tribunal Service appoints a clerk for the valuation tribunal, who acts as an adviser at hearings but has no role in the decision-making process. Normally three members form the tribunal, the judicial body which hears the appeals, although two members can act as a tribunal with the parties' approval.

The main obligations on the tribunal are to conduct the hearing in a way that is most suitable to clarify the issues; to ensure the fair and just handling of the matter; and to avoid formality. The hearing is usually informal and open to the public (unless a party has a specific reason to ask the tribunal to hear his case privately).

The appellant generally presents his case first, though the tribunal is able to adopt the procedure it thinks most appropriate in the circumstances. Parties can call witnesses, and can cross-examine the other parties' witnesses. The tribunal can require any witness to give evidence under oath or affirmation, although this is rare. Parties are usually offered the opportunity to sum up their case. The tribunal is not a formal court of law, and therefore is not bound by rules regarding the admissibility of evidence, ie it can consider hearsay statements, although it is up to the tribunal how much weight to apply to the evidence it hears.

The decision must be communicated in writing to the parties (although some valuation tribunals give an oral decision on the day prior to the formal decision being issued), with the reasons for the decision.

The valuation tribunal can decide on whatever level of RV is appropriate, even if this is higher than the value contended in the proposal. However, if the decided figure is higher than both the figure in the list and the figure in the proposal, the decided figure takes effect only from the date of the valuation tribunal decision. Recent case law on whether the valuation tribunal is bound by the scope of the proposal has been decided both ways, so this is an emerging area of law.

The valuation tribunal does not have the power to award costs, so each party bears their own.

Parties can apply for the review or setting aside of a tribunal decision, on very limited grounds:

- the decison is wrong due to a clerical error
- a party did not attend the hearing but can show reasonable cause why and
- the Lands Tribunal or High Court has dealt with another appeal on the same property, which affects the valuation tribunal 's decision.

Applications for a review or setting aside should be made within four weeks of the decision, otherwise the tribunal may (but does not have to) dismiss the application.

The main right of appeal against a valuation tribunal 's decision is to the Lands Tribunal, within four weeks of the valuation tribunal decision. The Lands Tribunal is a more formal body than the valuation tribunal, with qualified members, and the power to award costs. The Lands Tribunal hears the case afresh and can confirm, vary, set aside or revoke the decision of the valuation tribunal.

An appeal against a Lands Tribunal decision lies to the Court of Appeal, but only on a point of law. Appeal from the Court of Appeal is to the House of Lords, but only with permission of either of those bodies.

(g) Programming

Although not a statutory process, valuation tribunals now operate a system of programming, in liaison with the VOs in their area. The Department for Communities and Local Government has endorsed the system. A local programme is published, showing when appeals will be dealt with by the VO and when, if unresolved, they may be heard by the valuation tribunal. Ratepayers have the opportunity to make representations if they believe their place in the timetable is unreasonable and they experience hardship.

There are three key dates in programming:

- "start" date — when the discussion period for an appeal will begin
- "target" date — when discussions are expected to be concluded
- "hearing" date — when the case will be heard by the valuation tribunal if the matter has not been resolved.

The start and target dates for each class and/or location of property are shown in the local programme. Appellants are contacted by the VO at the identified start date, to begin discussions.

(h) Scottish appeal mechanism

New owners, tenants or occupiers may appeal within six months of acquiring an interest in a property. Where the assessor alters the valuation roll (by making a new entry or changing the value) the appeal must be lodged within six months of the date of the valuation notice. Appeals may be lodged at any time on the grounds of error or in the event of a material change of circumstances. The assessor contacts the appellant to confirm if the appeal is valid or if further information is required (eg evidence to confirm that a new occupier complies with the six month rule).

Error and material change of circumstances are defined in the Local Government (Scotland) Act 1975. An error relates to "any error of measurement, survey or classification or any clerical or arithmetical error in any entry". A "material change of circumstances means ... a change of circumstances affecting their value and, without prejudice to the foregoing generality, includes any alteration in such lands and heritages, any relevant decision of the Lands Valuation Appeal Court or a valuation appeal committee ... or the Lands Tribunal for Scotland ... and any decision of that Court, committee or Tribunal which alters net annual value or rateable value of any comparable lands and heritages."

Appeals against the valuation roll and council tax valuation list are initially to the relevant assessor. If a settlement is not reached, the appeal is heard by an independent valuation appeal committee. Members are drawn from an appeal panel appointed by the Sheriff Principal and are independent of the assessor, local council or valuation joint board.

Committees are assisted by a secretary, usually a qualified solicitor (also appointed by the Sheriff), who provides advice on eg the law or procedure, but who takes no part in the decision making process. Members of the panel are usually unqualified in valuation or legal matters, but are appointed because they are local ratepayers or council taxpayers.

The proceedings are relatively informal, but witnesses can be called and are normally placed on oath by the committee chairman. A committee usually comprises between three and eight members. Sometimes the assessor will conduct his own case or be represented by a senior member of his staff and at other times he will be represented by an advocate or counsel. Ratepayers are entitled to appear in person to present their own case or can be legally represented. A surveyor, family member or friend can also appear on behalf of the appellant. The chairman is responsible for the conduct of the hearing and will give advice and guidance as to the procedure to be followed at any time during the hearing.

The valuation appeal committees have a statutory timetable, and are required to deal with appeals from the 2005 revaluation by December 2008. As in England and Wales, very few of the appeals lodged will actually be heard by committees, the majority being resolved in discussion between ratepayers or their agents, and assessors. Valuation appeal committees give a minimum of 70 days notice of the place and time of the hearing.

If the appeal is particularly complex or technical, or raises major issues of law, the appellant can request that the valuation appeal committee passes the appeal straight to the Lands Tribunal for Scotland, instead of the VAC dealing with it first.

(i) Appeal procedures in Northern Ireland and the Channel Isles

In Northern Ireland, the valuation list is drawn up by the Valuation and Lands Agency, which is an Executive Agency of the Department of Finance and Personnel in Northern Ireland. The valuation list was revalued and came into force on 1 April 2003, (antecedent valuation date 1 April 2001), the previous revaluation in Northern Ireland having been in 1977.

To challenge an entry in the valuation list, an application may be made at any time to the local District Valuer of the Valuation and Lands Agency seeking a review. There are subsequent rights of appeal to the Commissioner of Valuation and then to the independent Lands Tribunal for Northern Ireland.

In the Channel Isles, (Jersey, Guernsey and Alderney), the draft Parish Rate lists are drawn up annually, based on an annual value — called quarters — for both the owner (the person with more than seven years interest) and the occupier. The Assessment Committee can be asked to review the rateable value if there is a significant difference (defined as 10% difference in value), between the assessment of the land/property concerned, and other similar land/property. The application for a review must be made within 14 days of the draft list being published. If the applicant is dissatisfied after the review, an appeal can be made to the Parish Rate Appeal Board.

6.7.2 Council tax

(a) Council tax — valuation appeals

The LGFA 1992 provides that the valuation list may be altered; the appeal mechanism is detailed in the Council Tax (Alteration of Lists and Appeals) Regulations 1993, as amended.

As for the VO in rating, the LO is under a duty to maintain the valuation list, and may alter it at any time to ensure its accuracy.

An interested person may challenge an entry in the list, but the circumstances in which they can propose an alteration are limited.

- A proposal querying the initial valuation band shown in the valuation list at 1 April 1993 had to be made not later than 30 November 1993. Given the passage of time, this appeal right is effectively extinct.
- A proposal may be made within six months of the person becoming liable to pay council tax at the dwelling.
- Where an alteration to the valuation list is made by the listing officer, or following a relevant valuation tribunal decision on another dwelling, an interested person may make a proposal within six months of the alteration.

However, alterations by the LO can only generally be made if, since the valuation band was first shown in the list:

a.   there has been a material increase in value of the dwelling and a relevant transaction. No alteration can be made unless there has been a "relevant transaction"
b.   there has been a material reduction in value (eg part demolished)
c.   the dwelling has become or ceased to be a composite hereditament
d.   there has been an increase or reduction in the domestic use of a composite hereditament.

A material increase is one which would make the dwelling fall into a higher council tax band. A relevant transaction includes a completed sale.

In general, an alteration has effect from the date on which the circumstances giving rise to the alteration occurred. However, where an alteration is due to (a) above, this has effect from the date of sale (or the creation of a new lease).

As for rating, the proposal to alter the valuation list entry is initially made to the Listing Officer, and then is automatically passed to the valuation tribunal as an appeal. The hearing procedures are the same as for rating. A valuation tribunal decision may be appealed to the High Court but only on a point of law.

At the time of writing, the DCLG is consulting on a proposal to implement an appeals direct system for council tax valuation appeals. This appeals direct system would require the LO to reach a decision on a proposal and advise the council taxpayer (or the person who made the proposal). The rejection of a proposal would not automatically be referred to the valuation tribunal as an appeal. Instead the proposer would have to separately make an appeal, direct to the valuation tribunal, if they were unhappy with the LO's decision. The consultation period ran until February 2007.

(b) Council tax — liability appeals

Although not a valuation matter, it is worth mentioning that a council taxpayer can appeal against certain aspects of his liability. An aggrieved person can initially make representations to the billing authority if they consider, for example: that a dwelling is not a chargeable dwelling; that they are not liable as the council taxpayer; that discounts or disregards have not been correctly applied; or that there is an incorrect calculation of the amount of tax charged. The billing authority has two months in which to consider these representations. If the person is still aggrieved, or if the authority fails to answer the representations within those two months, an appeal can be made directly to the valuation tribunal.

6.7.3 Appeals concerning penalty notices

In order to gather information to provide the rental evidence to draw up the rating list, the VO has the power, under the LGFA 1988, amended by the Local Government Act 2003, to issue forms of return to ratepayers, asking for details of rents, lease terms, etc. Enforcement of the VO's request for information is now a civil penalty.

If the information is not provided within 56 days, the VO can raise a penalty charge of £100 against the ratepayer. The penalty rises to £200 if the information is not provided in the next 21 days, and then £20 can be added to the penalty for each day after that. The penalty can rise to a maximum of the same amount as the rateable value.

If a person feels that the penalty has been wrongly imposed, he can appeal directly to the valuation tribunal. The only defenses against the penalty are that

- the appellant had a reasonable excuse for not providing the information or
- that the appellant did not have the information in their possession or control.

The valuation tribunal can order the VO to mitigate or remit the penalty.

6.7.4 Appeals concerning completion notices

Local authorities serve a completion notice where they feel that a newly built property is complete, is substantially complete, or the remaining work could be completed within three months. The VO then enters the new property into the rating list if it is non-domestic. The LO enters the property into the valuation list if it is domestic. Liability arises with effect from the date stated in the completion notice, (subject to any void periods). This has the effect of starting the period for calculating rating or council tax liability.

If the person on whom the completion notice is served feels that either the building is not complete (if the completion date has passed), or cannot reasonably be expected to be completed by the date stated in the notice, they have four weeks to appeal directly to the valuation tribunal.

The valuation tribunal can decide that the completion notice is correct; that a different date should be given (up to three months ahead); or that it is not reasonable to expect that the building could be completed within the next three months.

6.8 Transitional relief scheme

Rateable values can decrease or increase fairly significantly between revaluations. This does not necessarily mean there will be a significant change in the rates bill, as the multiplier, or Uniform Business Rate (UBR) is adjusted downwards following a revaluation to offset the overall rise in values. Some ratepayers, however, could see significant changes in their rates bill. Transitional arrangements are designed to soften the impact of revaluation by phasing in the changes to the rates bill over a period of time. Note that transition is not dependant on the rateable value change, but on the change in the rate bill following revaluations.

The government limits the percentage by which bills may increase or decrease each year. These transitional arrangements will operate until 31 March 2009 (the government has decided that this scheme will only run for four out of the five years of the rating list); so that by the end of the life of the current rating list (if not earlier), every ratepayer will be paying their correct liabilty of RV × UBR. (Under previous rating lists, the transitional arrangements continued for the whole life of the list. This created the anomaly in some cases that a ratepayer went forward into the next list, never paying their full tax liability, and having transition applied to transition.)

This section is not intended to provide full details of calculations — assistance can be received from individual local authorities.

Note that there is no transitional relief scheme for council tax; this applies only to non-domestic properties.

6.8.1 Transitional arrangements

Different transitional arrangements will apply, depending on whether the rate bill has increased or decreased, and whether the hereditament is classed as small or large.

A "small" hereditament is defined as a property with a rateable value of under £15,000 outside London and under £21,500 within London.

Current transitional arrangements are based on the change in the rates bill from 2004/05 to 2005/06.

Table 6.1 show the proposed limits by which a rates bill can increase in a single year before transitional arrangements apply:

Table 6.1

Year	Small business	Large business
2005/06	5%	12.5%
2006/07	7.5%	17.5%
2007/08	10%	20%
2008/09	15%	25%
2009/10	n/a	n/a

Slightly different arrangements apply for the proposed limits by which a rates bill can decrease in a single year, shown in Table 6.2 below.

Table 6.2

Year	Small business	Large business
2005/06	30%	12.5%
2006/07	30%	12.5%
2007/08	35%	14%
2008/09	60%	25%
2009/10	n/a	n/a

6.9 Business improvement districts

A Business Improvement District (BID) is not part of rating as such, but is closely allied to the rating system, and being a new local tax, is worth discussing briefly. BIDs came into effect under the Local Government Act 2003 which is applicable in England and Wales, though each will have its own regulations. The Scottish Executive is considering similar legislation, although there are not understood to be any similar plans in Northern Ireland.

A BID is a partnership between a local authority and the local business community, to develop projects and services that benefit the trading environment and public spaces within a defined area. They are designed to provide additional services, not to replace those already provided by a local authority.

The main aim of a BID is to support the long-term sustainability of town and city centres, partly due to pressures from out of town shopping which has increased since the 1980s. A BID can be developed in any business location where there are non-domestic rate payers. BIDs can also work, for example, on industrial estates, to improve the local environment.

Key features of the BID system are as follows:

- A BID can be proposed by any non-domestic ratepayer, property owner, local authority or other key stakeholder with an interest in the BID area. The BID proposal identifies the area and the issues, and includes a management structure, performance indicators and delivery guarantees.
- Non-domestic rate payers within a BID area pay for the BID through a supplement on their rates bill. The supplement can be up to 5% of the rateable value of a property, although most BIDs have opted for lower percentages, around 1 to 3%, in order to be acceptable locally.
- A BID proposal will state whether all rate payers will be charged, or if the charge only applies to a particular group. For example, some BID schemes have excluded hereditaments where the ratepayer receives charitable relief; others have excluded none. There is some concern that as the BID levy is charged to rate-payers; major (non-occupying) investors in property gain the benefit of improvements to the area, without having to pay the levy. Therefore, many BID schemes seek voluntary contributions from owners, to rectify a perceived unfairness.
- The local rate-payers who will be liable to pay the BID levy vote on the proposals. The postal ballot is conducted by the local authority. A successful vote for a BID must meet two tests: more than 50% of votes cast must be in favour of the BID, and the positive vote must represent more than 50% of the rateable value of the votes cast.

There is some concern that there is not a minimum time-period between ballots, i.e. where a ballot is unsuccessful, a second ballot can be held as soon as the BID body wishes. There is also concern at relatively low turn-outs. The legislation does not require a minimum percentage of rate-payers to have voted, only certain minimum results from those who have voted. Hence, ignorance or apathy can lead to a levy which the majority are not in favour of.

The BID levy is collected by the local authority on behalf of the BID body. Enforcement procedures are the same as for rating.

A BID's mandate is for a maximum of five years. A BID wishing to continue beyond five years must reaffirm their mandate through another ballot, based on a further proposal. The latest data on BIDs can be found at *www.ukbids.org*.

6.10 Future of local government funding

6.10.1 The Lyons inquiry

In July 2004, Sir Michael Lyons was commissioned to undertake an independent inquiry to consider whether the present system of local government funding in England should be changed, specifically including the reform of council tax. In September 2005, the Government announced an extension to the Inquiry's terms of reference to cover questions relating to the function of local government and its future role, as well as how it is funded.

The inquiry's initial terms of reference were to:

- consider the detailed case for changes to the present system of local government funding
- make recommendations on any changes and their implementation
- make recommendations on how to reform council tax
- assess whether local authorities should have increased flexibility to raise additional revenue and for making a significant shift in the current balance of funding
- consider options other than council tax for local authorities to raise supplementary revenue, including local income tax, reform of non-domestic rates and other possible local taxes and charges and
- consider the implications for the financing of possible elected regional assemblies.

The inquiry will also consider, as appropriate, any implications that its recommendations have for other parts of the United Kingdom. The additional terms of reference introduced in September 2005 were to:

- consider the strategic role of local government in the context of national and local priorities for local services, and the implications of this for accountability
- assess the link between the Government's agenda for devolution and decentralisation, together with changes in decision making and funding, and improvements to local services
- consider whether improved accountability, clearer central-local relationships, or other interventions could help to manage pressures on local services, and changes to the funding system which will support improved local services.

In his interim report, Sir Michael Lyons drew links between local government functions, its funding, and its accountability to local people. His interim report included the following views:

Local government is not just about the provision of services. It has a major and unique role to play in helping to develop and deliver vision for its communities, making decisions and trade-offs on their behalf, and shaping a strategic view of the area and its future — a role we might refer to as "place-shaping".

This role requires local government to have the trust of its citizens, and effective structures and processes to support it. This document asks how we can ensure that local government has the capability to perform this role effectively into the future.

There is a strong rationale for devolution and decentralisation to ensure a better fit with local needs and to allow local communities to exercise choice over priorities ... there is an ongoing debate about the appropriate balance between the need for devolution (which will produce diverse results and local choices), and the desire to see national standards for some services to avoid pronounced variations in service standards across the country.

... the current system of delivering to national standards, driven by central government in a variety of ways including targets, inspection and specific grants, appears to have some drawbacks in terms of confusion and complexity. This might hinder effective service delivery and choice at the local level, as well as producing inefficiencies. Such pressures may also divert local government from its strategic place-shaping role.

Discussion ... cannot separate local government priorities from central government's concerns about the management and performance of the public sector as a whole, or from the contribution of other agencies involved in public service delivery, or from local people as service users and taxpayers. Changes to the local government system of funding, activities and structures, as well as those of other parts of the public sector such as the police and health, may all affect council tax bills ...

Growing pressures and expectations on local government from a range of sources are likely to be unsustainable, and I intend to explore ways of managing these better as part of the foundation for a more sustainable base for the future. However, this must include a debate about what people are willing to pay for.

This report also sets out the issues raised by my work to date on the system of local government funding...

Lyons asked fundamental questions, such as whether it is important that local government is elected; what the strategic role of local government should be; whether there should be standard services provided across the country; and the extent to which devolution is appropriate.

In his "National prosperity, local choice and civic engagement: a new partnership between central and local government for the 21st century" report, Sir Michael Lyons went on to consider (among other issues):

- local choice
- "fairness" of local taxation and charges
- whether local revenue systems could be adapted to create greater public trust in the local and national government system
- whether local government functions and finance can be used to promote local prosperity
- whether any local government funding system could provide incentives for its functions, such as enabling local authorities to share in the financial benefits of housing growth and economic growth and
- strengthening local accountability.

A final report from the Lyons Inquiry will follow. It will include recommendations for local government funding, with implications for the current national non-domestic rate and council tax.

6.11 Case studies

6.11.1 Rating valuation case study — industrial evidence

A basket of evidence has been gathered from the Westgate Industrial Estate on the outskirts of a sub-regional town in Lancashire, ready for assessing rateable values for industrial units on the estate. The raw information is as follows: you will note that not all lease details are available, which is a problem the valuer often has to contend with. Also, all of the rental transactions post-date the antecedent valuation date (AVD); it so happened in this case that the pre-AVD rents were so much earlier to be of little usefulness.

Address	Built	Floors	GIA GF (m²)	GIA 1st (m²)	Total GIA (m²)	Rent £	Date	Lease terms	Analysis Rent/ Total GIA £
Unit A Great Barr	1999	GF, 1st	941.2	66.5	1007.7	46,800	Sep-04		46.44
Unit B Great Barr	1999	GF	1276.3		1276.3	63,200	Oct-04		49.52
Units 2–3 West Hill	1978	GF, 1st	1088.0	39.9	1127.9	42,500	Oct-03		37.68
Unit 11 West Hill	1978	GF	936.8		936.8	32,750	Oct-03		34.96
Unit 14 West Hill	1978	GF, 1st	2352.0	249.5	2601.5	106,000	Mar-06	5 years	40.75
Unit 31 West Hill	1994	GF	797.0		797.0	35,200	Aug-05	rent review	44.17
Unit 31 Extension, West Hill	1999	GF	3249.0		3249.0	133,150	Jan-04	10 years	40.98
Unit 32 West Hill	1994	GF	797.0		797.0	35,200	Aug-05	rent review	44.17
Unit 4 Great Barr	1978	GF, 1st	504.3	119.0	623.3	20,350	Sep-03	5 years 6 months rent free	32.65
Unit 15 Great Barr	1988	GF	2107.0		2107.0	66,900	Dec-03	rent review	31.75
3a and 3b, Great Barr	1995	GF	797.0		797.0	35,000	Jan-04	9 months rent free	43.91
Unit 35 Great Barr	1978	GF	791.0		791.0	35,000	Mar-04	lease renewal	44.25

When shown in address order, the basic analysis of rent per square metre shows no clear pattern (in this case, the valuer had already established that there was no impact on rental value of being positioned on the different roads in the estate; the vehicular access on each road was similar.) So, adjustments to the rent, and more detailed analysis needs to be carried out.

First, it is common practice when valuing industrial properties for rateable value to increase the rent weighting for office elements, which in this case are those units with first floors. The first floors are weighted at 125% of the ground floor, giving an adjusted area (often called in terms of main space).

Then, where rental details are known, adjustments to the rents can be made as appropriate. In this case, the properties where six months and nine months rent-free periods have been granted, are adjusted (as rental details are not known in every case, a simple calculation spreading the rent-free period over the period to the first review has been used, rather than the more subtle application of YP.)

The adjusted rents can then be divided by the adjusted areas, giving a slightly different analysis:

Address	Age	GIA GF (m²)	GIA FF (m²)	Adjusted area(m²)	Rent "ITMS" £	Date	Lease terms	Adjusted rent £	Analysis Adj Rent/ Adj Area £
Unit A Great Barr	1999	941.2	66.5	1024.3	46,800	Sep-04		46,800	45.69
Unit B Great Barr	1999	1276.3		1276.3	63,200	Oct-04		63,200	49.52
Units 2–3 West Hill	1978	1088.0	39.9	1137.9	42,500	Oct-03		42,500	37.35
Unit 11 West Hill	1978	936.8		936.8	32,750	Sep-03		32,750	34.96
Unit 14 West Hill	1978	2352.0	249.5	2663.9	106,000	Mar-06	5 years	106,000	39.79
Unit 31 West Hill	1994	797.0		797.0	35,200	Aug-05	rent review	35,200	44.17
Unit 31 Extension, West Hill	1999	3249.0		3249.0	133,150	Jan-04	10 years	133,150	40.98
Unit 32 West Hill	1994	797.0		797.0	35,200	Aug-05	rent review	35,200	44.17
Unit 4 Great Barr	1978	504.3	119.0	653.1	20,350	Sep-03	5 years, 6 months rent free	19,333	29.60
Unit 15 Great Barr	1988	2107.0		2107.0	66,900	Dec-03	rent review	66,900	31.75
3a and 3b, Great Barr	1995	797.0		797.0	35,000	Jan-04	9 months rent free	31,500	39.52
Unit 35 Great Barr	1978	791.0		791.0	35,000	Mar-04	lease renewal	35,000	44.25

Although the analysis now allows for greater comparability, the data is still not in a format that is easy to apply when valuing. Grouping the evidence now helps. In the following table, the data has not been altered, but has instead been arranged:

- first by the age of the building
- then by date of the rental evidence
- and then by the size of the unit.

Address	Age	Adjusted area (m²)	Rent £	Date	Lease terms	Adjusted rent £	Analysis Adj Rent/Adj area £
Unit 4 Great Barr	1978	653.1	20,350	Sep-03	5 years, 6 months rent free	19,333	29.60
Unit 11 West Hill	1978	936.8	32,750	Sep-03		32,750	34.96
Units 2–3 West Hill	1978	1137.9	42,500	Oct-03		42,500	37.35
Unit 35 Great Barr	1978	791.0	35,000	Mar-04	lease renewal	35,000	44.25
Unit 14 West Hill	1978	2663.9	106,000	Mar-06	5 years	106,000	39.79
Unit 15 Great Barr	1988	2107.0	66,900	Dec-03	rent review	66,900	31.75
3a and 3b, Great Barr	1995	797.0	35,000	Jan-04	9 months rent free	31,500	39.52
Unit 31 West Hill	1994	797.0	35,200	Aug-05	rent review	35,200	44.17
Unit 32 West Hill	1994	797.0	35,200	Aug-05	rent review	35,200	44.17
Unit 31 Extension, West Hill	1999	3249.0	133,150	Jan-04	10 years	133,150	40.98
Unit A Great Barr	1999	1024.3	46,800	Sep-04		46,800	45.69
Unit B Great Barr	1999	1276.3	63,200	Oct-04		63,200	49.52

The valuer now has a set of data that is much more useful when valuing properties on the industrial estate for rateable value.

Dealing first with the group of units built in 1978. There is a clear trend of increasing value over time. This will help to ascribe a value for the antecedent valuation date, (1 April 2003 for the April 2005 List), even though the valuer must draw the graph backwards, earlier than the available evidence. What is interesting is that Units 4 and 11 Great Barr, both built in 1978, and both with rental dates of September 2003, have analyses more than £5 m² apart. Not finding any lease differences, the valuer drew the conclusion that for this industrial estate, and for this location, there was a certain size beneath which demand in the open market was limited, hence giving a lower rental value. Conversely, Unit 14 shows a lower rental analysis despite being of later date. Here the valuer concluded that the much greater size restricted demand, rather the rental market generally having decreased from 2004 to 2006.

Unit 15 is the only unit built in the late 1980s where evidence is available. The valuer could apply the trends found in the late 1970s group to reach a rate as at the AVD.

The units built in 1994 and 1995 have been grouped together, as the effect of their ages on rental value should be insignificant. By coincidence they are of identical sizes and therefore show clearly the movement in rental value over time.

The late 1990s units are interesting. The extension to unit 31, being particularly large, fits the trend shown in the late 1970s group, that above a certain size, rental value is reduced. However, Units A and B, at almost the same rental date, and in a similar size bracket, give conflicting analyses. The valuer took the cautious approach and concluded that there was something off key about the Unit B rent, possibly a poorly advised tenant, and hence treated this rent as unreliable.

From these considerations, and erring slightly on the cautious side, the valuer drew the following conclusions for rateable value, based on estimated rental values as at April 2003:

Late 1970s units	less than 700 sq m, £24/m²
	between 700 and 1500 m², £29/m²
	above 1500 m², £19/m²
Late 1980s units	above 1500 m², £29/m²
Mid 1990s units	between 700 and 1500 m², £37/m²
Late 1990s units	between 700 and 1500 m², £41/m²
	above 1500 m², £36/m².

This set "bench mark" rates that could then be applied to value other premises on the industrial estate.

6.11.2 Rating valuation case study — retail evidence

This brief case study gives an extract from rental evidence, gathered from shop units around the central mall of a covered shopping centre in a prosperous town in Surrey. The sketch plan (p 000) shows the layout of the shops.

The raw data is given below: as the shopping centre is owned and run by an institutional investor, the lease terms are uniform and therefore play little part in rental value. The shops have all been measured to ITZA (in terms of Zone A).

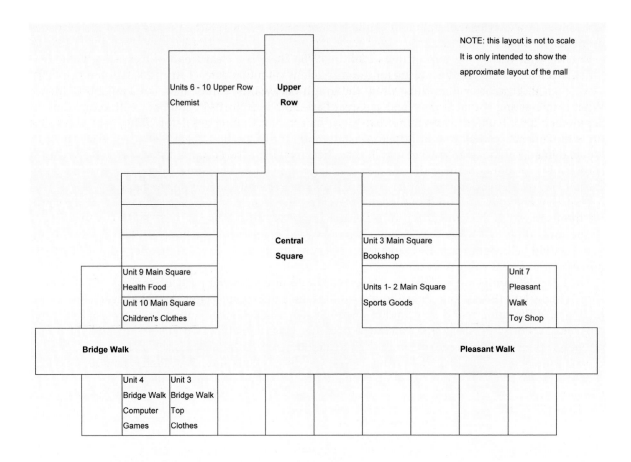

NOTE: this layout is not to scale
It is only intended to show the
approximate layout of the mall

Units 6 - 10 Upper Row
Chemist

Upper
Row

Central
Square

Unit 3 Main Square
Bookshop

Unit 9 Main Square
Health Food

Unit 10 Main Square
Children's Clothes

Units 1- 2 Main Square
Sports Goods

Unit 7
Pleasant
Walk
Toy Shop

Bridge Walk

Pleasant Walk

Unit 4
Bridge Walk
Computer
Games

Unit 3
Bridge Walk
Top
Clothes

Address & Occupier	Date	Rent £	Size (ITZA m^2)	Unadjusted analysis (£/m^2 ITZA)
1–2 Main Square, Sports Goods	Oct 99	157,900	140.4	1,124.64
1–2 Main Square, Sports Goods	Oct 04	157,900	140.4	1,361.47
9 Main Square; Health Foods	Sept 02	85,150	70.0	1,216.43
10 Main Square; Children's Clothes	Sept 03	91,150	68.2	1,336.51
3 Main Square; Bookshop	July 03	82,950	67.0	1,238.06
6,7& 8 Upper Row, Chemist	Apr 03	226,300	212.4	1,065.44
2 Upper Row, Handbags	Mar 03	82,150	69.4	1,183.72
7 Pleasant Walk, Toy Shop	Oct 03	71,050	66.0	1,076.52
3 Bridge Walk, Clothes	Apr 03	93,800	72.6	1,292.01
4 Bridge Walk, Computer Games	Jul 02	93,650	73.1	1,282.12

The analysed rents range from £1,065 to £1,336/m² ITZA, which does not initially give the valuer much assistance. However, various issues are apparent, for which the rental data needs adjusting. The rents are either side of the AVD (1 April 2003 for the April 2005 Rating List), from October 1999 to October 2004. Location is an important factor for retail property, with only small differences in position having significant rental impact. Two of the shops in the data set have return frontages, which is also significant to rental value. Finally, the shops vary in size. The majority are in a standard range, but two are much larger. Dependant upon the local market, quantum can be an issue, either because there is demand for large shop units, or the reverse.

Grouping the rental evidence into the separate malls, and then in date order of the rental transaction, gives the following:

Address & Occupier	Date	Rent £	Size (ITZA m²)	Unadjusted analysis (£/m² ITZA)	Adjustment required
1–2 Main Square, Sports Goods	Oct 99	157,900	140.4	1,124.64	Return frontage; size
9 Main Square; Health Foods	Sept 02	85,150	70.0	1,216.43	
3 Main Square; Bookshop	July 03	82,950	67.0	1,238.06	
10 Main Square; Children's Clothes	Sept 03	91,150	68.2	1,336.51	Return frontage
1–2 Main Square, Sports Goods	Oct 04	191,150	140.4	1,361.47	Return frontage; size
2 Upper Row, Handbags	Mar 03	82,150	69.4	1,183.72	
6,7& 8 Upper Row, Chemist	Apr 03	226,300	212.4	1,065.44	Size
7 Pleasant Walk, Toy Shop	Oct 03	71,050	66.0	1,076.52	
4 Bridge Walk, Computer Games	Jul 02	93,650	73.1	1,282.12	
3 Bridge Walk, Clothes	Apr 03	93,800	72.6	1,292.01	

Looking first at the shops around Main Square, these show a clear trend of increasing rent over time, even setting aside the need for adjusting Units 1–2's rent for return frontage and size.

The adjustment for return frontage can be demonstrated by comparing 3 and 10 Main Square. These shops are opposite one another, and although their rents are only two months apart, their ITZA £/m² is approximately 8% apart. Allowing for the time factor, the valuer chose to use a 7.5% adjustment, to apply to the two shops with return frontages, 1–2 and 10 Main Square. (As the occupier is paying more for a property with the benefit of a return frontage, the rent would have to be reduced when analysing, in order to equate to a standard shop.)

The adjustment for size/quantum can be deduced by looking at the shops in Upper Row. Although only a month apart in rental date, the analyses of 2 and 6, 7 & 8 Upper Row are 10% apart. The valuer can assume, (as there are no lease differences), that this rental difference is due to size. 6, 7 & 8 is a triple unit; the valuer then made the assumption that a double unit would warrant a lower adjustment, and adopted 5% for 1–2 Main Square. (This evidence demonstrates that in this town, rental values reduce pro rata with increasing size. Therefore a tenant of a larger shop is paying less ITZA than for a standard shop, and the rental analysis will have to be increased in order to equate it to the standard size range.)

Building these adjustments into the data provides the following:

Address & Occupier	Date	Rent £	Size (ITZA m^2)	Unadjused analysis (£/m^2 ITZA)	Adjustment required	Adjusted analysis (£/m^2 ITZA)
1–2 Main Square, Sports Goods	Oct 99	157,900	140.4	1,124.64	Return frontage −7.5%; size +5%	1,092.31
9 Main Square; Health Foods	Sept 02	85,150	70.0	1,216.43		1,216.43
3 Main Square; Bookshop	July 03	82,950	67.0	1,238.06		1,238.06
10 Main Square; Children's Clothes	Sept 03	91,150	68.2	1,336.51	Return frontage −7.5%	1,236.27
1–2 Main Square, Sports Goods	Oct 04	191,150	140.4	1,361.47	Return frontage −7.5% size +5%	1,322.33
2 Upper Row, Handbags	Mar 03	82,150	69.4	1,183.72		1,183.72
6,7& 8 Upper Row, Chemist	Apr 03	226,300	212.4	1,065.44	Size +10%	1,171.98
7 Pleasant Walk, Toy Shop	Oct 03	71,050	66.0	1,076.52		1,076.52
4 Bridge Walk, Computer Games	Jul 02	93,650	73.1	1,282.12		1,282.12
3 Bridge Walk, Clothes	Apr 03	93,800	72.6	1,292.01		1,292.01

The re-analysed rents now show more clearly the trend over time, allowing the valuer to extrapolate the reasonably expected ITZA £/m^2 as at April 2003. His view of the reasonably expected rent at April 2003 for the locations considered is:

Main Square, £1,230/m^2 ITZA
Upper Row, £1,180/m^2 ITZA
Pleasant Walk, £1,065/m^2 ITZA
Bridge Walk, £1,290/m^2 ITZA

The valuer, of course, notes that the analyses are location specific, and in order to carry out a rateable valuation across the whole shopping centre, more rental evidence in each mall would be required.

Part 3

Compulsory Purchase and Planning Compensation

Introduction and Compensation for Land Taken

7.1 The Slough of Despond

This issue involves a consideration of various Acts and orders which Harman LJ likened to a "Slough of Despond". The arguments and the evidence in regard to it were quite properly exhaustive; they left us exhausted and we cannot hope in a decision of reasonable compass to cover the whole range of either. Therefore we think it best to express our views as best we can within a reasonable space, drawing attention only to those points of argument and evidence which appeared as sign-posts through the bog.

(Sir Michael Rowe, QC, and John Watson, FRICS, Members of the Lands Tribunal, in *Kaye v Basingstoke Development Corporation* (1967))

This issue is the assessment of compensation upon the compulsory acquisition of an interest in land. The valuer's task in this particular subject area has been complicated over recent years by new statutes and cases which have added to an already confused juxtapositioning of statute law and common law principles.

The author of "Legal Notes" in *Estates Gazette* has referred to yet another confusing decision (where the Court of Appeal, reversing the decision of Lands Tribunal, was overruled in turn by the House of Lords) as the House giving "another twitch to the tangled mass of compulsory purchase compensation rules".

Between 2000 and 2003, a consultation process was put into action with a view to a complete reform of the whole body of statue and case law which makes up the current compensation code. This resulted in December 2003 in a Law Commission report proposing a new statute which would reform and simplify the present confused and outdated system. Unfortunately, it only succeeded in convincing the government of the enormity of the changes that are required, and in December 2005 the reform process was quietly shelved. In a written statement the Office of the Deputy Prime Minister announced that while it accepted that "the convoluted state of the compulsory purchase legislation as it currently stands does not make it readily accessible", it did not consider that "the amount of effort and resources which would be required can be justified at the present time in the light of the government's overall priorities".

Some minor changes were made in the Planning and Compulsory Purchase Act 2004, but for the most part our task in the following four Chapters remains to try and untangle a set of rules which has been developed gradually over a period of 160 years, so that a logical valuation methodology may be applied to the often confusing problem of establishing the correct quantum of compensation.

Figure 7.1 Parties involved in compulsory purchase

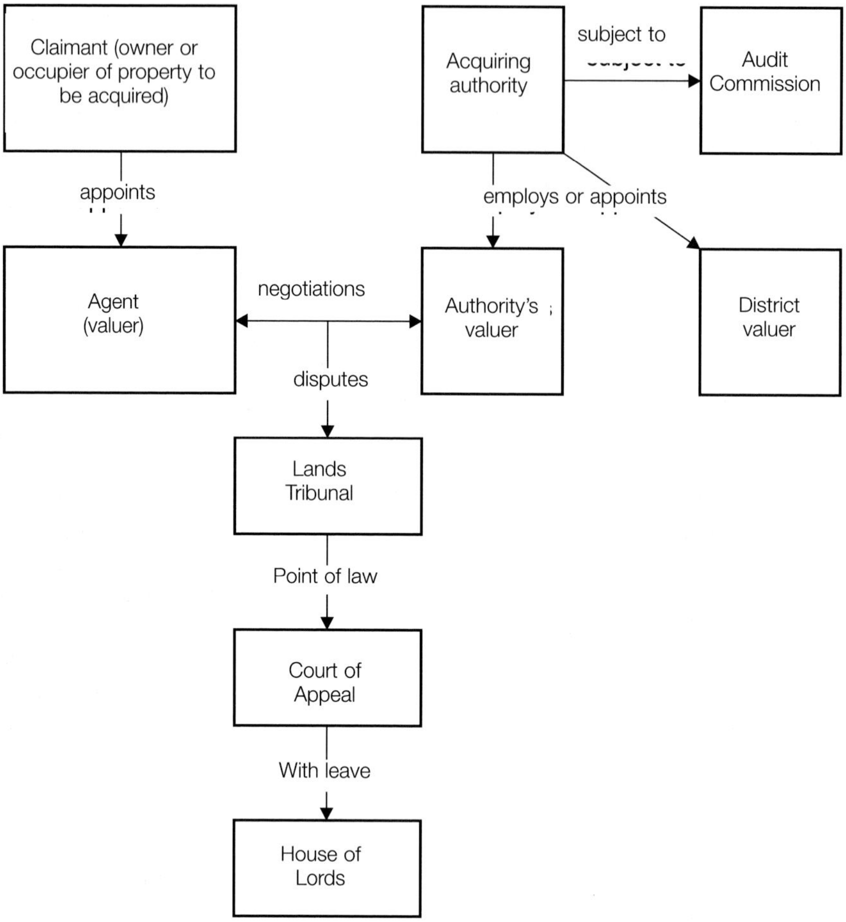

7.2 The parties involved

Interests in property may be compulsorily acquired by local authorities, statutory undertakers or other public bodies, and central government, provided that the authority is given powers of compulsory acquisition by an authorising Act. With the privatisation of many statutory undertakers in recent years the number of private, as well as public, bodies with statutory powers has increased dramatically.

The claimant may be an owner or occupier with any legal interest in the property. Consequently, the acquisition of a single property may give rise to more than one claim. Even persons without compensatable interests, for example assured tenants, may receive certain payments and a preliminary task for the valuer is to establish the number of claims he will be required to quantify.

The Audit Commission exercises a check on the expenditure of all local authorities. It is a government body whose function is to audit the annual accounts and check both the accuracy and legality of items of expenditure. If the legality of a compulsory acquisition is in question due to the *ultra vires* rule then the members of the council which voted for the expenditure may be personally surcharged.

The fees of the claimant in employing a solicitor, valuer, and (perhaps) an accountant to transfer the title of his interest and to quantify and negotiate his claim for compensation will be paid as of right by the acquiring authority (see *Tobin v London County Council* (1959)). Consequently, any claimant should appoint a valuer to act on his behalf and exploit the free benefits of his expert advice. Claimants who have represented themselves without a valuer's advice in the Lands Tribunal have regularly been the unwilling recipients of complete disregard from members of the Tribunal (admittedly often valuers) and have usually lost the argument comprehensively. This is particularly dangerous in the Lands Tribunal where, if good comparable evidence or logical argument is not produced in defence of a claim, the authority's valuation will usually be relied upon without argument, even though this figure might have represented a starting point and not the highest figure the authority were happy to expend. This had particularly embarrassing ramifications in *Fawcett v Newcastle-upon-Tyne MDC* (1980) where the sealed offer of compensation by the acquiring authority exceeded the actual award of the tribunal by 20%.

Most of the larger acquiring authorities employ chartered surveyors who act as valuers on their behalf in the quantification and negotiation of claims for compensation with the claimant's valuer. The recommendations made by the authority's valuer will usually be subject to ratification by a committee of the council, which is by no means obligatory, and approval by the full council (albeit less subject to argument) is also required in the majority of cases.

Central government and smaller local authorities and statutory undertakers who employ no valuers may instead appoint the district valuer, the supervisor of a branch of the valuation office of the Board of Inland Revenue, to act for them. Each branch of the valuation office is likely to include specialists in compensation matters. Alternatively, and more commonly in recent years, they may appoint valuation firms in private practice to act on their behalf.

Disputes over the quantum of compensation where the claimant's valuer and the authority's valuer fail to reach agreement may be considered by the Lands Tribunal, whether the dispute concerns a point of law, or a simple difference of opinion concerning the value of the property. Dissatisfaction with the award of the Lands Tribunal may prompt either party to seek satisfaction in the Court of Appeal, but the dispute will only be considered by this body if it centres upon a point of law.

Appeal to the House of Lords is the final recourse of the disenchanted party, provided leave is given by the Court of Appeal.

Compulsory purchase compensation may be conveniently divided into three subject areas. Any claim for compensation may involve one or more of the following.

(a) Compensation for land taken, where land is acquired from someone with a compensatable interest, in the general case (Chapter 7).
(b) Compensation for severance and injurious affection (or a reduction of the compensation figure for betterment) where other land in the claimant's ownership is increased or reduced in value as a result of the acquisition (Chapter 8).
(c) Compensation for disturbance or other losses suffered by the claimant as a result of the acquisition not based on the value of land. Technically, this forms part of the compensation for land taken but is dealt with as a separate item in Chapter 9.

It is good practice always to set out a compensation claim under these three heads, even though nil is likely to be entered under one or more head, as it is quite unusual for all three to apply.

The following chapters explore these broad subject areas. We concentrate on the assessment of compensation, rather than the procedural issues relating to the making and implementation of compulsory purchase orders.

7.3 The background

It is not the purpose of this text to dwell upon the archaeology of statutory valuations. However, a concise consideration of the way compensation for land taken in a compulsory acquisition has been assessed over the last century and a half is helpful, not only because many of the old rules continue to apply, but also to explain the introduction of some recent measures.

An examination of the history of compensation for land taken runs into difficulties as one attempts to go back before 1845. Blackstone had something to say in 1765, but it was in the 19th century that the need for powers of compulsory acquisition became highlighted. At the start of the 1830s private Acts of Parliament allowed the cutting of canals and the construction of railways. These private Acts had to include certain standard clauses, which were eventually collected in the 1845 Lands Clauses Consolidation Act. This Act is the natural springboard for a study of compensation, yet it omits to refer directly to the assessment of the compensation, concentrating instead upon procedural matters. The legislature at the time apparently considered that the assessment of compensation was a practical matter which had to be left to the discretion of the professional valuer. One must imagine the attitude of arbitrators of the time, who must have been guided by Blackstone's words:

> If a new road, for instance, were to be made through the ground of a private person, it might perhaps be extensively beneficial to the public ... In this and similar cases the legislature...can, and indeed frequently does, interpose and compel the individual to acquiesce. But how does it interpose and compel?
>
> Not by absolutely stripping the subject of his property in a arbitrary manner; but by giving him a full indemnification and equivalent for the injury thereby sustained. The public is now considered as an individual, treating with an individual for an exchange. All that the legislation does is to oblige the owner to alienate his possessions for a reasonable price.

Interpretation of this 18th-century guide in the mid-19th century is illustrated in *Penny* v *Penny* (1867). The value to be assessed in this case was interpreted as being the value to the owner and "any difference in the result which is due to the accident of the property being taken by a public body is not to be thrown into the compensation fund". In *Stebbing* v *Metropolitan Board of Works* (1870), it was held that the owner's loss shall be tested by what is the value of the thing to him, not by what will be its value to those persons acquiring it.

This basis of compensation for land taken was changed in 1919 by the Acquisition of Land (Assessment of Compensation) Act. The Act laid down, for the first time, statutory rules governing the assessment of compensation. Section 2 of the Act contained the major provision which remains in force to this day (with minor amendments), having been consolidated in the Land Compensation Act of 1961. It lays down six rules for the assessment of compensation which form the single most important guide for the valuer concerned with compulsory acquisition. These rules are fully considered at 7.6 below and onwards. The other most important compensation provisions are set out in the Compulsory Purchase Act 1965. Three subsequent enactments promised major reformation of the compensation code, but delivered only minor reforms. These are the Land Compensation Act 1973, the Planning and Compensation Act 1991 and the Planning and Compulsory Purchase Act 2004.

Before considering the assessment of compensation, however, two questions of importance to the valuer must be considered. First, of great relevance in times of inflation, what is the date at which the interest should be valued? Second, as a single property may be the subject of a changing pattern of tenure and ownership, what interests are to be the subject of compensation? The considerable time-scale of a compulsory acquisition requires the identification in specific terms of these two factors.

7.4 Date of valuation

In the 1968 *West Midlands* case (*West Midlands Baptist (Trust) Association (Incorporated) v Birmingham Corporation*) a spurious rule concerning the date of valuation, gleaned from *Penny v Penny* (1867), referred to above, was condemned to impotency and replaced by new rules derived from the words of Lord Reid in the later case.

After *Penny v Penny*, the rule that the date of valuation was to be the date at which notice to treat was given became accepted without question. In times of rising prices and costs it became inevitable that a claimant threatened with a substantial loss in compensation would challenge the rule in the High Court due to a delay on behalf of the acquiring authority in taking possession after serving notice to treat. These very facts kindled the *West Midlands* case, where notice to treat for the acquisition of a chapel was deemed to have been served in August 1947, when compensation would have been awarded at a figure of £50,025, but the actual date of removal of the occupier was not to begin until 1961, by which time the relevant figure of compensation had risen to £89,575. The rule in *Penny v Penny* would have ensured payment of the lower figure which was patently inequitable to the claimant.

Lord Reid described what he saw as the rationale behind a new correct rule as follows:

> No stage can be singled out as the date of expropriation in every case. Sometimes possession is taken before compensation is assessed. Then it would seem logical to fix the market value of the land as at that date. But if compensation is assessed before possession is taken, taking the date of assessment can, I think, be justified because then either party can sue for specific performance and the promoters obtain a right to the land, as if there had been a contract of sale at that date.

In summary, the result of the *West Midlands* case and subsequent interpretation of it is that the date of valuation shall be as follows.

(a) In rule 2 cases, where market value is the basis of compensation (see 7.6.2), the date of valuation shall be the earlier of:
(i) the date when compensation is being agreed or, in the case of a reference to the Lands Tribunal, the date of the award and
(ii) the date of physical possession. (For a dispute over the date of taking possession, see *Courage Ltd v Kingswood DC* (1978). For a dispute over the validity of an agreement of compensation, see *Duttons Brewery v Leeds City Council* (1982).) The situation where possession is taken in stages was considered in *Chilton v Telford Development Corporation* (1986). Where possession of eight small parcels of land was taken on different dates the Lands Tribunal held there should be eight separate valuation dates even though there was only one notice of entry. This decision was overturned by the Court of Appeal which held that the authority would be treated as taking possession of the whole at the date of entry on to the first parcel. In the absence of firm precedent

the court took the line most favourable to the claimant. It may be, however, that the decision will work against claimants in the majority of cases where it is applied, for example where possession is taken of a small part of a large holding some years before entry to the remainder. In such cases compensation will be assessed at a level of values applicable some years before the compensation is paid, exactly the situation which the *West Midlands* decision was intended to avoid.

(b) In rule 5 cases, where equivalent reinstatement is the basis of compensation (see section 7.6.5), the date of valuation shall be the date at which reinstatement becomes reasonably practicable or the date of physical possession, whichever is the earlier. In the *West Midlands* case the date at which reinstatement became practicable was found to be the date at which building of the new chapel was to begin.

The *West Midlands* principle applies whenever the acquiring authority uses the traditional route to obtaining possession: the service of notice of entry followed by notice to treat.

There is, however a more modern alternative. The Compulsory Purchase (Vesting Declarations) Act 1981 introduced a simplified process under which a single notice transfers legal title to the acquiring authority, and empowers them to take possession of the land. Under this procedure the valuation date will be the date of agreeing compensation, or the date of vesting, whichever is the earlier. The main benefit of the general vesting declaration (GVD) to the acquiring authority is the transfer of title at the date of the notice, rather than the date of conveyance following agreement of compensation, which applies under the old system. This is of little practical significance where the compulsory purchase is for public works, such as road widening, but is important where they may wish to sell or lease the property later, for example land within a general development area. In spite of the benefits of the GVD procedure to acquiring authorities, the old system remains in common use.

It is important to note that the statute of limitations applies to the right to compensation. In *Co-operative Wholesale Society* v *Chester-le-Street* (1996) the Lands Tribunal confirmed that a claim could only be referred to the tribunal within a period of six years from notice to treat. On the facts of that case, however, it decided that the acquiring authority had waived its right to rely on the limitation period by continuing negotiations, as if the claimant still had a valid claim, after the six year period had elapsed. This decision, subsequently confirmed by a number of others, came as something as a revelation at the time. Compulsory purchase tends to be a lengthy process, and many of the most important cases in this field would not even have been considered by the Lands Tribunal if it had been known that there was a time-limit on referrals.

7.5 The interests to be acquired

Over the period of execution of the compulsory acquisition, even if this is not especially lengthy, changes in the pattern of tenure attaching to the subject property may be engineered. In addition, there may be both legal and physical amendments to that property and the problem of deciding upon the precise subject matter of the valuation exercise will often be encountered.

In *Mercer* v *Liverpool, St Helens and South Lancashire Rail Co* (1904) a preliminary rule was established: the notice to treat was seen in that case as a move of significance as it "binds the (acquiring) authority to purchase and the owner to sell at a price to be ascertained". It is also well settled that the owner served with a notice to treat may not, by altering the land or creating new interests in it, increase the burden on the acquiring authority as regards the compensation to be paid (a principle evinced in the *West Midland* case, having been given statutory weight by schedule 2, para 8 of the Acquisition of Land

(Authorisation Procedure) Act of 1946). Hence, the owner of a freehold interest who grants a leasehold interest out of it after the service of notice to treat and obtains a premium for that lease has saddled the leaseholder with a worthless interest and a probable action against his conveyancer.

The general principle that the notice to treat fixes the interests to be acquired has to be reconciled with the rules concerning the date of valuation. In *Holloway v Dover Corporation* (1960), a bakery had been let on a lease due to expire in 1954 and notice to treat was deemed to have been served in 1949. At the expiry of the lease, the acquiring authority had made no attempt to obtain possession and the tenants continued in occupation under a statutory tenancy. The date of valuation being 1957, the compensation to be paid was simply disturbance compensation under the Landlord and Tenant Act 1954, as the premises were to be redeveloped. At the date of valuation the tenants no longer had a valuable leasehold interest although at the date of service of the notice to treat they certainly had. This case leads to the conclusion that the rule for the date of valuation of leaseholds which expire over the period of compulsory acquisition is exactly the same as the rule summarised at 7.4.

This interpretation may result in the application of section 20 of the Compulsory Purchase Act 1965 which grants a right to compensation to the holder of a lease with less than a year to run at the date of valuation, for "any just allowance which ought to be made to him by an incoming tenant, and for any loss or injury he may sustain" (see *DHN Food Distributors Ltd v London Borough of Tower Hamlets* (1976)).

In *Babij v Rochdale Metropolitan Borough* (1976), the claimant owned a long leasehold interest in a terraced house which was compulsorily acquired. The value of the owner-occupied interest was agreed at £3,500. Before the date of valuation, however, the claimant, having found alternative accommodation, had let the house on what he considered to be a temporary basis pending notice of entry being served. Unfortunately, he inadvertently had created a statutory protected tenancy, which reduced the compensation figure to £900.

It is now possible to expand upon the position of the imaginary freeholder who granted a leasehold interest after the service of notice to treat. In addition to providing the leaseholder with a worthless interest, he may have considerably reduced the value of his own interest. The only party to gain from such a move is the acquiring authority.

The decisions in *Ali v Southwark London Borough* and *David v London Borough of Lewisham* (both 1977) may be contrasted. In the former case, a freehold dwelling house was found to be encumbered by statutory tenancies and, despite the fact that the tenants were close friends of the landlord with whom they shared a understanding that the premises would be vacated if he wished to re-occupy, compensation was awarded at a low figure reflecting the encumbrance. In the latter case, on the other hand, a low compensation award was avoided in similar circumstances as occupation by the claimant's daughter was held to be by virtue of a licence, rather than a tenancy, which enabled the valuation to be made on the basis that the property was vacant.

Where the acquiring authority rehouses a protected tenant between the date of notice to treat and notice of entry, it has been successfully argued that compensation should be paid on the basis of the freehold interest being vacant and unencumbered (see *Banham v London Borough of Hackney* (1970)). Since the 1973 Land Compensation Act, however, the prospect of rehousing the occupier of residential property is irrelevant to the question of compensation and must be disregarded, whether as an argument for increasing the landlord's compensation or for decreasing the tenant's, in those rare circumstances where the tenant of residential property has a valuable interest.

The property should be valued *rebus sic stantibus*: in its actual physical condition at the valuation date. It is widely accepted that any damage caused to the premises, after the date of notice to treat but before the date of valuation, is the responsibility of the claimant, in accordance with the rules laid down in the *West Midlands* case (see *Lewars* v *Greater London Council* (1982)). An exception to this was *Gateley* v *Central Lancs New Town Development Corporation* (1984), where the Lands Tribunal held that the scheme underlying the acquisition encouraged vandalism and the authority should bear part of the resulting diminution in value of the property. In that case the acquiring authority had taken possession of most of the other properties in the scheme, and had been so careless in its maintenance of them that it was considered to have made a major contribution to the vandalism and disrepair which had affected the subject property by the valuation date. It is clear, however, that the general principle of *rebus sic stantibus* will be followed in all but the most exceptional cases.

Finally, it is noteworthy that a strange anomaly in compulsory purchase procedure is removed by section 67 of the 1991 Planning and Compensation Act. Whereas a notice to treat must be served within three years of confirmation of a CPO, there was previously no time-limit once a notice to treat has been served. As a result of section 67 a notice to treat will be rendered invalid unless within three years:

(a) compensation has been agreed or
(b) physical possession has been taken or a general vesting declaration served or
(c) the question of compensation has been referred to the Lands Tribunal.

Having established both the precise subject matter of the valuation and the date at which that valuation is to be made, it is now proposed that the principles governing the valuation shall be considered. The springboard for such a consideration is section 5 of the Land Compensation Act 1961, which sets out what are generally known as the six main rules of compulsory purchase. While there may have been some validity to this statement in some long-forgotten age, only three of the rules (rules 2, 5 and 6) can now be considered important, the others being of marginal significance at best. In any modern recital of the main rules, those relating to the no-scheme world and planning assumptions would certainly be promoted to the premier league at the cost of some of the section 5 rules. We will, however, consider the rules in their traditional order.

7.6 The six rules (Land Compensation Act 1961, section 5)

7.6.1 Rule 1

No allowance shall be made on account of the acquisition being compulsory.

This rule was introduced by the 1919 Acquisition of Land Act. Prior to 1919 it had become accepted practice to add 10% to the agreed quantum of compensation as a consolation for compulsory expropriation. This is no longer permitted.

In recent years, the principle that at least some claimants should receive additional compensation for the inconvenience of compulsory purchase has regained acceptance. The 1973 and 2004 Acts provide for home loss, basic loss and occupier's loss payments, up to a maximum of 10% of the compensation payable (see Chapter 9.9).

7.6.2 Rule 2

The value of land shall, subject as hereinafter provided, be taken to be the amount which the land if sold in the open market by a willing seller might be expected to realise.

This, the willing seller rule, is the point of reference for the valuer concerned with compulsory acquisition, against which all other rules derived from statutes and cases may be seen as either amplifications of, or exceptions to, the general basis of compensation: the market value. The valuer is uniquely qualified to assess compensation on this basis, and few problems in assessing compensation would arise if it were not for the necessary presumption of certain artificialities which the valuer must envisage, an example of which is the concept of the "willing seller".

This is the major essential contradiction in the compensation basis. Coincidences apart, the seller is not willing: he is being compulsorily expropriated. Yet the valuer has to assume that the vendor is happy to part with the property. This must be qualified by the proviso that the seller is only willing at a certain price, that is market value, and it is not reasonable for the acquiring authority to argue that a willing seller might accept less than this figure as a result of his eagerness to sell.

In addition, the Act does not require the valuer to assume that the actual seller is willing, but requires that the value of the land is fixed at the price it could be expected to realise on a hypothetical sale in the open market by a hypothetical willing seller. This assumption enables the actual seller to be regarded as a possible bidder for his own property.

In *Solarin* v *Wandsworth London Borough* (1970), the claimant had paid an exceptionally high price for his residence, which was later compulsorily acquired from him. The acquiring authority argued that he was a special purchaser and that rule 3, as it then stood, required his bid to be left out of account. However, as a notional bidder in the hypothetical market to be assumed under rule 2 he would, it was assumed by the Lands Tribunal, again be prepared to offer a higher price than other bidders and compensation was awarded at a figure which reflected this.

Where part of a large holding is being acquired, the hypothetical willing seller rule means that it is necessary to assess the market value of the land taken in isolation even though it may well have virtually no value except as part of the whole holding (see *Hoveringham Gravels Ltd* v *Chiltern District Council* (1977), and Chapter 8).

The test of the market value of the interest under rule 2 is good comparable evidence. Considerable reliance will be placed, both in negotiations between the parties' valuers and in the Lands Tribunal, upon first-hand experience of sales of similar properties in the open market or, failing such experience, well-kept and reliable records. Such comparable evidence may often be difficult to collect, particularly in run-down residential areas where the local authority, using its compulsory purchase powers, may be the only recent purchaser. In these circumstances the authority may build up a vast knowledge base of agreed prices, or settlements, none of which have been tested in the open market. These can become self-perpetuating, as they are able to use this body of evidence to justify continuing to buy properties at a similar price, even though underlying values may be rising in the surrounding area. Such a problem arose in *Zarraga* v *City and County of Newcastle-upon-Tyne* (1968), where a single comparable sale in the open market produced by the claimant's valuer held sway over a list of accepted compensation awards under the same scheme produced by the authority. The member said: "I am in no doubt myself that this open market sale, although a solitary transaction, must be given greater weight than all the corporation's settlements. The sale blows like fresh air through the whole debate."

In *Fairview Estates (Barnet) Ltd* v *Hertfordshire County Council* (1976), the date of valuation of 6.25 acres of residential building land was February 1974, at which time it was discovered that no comparable transactions had taken place for a considerable period. It was also apparent that property values were changing at that time and the tribunal was forced into the solution of taking a 1973 comparable and deducting a percentage from the value derived from this evidence to account for an apparent fall in values between 1973 and 1974. Such a course is to be avoided, as future tribunals will only adopt such procedures in desperation.

However unreal the artificial world created in a compulsory purchase situation, it should never be forgotten that, having established the parameters of that world, the normal principles of valuation apply. The appropriate method of valuation should be used, and sound comparable evidence should always be gathered and analysed. In effect, the general basis of compensation for land taken can be summarised as open market value, as if offered for sale by a hypothetical willing seller, in its actual physical condition at the valuation date.

7.6.3 Rule 3

The special suitability of adaptability of the land for any purpose shall not be taken into account if that purpose is a purpose to which it could be applied only in pursuance of statutory powers, or for which there is no market apart from the requirements of any authority possessing compulsory purchase powers.

It may be argued that, in part at least, this rule is a second departure from reality encapsulated in the compensation code, as it appears to place the acquiring authority in a special position. It has, however, been much weakened by the provisions of the 1991 Act, which removed a further obligation to leave out of account any value arising from the needs of a special purchaser.

The rule should be split into two parts in order to examine its effects.

(a) The special suitability or adaptability of the land for any purpose shall not be taken into account if the purpose is one to which the land could only be applied in pursuance of statutory powers.

This part of the rule is relatively uncontroversial and rarely applied due to the rarity of circumstances required for it to be relevant.

Livesey v *Central Electricity Generating Board* (1965) illustrates the classic operation of this part of the rule. Agricultural land was being acquired for use as a power station and the claimant hoped for a large figure of compensation to reflect the proposed industrial use. However, statutory authority being necessary in order to set up a power station, this claim was discounted and compensation on the basis of agricultural value was awarded due to the operation of rule 3.

As with many of the leading cases in this field, *Chapman Lowry Puttick Ltd* v *Chichester District Council* (1984) concerned a small area of waste land which was acquired to provide access to a development site owned by the acquiring authority. The perennial question is whether the basis of compensation should be the price which a developer might be expected to pay to acquire access and unlock the development potential, the ransom value; or whether rule 3 requires the additional value over and above existing use value to be left out of account. In this instance the Lands Tribunal held that the land could be used for access even without statutory powers so the first part of rule 3 did not apply. Compensation was awarded at its ransom value of £25,000, not its existing use value of £200.

(b) The special suitability or adaptability of the land for any purpose shall not be taken into account if the purpose is one for which there is no market apart from the requirements of any authority possessing compulsory purchase powers.

In addition to its application to cases such as that which formed the subject of dispute in *Livesey* v *Central Electricity Generating Board* (1965), this part of the rule is widely considered to prevent competition between more than one statutory or local authority, for example water board and district council, being used as an argument to force up the compensation to be paid upon compulsory acquisition by one of these authorities. Again, this part of the rule is uncontroversial and rarely used.

A much more controversial second limb to this rule was removed by schedule 15 of the 1991 Act which removed the words "the special needs of a particular purchaser or" before "the requirements of...".

This part of the rule caused particular difficulties and it appeared to confirm what would happen in an auction. Suppose, for example, that land had a value of £10,000 as a car park, but the acquiring authority, owning similar adjacent land, could exploit marriage value by acquiring it. Acquisition of the subject land might increase the value of the acquiring authority's holding by £15,000: but the authority would have no need to bid up to that figure to secure the land, and a bid marginally over £10,000 would be sufficient. This part of the rule appeared to confirm the auction principle.

But what would be the position if the land were worthless in its existing state, yet it increased the value of the authority's holding by £15,000? It is not reasonable to suppose that a nominal bid would secure the purchase as the owner would probably not sell. Although a hypothetical willing seller is to be assumed, he is not assumed to be willing to sell at any price: *Inland Revenue Commissioners* v *Clay and Buchanan* (1914).

The *Indian* case (*Raja Vyricherla Narayana Gajapatiraju* v *The Revenue Divisional Officer, Vizagapatam* (1939)) resulted in a Privy Council decision which appears to have influenced the Lands Tribunal. This case concerned the acquisition of land in India required for the construction of a harbour which was required to provide a water supply essential for the suppression of a malaria epidemic. The land was virtually worthless apart from the needs of the acquiring authority which was the only potential purchaser. Rather than award a nominal figure of compensation, the Privy Council argued that the valuer must envisage a friendly negotiation between a willing seller who expects a reasonable price for his land and a willing purchaser, rather than an auction. Therefore, although only a willing seller is referred to in rule 2, it seems that a willing purchaser can also be assumed to exist.

In *Britton* v *Hillingdon London Borough* (1977), 27 m² of land were being acquired. The land separated a 0.8 ha development site, which was also being acquired, from a road. A cost saving of £10,000 would be made by the acquiring authority if it were to use the subject land as an access. The award of compensation was based on the existing use value of the land plus 25% of the cost saving. Despite the obvious relevance of rule 3 as originally drafted, that rule does not appear to have been considered.

Rathgar v *Haringey London Borough* (1978) concerned very similar facts. In this case the purchase of land as an access to a block of flats in the authority's ownership enabled the acquiring authority to make a cost saving of £19,000. The acquiring authority attempted to rely upon rule 3, which was held not to apply. The Lands Tribunal envisaged that there was more than one prospective purchaser of the land for the purpose of gaining an access to the flats, including a speculator interested in reselling to the authority at a profit. *Blandrent Investment Developments* v *British Gas Corporation* (1978) was of similar effect.

The logic of the concept of the interposing speculator seems to be designed to prevent the acquiring authority from relying upon rule 3 both in the purchase of the back land without an access and the access strip, thereby making a considerable profit, and this may explain the loss of power of this second part of rule 3.

The enormous significance of the avoidance of rule 3 in simple cash terms is illustrated by *Ozanne* v *Hertfordshire County Council* (1991). In this case the land had an agricultural value of £5,500. However, it was required to carry out a highway improvement without which a comprehensive development scheme could not proceed. The Lands Tribunal held that the second part of rule 3 did not apply as there was more than one developer involved and their decision was upheld by the

House of Lords. Compensation was assessed on the basis of 50% of the development value released. The sum awarded was £1.24 m!

Happily, the courts need no longer devise ingenious arguments to avoid this previously much criticised rule and in its amended form it is unlikely to cause too many problems.

For further details of ransom value see p 258.

Finally, it is apparent that the special suitability rule in toto applies to the land and it is not to be interpreted as applying to the special suitability of an interest in the land. This is made clear by *Lambe* v *Secretary of State for War* (1955). In this case the acquiring authority was the sitting tenant of a property, the freehold reversion to which was the subject of the compulsory acquisition. Marriage value would be exploited by the sitting tenant who could logically afford to pay more for the reversionary interest. The acquiring authority attempted to rely upon rule 3 in order to exclude that extra value from the compensation to be paid, but found the Court of Appeal to be unsympathetic to this argument for the reason stated at the beginning of this paragraph.

In the House of Lords decision of *Waters* v *Welsh Development Agency* (2004), Lord Nicholls expressed the view that rule 3 should be interpreted so narrowly that it has effectively become redundant; its main role having been taken over by the *Pointe Gourde* rule.

7.6.4 Rule 4

Where the value of land is increased by reason of the use thereof or of any premises thereon in a manner which could be restrained by any court, or is contrary to law, or is detrimental to the health of the occupants of the premises or to the public health, the amount of that increase shall not be taken into account.

This rule can be summarised as — leave out of account any value attributable to an illegal or immoral use.

Few disputes appear to have been considered by the Lands Tribunal concerning this rule, which is either beyond dispute in its clarity of application or rarely relevant. The latter is more likely to be the case. A house which is split into an unhealthily large number of bed-sitting rooms and is subject to the overcrowding provisions of the Housing Acts should be valued on the basis of a legally acceptable concentration of occupation less the costs of conversion necessary to implement this (say by converting the property to two or three flats).

Residential property used without planning permission as offices is also a use which is contrary to law, and upon compulsory acquisition of the property compensation must be limited to a value which ignores the office use. It is incumbent upon the valuer to check upon the existence of planning permission, or an established use right, for the current use of the property before he values it on that basis, in exactly the same way as for a valuation for any other purpose.

The Royal Institution of Chartered Surveyors identified a potential shortcoming of rule 4 in its first discussion paper suggesting improvements to the compensation code in 1983. Under planning law, building works completed more than four years ago, and certain changes of use, were immune from enforcement action by the planning authority. However, they were technically illegal but unenforceable, and so should be left out of account in assessing compensation, even though the open market value of such works in the absence of a scheme would be as high as for a legal use. This point was taken to heart by the government and under 3.4 of the 1991 Act breaches of planning consent which are outside the limitation period for enforcement action become lawful. That limitation period is, however, altered to 10 years except for building operations and changes of use to a single dwelling house where the four-year rule is retained.

Unfortunately, the amendment arrived a little too late as the interpretation of rule 4 had already been clarified by the House of Lords. In *Hughes* v *Doncaster* (1990) it was held that, even under the old system, a planning use which had become immune from enforcement action could not be said to be illegal within the meaning of rule 4.

7.6.5 Rule 5

Where land is, and but for the compulsory acquisition would continue to be, devoted to a purpose of such a nature that there is no general demand or market for land for that purpose, the compensation may, if the Lands Tribunal is satisfied that reinstatement in some other place is bona fide intended, be assessed on the basis of the reasonable cost of equivalent reinstatement.

This rule encapsulates the major exception to the market value rule and represents the continuation, in certain circumstances where the new rule is inappropriate, of the old value to the owner rule. It is born of necessity as it is impossible to fulfill the requirements of rule 2 when attempting to value certain properties. It allows compensation to be assessed on the basis of replacement cost, as an alternative to market value, where the qualifying rules are met.

An example of a case where this equivalent reinstatement rule was applied instead of rule 2 is *Sparks* v *Leeds City Council* (1977), the report of which is useful both because it examines four requirements made by the rule and because a valuation under the rule 5 basis is set out.

The four requirements are as follows:

(a) The subject land is devoted to the purpose, and, but for the compulsory acquisition, would continue to be so devoted.

Assessment of devotion to the purpose is a matter of fact at the date of notice to treat. It is the valuer's responsibility to check this requirement both by ensuring that the premises are wholly or mainly used for the purpose in question, and that no imminent removal, before the likelihood of compulsory acquisition was made public, was being considered.

(b) There is no general demand or market for the land for that purpose.

A study of previous Lands Tribunal decisions on this point provides little guidance as to how they have arrived at their decision, and what they are likely to decide in the future. Again this seems to be a matter for factual evidence presented by the valuer. The existence of a market would allow the use of rule 2 and obviate the need for rule 5. Uses of the land which have been suggested as qualifying under this requirement include religious purposes (*Zoar Independent Church Trustees* v *Rochester Corporation* (1974)), working men's clubs (the *Sparks* case), charitable purposes (*Aston Charities Trust* v *Stepney Borough Council* (1952)), hospitals, etc. (Trustees of the *Manchester Homeopathic Clinic* v *Manchester Corporation* (1970)), theatres (*Trustees of the Nonentities Society* v *Kidderminster Borough Council* (1970)), and a private railway (*Ffestiniog Railway Soc* v *Central Electricity Generating Board* (1962)). The list is not finite.

In *Wilkinson* v *Middlesbrough Borough Council* (1979), confirmed on appeal, 1981, the claimants failed to establish that rule 5 should apply because there was held to be a general demand or market for the land for the purpose of a veterinary practice, even though such practices were never marketed in the usual way.

However, the House of Lords decided in favour of the claimant in the case of *Harrison & Heatherington Ltd* v *Cumbria County Council* (1985) which concerned a livestock market. Although such businesses are occasionally sold it was held that this did not happen frequently enough to constitute a market. This decision is difficult to reconcile with the *Wilkinson* case above, and it may be that a less restrictive view of rule 5 will be taken in the future. Clearly, each case must be treated on its own merits and the decision will depend upon the extent to which properties of the type in question are sold on the open market.

(c) There is a bona fide intention to reinstate on another site.

This requirement is intended to prevent a claimant from claiming compensation on a reinstatement basis, but not actually undertaking the reinstatement. It may actually depend upon the receipt of rule 5 compensation, which fact will not necessarily disqualify the claim under this requirement (the *Zoar* case). Again, the intention will be established by factual evidence and the actual purchase of an alternative property is excellent evidence of such an intention. In the *Sparks* case, the commissioning of a feasibility study coupled with acceptance of the financial ability of the club to reinstate combined to persuade the tribunal.

It was held in *Edge Hill Light Railway* v *Secretary of State for War* (1956) that no bona fide intention to reinstate existed, as the claimants had failed to draw up plans for the alternative route of the railway and had failed to demonstrate sufficient directness of purpose during the process of expropriation.

Reinstatement can be bona fide intended even by a claimant with a leasehold interest of less than a year who is claiming under section 20 of the 1965 Compulsory Purchase Act (*Runcorn AFC Ltd* v *Warrington and Runcorn Development Corporation* (1982)).

(d) The tribunal's reasonable discretion should be exercised in the claimant's favour.

Even if the claimant has satisfied the tribunal on the above three points, it remains in the tribunal's jurisdiction to award compensation under rule 2 rather than rule 5, if it does not consider that the cost of reinstatement is reasonable in relation to any financial or even social value of the undertaking. This conclusion remains despite the decision in the *Zoar* case where the tribunal appears to have waived its right to exercise its discretion despite a high reinstatement cost to benefit a current church congregation which numbered 12. However, it is the purpose not the building which is to be reinstated, so the size of the replacement building would reflect the size of the congregation, rather than the size of the church to be replaced.

The tribunal did operate its discretion to the claimant's disadvantage in the *Ffestiniog Railway* case, where it was held that the value of the railway was less than the cost of reinstatement and that compensation should be confined to market value of the land taken. In the *Sparks* case, on the other hand, rule 5 compensation was awarded even though the value of the club's assets did not approach the cost of reinstatement. In this latter case, the claimants' valuer's figure of claim was accepted in full and his valuation, slightly simplified and modified, is laid out below. The basis of compensation under rule 5 is "the reasonable cost of equivalent reinstatement", requiring the valuer to carry out a contractor's test valuation. The basis of such a claim is the cost of rebuilding the existing premises (less an allowance for disrepair and obsolescence) plus the cost of a suitable site and fees and expenses. A deduction for disrepair and obsolescence is necessary to ensure that the claimant is not established in a vastly improved position as a result of the acquisition, while the cost of an alternative site will often only be estimable by reference to actual sites which are suitable, and available for purchase.

The *Sparks* case valuation was complicated by the fact that the new premises that were to be constructed were larger than the old, and that a planning condition committed the claimant to providing a car park for public use on the new site.

Valuation under rule 5: (from *Sparks* v *Leeds City Council*)

Net cost of reinstatement adjusted for items not equivalent and for extras	£143,611
To be reduced for comparative size:	
4,880/8,682 sq ft = 56.2%	£80,709
Add for architect's fees @ 12.5%	£10,088
Total new cost	£90,797
Less dilapidations to existing premises and to bring up to by-law standards	£7,158
	£83,639
Add for fixed seating as in existing premises	£1,750
	£85,389
Add cost of alternative site, adjusted for size	£6,750
	£92,139
Add for obligation to provide new car park	£4,200
	£96,339
Add legal fees for site purchase, valuation fees on new site, removal expenses, survey fees on alternative site, and car park supervision fee	£1,493
Total compensation claimed	£97,832

This (the awarded quantum of compensation) may be compared with the district valuer's valuation for the acquiring authority, assessed under rule 2, of £7,842.

The difference between the valuations for claimant and authority is illustrative of the importance of establishing rule 5 as the relevant basis of compensation in certain cases. It should not, however, be assumed that it will always be advantageous to do so and it must be remembered that, even if the claimant has a putative right to rule 5 compensation, he may still claim under rule 2 if this would increase his compensation claim. Two valuations are therefore always necessary in possible rule 5 cases (see, for example, *Essex Incorporated Congregational Union* v *Colchester Borough Council* (1982)).

7.6.6 Rule 6

> The provisions of rule (2) shall not affect the assessment of compensation for disturbance or any matter not directly based on the value of land.

The first point to make about this rule is that, while it is sometimes described as the statutory basis of the disturbance element of the claim, it is clearly no such thing. The wording assumes that disturbance already exists, and simply gives guidance as to how it will be assessed. The principle of disturbance actually arises from common practice and case law developed from the 1845 Act, rather than from any specific statute.

This rule is necessary in order to avoid a strict interpretation of rule 2 disqualifying expropriated landowners from their hitherto well-established right to disturbance compensation. Other matters "not directly based on the value of land" which do not constitute disturbance compensation include the professional fees of a surveyor or valuer, a solicitor, and perhaps an accountant in conveying the title of the claimant's interest and establishing his claim to compensation. These fees are payable whether or not the claimant is entitled to disturbance compensation (see Chapter 9).

To conclude, market value is the basis of the assessment of compensation for land taken. No extra allowance is to be made on top of this as a consolation, but the right to disturbance compensation and certain other payments remains unaffected. Rules 3 and 4 qualify the market value basis to a certain extent, while equivalent reinstatement compensation under rule 5 remains the sole exception to the market value provision, which is assumed to constitute the basis of compensation throughout this chapter.

7.7 Other rules

In addition to the six rules laid down in section 5 of the Land Compensation Act 1961, other rules may be appended to establish a compensation code by which the valuer should work. Section 7 of the 1961 Act, which deals with the deduction from compensation of an increase in value of contiguous land in the same ownership, is dealt with in Chapter 8. Other rules which are directly relevant to compensation for the subject land itself and which may be regarded as addenda to the six rules in section 5 are sections 6 and 9 of the same Act and a common law rule known as the *Pointe Gourde* rule. These, taken together, require the impact on value of the scheme underlying the acquisition to be left out of account.

Consideration is than given to the planning assumption rules contained in sections 14–17 of the 1961 Act.

7.7.1 Section 6, Land Compensation Act 1961

(As amended by section 145 of the Local Government, Planning and Land Act 1980.)

No account shall be taken of any increase or diminution in the value of the relevant interest which, in the circumstances described in any of the paragraphs in the first column of Part 1 of the First Schedule to this Act, is attributable to the carrying out or the prospect of so much of the development mentioned in relation thereto in the second column of that Part as would not have been likely to be carried out if —

(a) (Where the acquisition is for purposes involving development of any of the land authorised to be acquired) the acquiring authority had not acquired and did not propose to acquire any of that land; and
(b) (Where the circumstances are those described in one or more of paragraphs 2 and 4 in the said first column) the area or areas referred to in that paragraph or those paragraphs had not been defined or designated as therein mentioned.

The effect of section 6 must be ascertained in conjunction with the first schedule, which defines in considerable detail the type of development which is to be left out of account.

The intention of rule 6 is to prevent the acquiring authority from paying a price which is increased, or reduced, by the nature of the proposed development. For example, if buying property for a sewage works it could argue that property values are falling locally as a result of the scheme. Alternatively, if the scheme is urban regeneration, the claimant could argue that values are rising as a result of the

Figure 7.2 Explanation of the rule in s.6 of the Land Compensation Act 1961

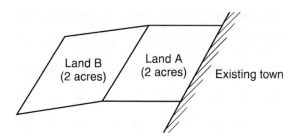

range of measures within the scheme, aimed at improving the local environment. In either case the intention is that compensation should be assessed as if the scheme underlying the acquisition did not exist. The necessity of making assumptions in the absence of the acquiring authority's scheme has produced the often quoted concept of the no-scheme world, in which the valuer should employ the unrestricted faculty of his imagination.

Where development falling under the first item is the purpose of the acquisition, no account shall be taken of any increase or diminution in the value of the relevant interest which, where the acquisition is for purposes involving development of any of the land authorised to be acquired, is attributable to the carrying out, or the prospect, of development of any of the land authorised to be acquired, other than the relevant land, being development for any of the purposes for which any part of the first-mentioned land (including any part of the relevant land) is to be acquired as would not have been likely to be carried out if the acquiring authority had not acquired, and did not propose to acquire, any of the land authorised to be acquired.

This rule obviously suffers from its length. While legal provisions are often criticised for being too vague, section 6 is unusual for having been thoroughly berated for being too detailed. By defining in such great detail the forms of development which are to be left out of account, it runs the risk that one party or another could defeat the intention by arguing that a particular development does not fall within any of the strict definitions.

In order to explain the operation of the rule it is helpful to refer to a diagram (Figure 7.2).

Assume that both lands A and B, in separate ownership, are to be acquired by the local authority for the expansion of the nearby town. Land A, being adjacent to the existing town boundary, is "ripe" for development, being close to existing roads and services, and would sell for £80,000 as a site for residential development. Land B, on the other hand, is virgin land, separated from the existing town's roads and services by land A, and would sell for only £40,000 as a site for residential development.

Assume that the local authority acquires land A and begins to develop it, having compensated the owner at its full market value of £80,000. Roads and services are constructed over and under land A. The value of land B rises to £80,000 as a result, as it has now become ripe for development.

Section 6 would prevent the payment of £80,000 as the compensation for land B, the correct quantum of compensation being £40,000 as long as the development which the acquiring authority commenced on land A would not have occurred but for the authority's development proposals. This is a matter for factual evidence: would a private developer have immediately acquired and developed land A? If so, £80,000 may be payable for land B.

Interpretation of section 6 was tested in the House of Lords in 1965 in *Davy* v *Leeds Corporation*. In this case, 10 houses in a clearance area were being acquired by Leeds Corporation. The houses were unfit for human habitation, so the basis of valuation was that of a site cleared of buildings. An individual cleared site, surrounded by standing buildings would be too small to be developed and would therefore have little value. A somewhat higher value would apply if it was surrounded by other cleared sites, with which it could be merged to create a developable area. Should the valuer appraise the site on the basis that the area surrounding and adjoining it had been, or was to be, cleared of slums, the reality? The conclusion of the House of Lords was that the effect of section 6 was that the valuer should assume that the site continued to be surrounded by slum property, as there would otherwise be an increase in value of the subject property which is to be discounted under section 6 and schedule 1, case 1 (bearing in mind that development, for planning purposes, embraces demolition).

A second 1965 case considered by the Lands Tribunal, *Halliwell* v *Skelmersdale Development Corporation*, included an examination of section 6. In this case, there was a dispute over the compensation payable for 34 acres of land to be acquired by Skelmersdale Development Corporation. Under section 6, no account should be taken of any increase in value of the subject land which was attributable to the development of the new town. The claimants argued that this carried with it an implication that such development as would have been likely apart from the new town scheme should be reflected in the valuation assumptions to be made. Upon this basis, there would eventually have been a demand for the development of the land and planning permission for the acquiring authority's proposals could be assumed under section 15 of the 1961 Act (see section 7.7.4). The claimant's valuer's valuation of the land was thus based on deferring the development value of the land in stages as demand for that land became known totally in ignorance of a new town scheme. It is submitted that this is a correct approach.

Yet the Lands Tribunal failed to accept that it was correct to envisage what would have happened in the absence of the new town scheme and ignored future development value. Later cases, for example *Myers* v *Milton Keynes Development Corporation* (1974), have discredited this decision and it is apparent that the valuer should adopt the approach of the claimant's valuer in the *Halliwell* case. Given that the development to be ignored by the valuer under section 6 is development which would not have been likely to be carried out in the absence of the acquiring authority's proposals, the valuer must envisage what would have happened in the absence of those proposals and value on that basis. Such an approach was suggested by the Lands Tribunal in *Crank* v *Winsford UDC* (1965), *Collins* v *Basildon Development Corporation* (1969), and *Marshall* v *Basildon Development Corporation* (1970).

The 1973 Land Compensation Act introduced a refinement to section 6. Section 51 of this later Act was introduced supposedly as a pre-emptive measure in the light of proposals for a third London airport and an attached new town (Maplin). As the twin proposals would constitute separate schemes of development under section 6 and schedule 1, and because the value of land forming the site of Maplin new town would increase due to the proximity of the airport, Parliament considered that central government should be protected from paying high land prices in this and similar cases. The Secretary of State may give a direction that the construction or extension of a new town may be in connection with a specific public development. Having given such a direction, then the increase or diminution to be left out of account by virtue of section 6 shall include those effects on value which are attributable to the carrying out of the specific public development.

It was in the case of *Waters* v *Welsh Development Agency* (2004), that section 6 received its most severe criticism from the House of Lords, and in the earlier Court of Appeal hearing where Carnwath LJ had commented that "there can be few stronger candidates for urgent reform, or simple repeal, than section 6". The main difficulty with section 6 was that it requires the scheme underlying the acquisition to be left out of account in so far as its purpose is implemented on land other than that which is the subject of the compensation claim. However, it makes no references to disregarding the scheme to the extent that it affects the subject land itself. Can parliament really have intended that, when valuing land for compensation purposes, the part of the underlying scheme taking place on other land should be ignored, but the part taking place on the land itself should be taken into account? In order to avoid the apparently anomalous result of such an interpretation, the courts have continued to apply the *Pointe Gourde* rule (considered below), which requires the whole scheme to be left out of account, but which was originally intended to be replaced by section 6.

7.7.2 Section 9, Land Compensation Act 1961

No account shall be taken of any depreciation of the value of the relevant interest which is attributable to the fact that (whether by way of designation, allocation or other particulars contained in the current development plan, or by any other means) an indication has been given that the relevant land is, or is likely, to be acquired by an authority possessing compulsory purchase powers.

The possibility that a property may be compulsorily acquired is likely to affect its value, even though owners will obtain market value compensation upon expropriation, due to the reluctance of prospective purchasers to invest in a property which may shortly be taken from them. For example, assume that the value of houses in a clearance area prior to the scheme being published is in the order of £80,000. After publication, a house changes hands at a price, reflecting the blight caused by the local authority's scheme, of £60,000. Upon subsequent acquisition by the authority, this open market sale may be used as the best available evidence of the market value of properties in the area and as a result of their scheme the authority purchases properties at 75% of their value.

Section 9 is designed to prevent this eventuality and ensures that £80,000 is the amount of compensation to be paid in such a case.

In *London Borough of Hackney v MacFarlane* (1970), six houses were the subject of a compulsory purchase order. Other houses in the area had been classified as unfit for human habitation: the acquiring authority contended that the subject properties were about to be similarly classified. The Court of Appeal confirmed the Lands Tribunal's decision that section 9 should be employed to ensure that the houses were valued as fit for human habitation.

Trocette Property Co Ltd v Greater London Council (1974) concerned the acquisition of a leasehold interest in a disused cinema of which the acquiring authority owed the freehold reversion. The GLC indicated that a new lease would not be granted to Trocette after expiry of the current lease due to the possibility that the property would be required in the future for road widening. Trocette applied for planning permission for demolition of the cinema and its conversion to shops, which was refused for the same reason, and served a purchase notice on the GLC which was forced to acquire the leasehold interest as if it were a compulsory acquisition (see section 7.9). It was held by the Court of Appeal that depreciation in the value of the subject interest, suffered as a result of the loss of development value by Trocette as leaseholders, should be disregarded under section 9.

Over recent years the courts have been troubled by the question of whether the section 9 principle is subservient to the rule that the land should be valued in its actual physical condition at the valuation date, or whether it overides this rule. A number of cases have concerned the situation where planning permission for development has been subject to conditions requiring the development to be set back to allow for future road widening. Some years later the land is compulsorily acquired for the road widening scheme, by which time it is a virtually valueless strip separating the public highway from the new development. Rule 2 requires the land to be valued in its actual physical condition at the valuation date, suggesting a nominal compensation payment. The claimant will, however, argue that the original planning condition was due to an indication having been given that the land was likely to be acquired by an authority possessing compulsory purchase powers. Section 9 requires this to be left out of account and therefore the land should be valued as if it were part a site, the whole of which is available for development, as it would have been but for the road widening proposal. Supporting this interpretation in *Myers v Milton Keynes Development Corporation* (1974) Lord Denning encouraged the valuer to "conjure up a land of make-believe and let his imagination take flight to the clouds".

The use of section 9 was perhaps stretched too far by the Court of Appeal who accepted the claimant's (and Lord Denning's) argument in *Jelson Ltd v Blaby District Council* (1977). Jelson, a firm of builders, owned land which was allocated for residential purposes on the outskirts of Leicester. Before the Second World War it was proposed to construct a ring road around the city which ran through the Jelson land. Jelson was granted planning permission to build housing estates either side of the route of the proposed road, and did so.

The construction of the M1 motorway nearby caused the abandonment of the ring road scheme whereupon Jelson applied for permission to develop what was now a 146 ft wide strip of land. This was refused and a purchase notice was served and accepted. Without permission for residential development, the land was worth around £6,000; with permission, it was worth around £60,000. Despite the lack of planning permission, and the fact that at the valuation date the land was an inaccessible strip in the middle of a housing estate, £60,000 was awarded, due to the use of section 9.

Abbey Homesteads Group v Secretary of State for Transport (1982) concerned similar circumstances, though somewhat less extreme as at the valuation date the adjacent land had not been developed, so in the no scheme world there was still the prospect of it being incorporated into a housing estate. In this case, however, the Court of Appeal confirmed that the property had to be valued as a strip of land, too narrow to be developed in isolation (see p 278). This decision directly contradicts Jelson and is of higher authority.

These circumstances were considered again by the Court of Appeal in *English Property Corporation plc v Royal Borough of Kingston Upon Thames* (1998), which concerned an area of land which, by the valuation date was a paved forecourt situated between the public highway and an office building which had been built some years earlier, and set back to allow for future road widening. Only part of the forecourt was being acquired, and an earlier lands tribunal had awarded £85,700 as compensation for the land taken, and a further £85,700 for depreciation in the land retained. It was common ground that the land taken was, "at the date of valuation, in its then condition, valueless". However, it was agreed between all the parties that the land should not be valued as it actually was, but as it would have been if the set back had not been required. In other words it was necessary to assume that the development, which had actually taken place 17 years previously, had not yet happened, and value the subject land at its value for inclusion in the development scheme. At full development value the land would be worth £171,400. However, the land did have to be valued as if being offered for sale by a hypothetical willing seller, and the subject site was too small to be developed in isolation. 50% of full development value, or £85,700 was therefore awarded as being the price a developer would be likely to pay to acquire the subject land for incorporation into a larger, developable unit. The tribunal then awarded the other 50% of the marriage value, a further £85,700 as compensation for depreciation in the value of the retained land, making total compensation of £171,400. The Court of Appeal rejected the second half of the payment on the grounds that section 9 applies only to the land taken, not to retained land. Therefore, even if it is necessary to imagine the subject land to be the condition it would have been in if the development had not yet taken place, no such assumption can be made in respect of land retained, which remains part of a useless forecourt. The £85,700 awarded for land taken was not subject to appeal, and therefore, the Court of Appeal could only confirm that the Lands Tribunal was entitled to award that sum. Morritt LJ did, however comment in relation to section 9 that "there is nothing to displace the normal rule that the land retained, like the land acquired, is to be valued as at the date of entry in its actual physical condition as at the date of notice to treat". Applying this rule to the subject land would clearly produce a nil or nominal compensation award, which suggests that the £85,700 payment in respect of land taken may also have been overturned if it had been considered by the Court of Appeal.

In *Fletcher Estates* v *Secretary of State* (2000) the House of Lords decided that any section 17 certificate (certificate of appropriate alternative development) should be issued on the basis that the scheme has been cancelled shortly before the certificate is issued, rather than trying to imagine what might have happened to the land and to planning policy if no scheme had ever existed.

The *English Properties* case confirms that a planning condition requiring a development to be set back for future road widening is an indication having been given that the land was likely to be acquired by an authority possessing compulsory purchase powers within the meaning of section 9, but that the land should then be valued in its actual condition at the valuation date, on the assumption that there is no threat of compulsory purchase.

Section 9 is often useful in tandem with blight notice procedure (see section 7.10) by which the owner of property which is to be acquired at some time in the future may engineer its purchase in advance of its requirement by the acquiring authority at its full market value by relying upon section 9, despite the fact that the property has become unsaleable at that price in the real world as a result of the blight.

For the sale of clarity it should be remembered that section 6 concerns itself with the development proposed under the scheme, while section 9 relates to the threat of acquisition itself.

7.7.3 *The* Pointe Gourde *rule*

The courts have often been faced with technical arguments concerning the wording of section 9, and more particularly, section 6. Whenever they have found that these arguments have the potential to create an unfair compensation award they have tended to revert to the *Pointe Gourde* rule, which predates section 6 and section 9, and was intended to have been replaced by them.

This so-called common law rule became fashionable in the 1950s and 1960s as a major part of the compensation code and remains in that role today. This is so in spite of its dubious origin and seductive simplicity.

In *Pointe Gourde Quarrying and Transport Company Ltd* v *Sub-Intendent of Crown Lands* (1947), the United States required, for the establishment of a naval base in Trinidad, certain lands owned by the claimant at Pointe Gourde on that island. The Crown acquired the land by the exercise of compulsory powers.

On part of the subject land there was a large quantity of good limestone. The land thus worked had access to smooth water which made for the easy transportation of the products of excavation which would be suitable for use in the construction of the naval base. $15,000 of the original award of the arbitrators was in respect of this particular advantage of the subject property.

An appeal from this award concerned the possible applicability of the special suitability rule (section 5(3) of the 1961 Land Compensation Act as it stood prior to amendment by the 1991 Planning and Compensation Act) which was held in the court of first instance not to apply as there was a market for the land for the particular purpose of the acquisition apart from the acquiring authority. The purpose was thought to be the quarrying of limestone.

The full court of Trinidad and Tobago reversed this decision, holding that the rule did apply because the purpose was more specific than the previous court had thought. The purpose was the building of a naval base, for which there was only one prospective purchaser. The $15,000 was thus disallowed.

The case was then considered by the Privy Council. Here it was held that rule 3 referred to land and not its products and did not, therefore, apply in this case. Even so, the Privy Council was not of the opinion that the $15,000 should necessarily be awarded and chose to disallow it by the resurrection of what was a thread of a principle used in certain pre-1918 cases, which had almost certainly been consolidated into rule 3, originally part of section 2 of the 1919 Act. Lord MacDermott said: "It is well settled that compensation for the compulsory acquisition of land cannot include an increase in value which is entirely due to the scheme underlying the acquisition."

It is this sentence which is currently often quoted with apparent authority as the *Pointe Gourde* rule. In this original form the rule states only that an increase in value which is entirely due to the scheme underlying the acquisition should be left out of account. It says nothing about the situation where a scheme causes values to decrease, nor does it refer to the effect on value (invariably downwards) of the threat of compulsory purchase. Subsequent cases have, however, widened the rule so that it is now applied so as to leave out of account any increase or decrease in value, caused either by the nature of the development proposed, or the threat of compulsory purchase: an all-encompassing rule which effectively renders sections 6 and 9 redundant.

(a) The conceptual problem

There has been considerable debate concerning the proposition, made by Viscount Dilhorne in *Davy* v *Leeds Corporation* (1965), that section 6 of the 1961 Act had "given statutory expression to the principle which Lord MacDermott stated was well settled". In other words, if rule 3 had failed to embrace the *Pointe Gourde* rule, it was intended that section 6 should do so.

This proposition was cast aside by Lord Denning in *Viscount Camrose* v *Basingstoke Development Corporation* (1966). It was argued for the claimants that section 6 operated to disregard increases in value of the relevant land caused by the prospect of development of land included in the scheme other than the relevant land and that, by implication, increases in value of the relevant land caused by the prospect of development of that land by the authority may be taken into account. Lord Denning plugged what he saw as a loophole by suggesting that the *Pointe Gourde* rule, operating in tandem with section 6, applied to increases in value of the relevant land caused by demand for that land created by the scheme. The result of this dictum was that the *Pointe Gourde* rule became ensconced in the law.

A classic example of the use of the rule was the Lands Tribunal decision in *Gateley* v *Central Lancs New Town Development Corporation* (1984). In this case the Lands Tribunal held that damage to the subject property was a direct result of the acquisition programme and so should be left out of account in assessing compensation in accordance with the *Pointe Gourde* rule. Section 9 does not appear to have been considered even though it exists for exactly these circumstances.

(b) What is a scheme?

The current interpretation of the *Pointe Gourde* rule is relatively uncontroversial. Any increase or decrease in value due to the scheme should be disregarded. The much more difficult question is what is the scheme? A "scheme", said Lord Denning in *Wilson* v *Liverpool Corporation* (1971), "is a progressive thing. It starts vague and known to few". The valuer cannot be satisfied with this. If *Pointe Gourde* is to be used, he must estimate the precise value effects of what he considers to be the scheme for the purposes of the rule.

Take, for example an industrial unit being bought under compulsory powers for a road junction improvement. The acquiring authority is a development corporation set up to regenerate a large industrial estate. The claimant argues that the scheme to be left out of account is the junction improvement and he should be compensated at open market value, leaving out of account only this junction scheme. However the authority could argue, and has done so, that the junction improvement is just part of a wider scheme, the whole of which should be left out of account. The junction improvement is part of a wider road improvement programme for the area. This, in turn is part of a range of infrastructure improvements including rail and light rapid transit systems. The infrastructure improvements are part of a wider programme of enhancements including environmental works, grants to encourage inward investment etc. In fact the scheme to be left out of account is not a junction improvement, it is the very existence of the development corporation and all its activities. It is very likely that all these activities will cause substantial rises in property values. Should, therefore, compensation reflect the enhanced value in the absence of the junction work, or the value which the unit would have had, if a development corporation had never been created. There is no fair answer to this dilemma. If the former, the authority will be forced to pay additional compensation which is entirely due to it performing its role well. If the latter, the claimant will receive insufficient compensation to enable him to purchase a nearby property on the estate which is identical to the one taken from him under compulsion. Similar problems will arise, but in reverse, where the scheme has a damaging effect on values — say a house bought for a junction improvement scheme as part of improved access arrangements to a new airport. In this case the acquiring authority will argue for a narrow definition of a scheme, whereas the claimant will argue that the whole concept of building an airport in the area should be left out of account.

In the Law Commission Consultation Paper (no 165), Towards a Compulsory Purchase Code the Commission described the question of identifying the scheme as "the most difficult issue we have had to address".

Identifying what comprises the scheme underlying the acquisition has taxed the courts on many occasions in recent years. Decisions have relied heavily on the facts of each case and general guidelines are difficult to identify. In *Waters* v *Welwyn Hatfield District Council* (1991) the Lands Tribunal was concerned with the compensation payable on the compulsory purchase of a petrol filling station for a by-pass which was to be built in a tunnel connecting the two sections of the A1(M) motorway which stopped immediately to the north and to the south of Hatfield. There was a large petrol throughput, and the claimant argued that, for the purpose of assessing compensation, it should be assumed that petrol sales would continue at a similar level indefinitely. While it was obvious that the two sections of motorway would eventually be linked, and that any such link would take traffic away from the filling station and reduce petrol throughput, the proposal to construct a by-pass was the scheme underlying the acquisition and should be ignored.

The Tribunal, however, supported the acquiring authority and held that:

> In the present case the initial overall aim to link the two motorways did not and could not have involved the compulsory purchase of the subject premises until the preferred route had been firmly established. Therefore it seems to us that the scheme underlying the compulsory purchase of the subject premises is the detailed scheme involving the tunnel route. That is the scheme which gave rise to the compulsory acquisition and that is the scheme that has to be disregarded for the purposes of assessing compensation.

The Tribunal reduced its valuation of the subject premises by 10% to reflect the risk that an alternative by-pass scheme might have been adopted in the no scheme world. The decision would have been different if this case had been decided following the passing of the 1991 Act. Under section 64 it is now necessary to disregard not only the road scheme underlying the acquisition but any other scheme designed to meet substantially the same purpose (see p 246). The principle established by this case will, however, still apply to acquisitions for other than highways purposes.

In the House of Lords decision of *Waters v Welsh Development Agency* (2004) the application of the *Point Gourde* rule was expanded and clarified. The case concerned the compulsory purchase of an area of natural wetland, to replace a similar area required for the Cardiff Bay Barrage. The claimant argued for a ransom value approach to the valuation of the land, on the basis that the barrage could not be built unless this replacement land was acquired. The authority's view was that any such value was entirely due to the underlying scheme and should be left out of account. The main difficulty facing the court was establishing the extent of the scheme which was to be left out of account. In the words of Lord Nicholls:

> This goes to the very fairness of the *Pointe Gourde* principle as currently applied. The wider the scheme, the greater the potential for inequality between those outside the area of acquisition, whose land values rise by virtue of the scheme, and landowners whose properties are acquired at a value which disregards the scheme. Conversely, the narrower the scheme, the greater the potential for an authority being called upon to pay compensation inflated by its own investment in improved infrastructure or other regeneration activities. Holding the balance between these conflicting interests is preeminently a subject for decision by Parliament. But, as matters stand, there are indications that in some cases the application of the *Pointe Gourde* principle has become too wide-ranging.

Lord Nichols set out six principles to be considered in applying the *Pointe Gourde* principle:

(1) the *Pointe Gourde* principle should not be pressed too far. It should be applied in a manner which achieves a fair and reasonable result.
(2) a result is not fair and reasonable where it requires the valuation exercise which is unreal or virtually impossible
(3) the valuation result should be viewed with caution when it would lead to a gross disparity between the amount of compensation payable and the market values of comparable adjoining properties which are not being acquired
(4) the *Pointe Gourde* principle should be applied by analogy with the provisions of the statutory code set out in section 6 of the 1961 Act. In the type of case covered by class 1 of section 6 the area of the scheme should be interpreted narrowly. In other cases Parliament has spread the disregard net more widely
(5) normally the scope of the intended works and their purpose will appear from the formal resolutions or documents of the acquiring authority. But this formulation should not be regarded as conclusive
(6) when in doubt a scheme should be identified in the narrower rather than the broader terms.

Lord Nicholls also made it clear that the *Pointe Gourde* rule should override sections 6 and 9. While the courts are generally reluctant to allow a common law principle to override a specific statutory provision, "this is less repugnant as an interpretation of the Act than the alternative".

The decision also appears to support the view that the scheme should be ignored by assuming that it has been abandoned immediately prior to the valuation date. This appears much more sensible than the approach adopted by Lord Denning in *Myers*, which requires the valuer to imagine what may have happened to the land over a period which may be many years, had no scheme ever been envisaged.

On the facts of the case the Law Lords decided the scheme to be left out of account was not just the nature reserve, as contended by the claimant, but the whole Cardiff Bay Barrage. Ransom value was not, therefore an appropriate basis of valuation.

The impact of *Waters* was considered in some detail by the lands tribunal in *RMC and Case* v *Greenwich* (2005). The case concerned two plots of land which were the subject of compulsory purchase for the Millennium Dome complex. The claimants argued that the scheme to be disregarded was simply the Millennium Dome, and the plot of land should be valued taken into account the benefit of the comprehensive infrastructure and development works taking place in the surrounding area. The acquiring authority argued that the Dome was only part of a wider scheme which had to be left out of account and which comprised the comprehensive development of the Greenwich peninsula.

In the opinion of the Tribunal:

> *Waters* gives rise to a fundamental change in the way that the *Pointe Gourde* principle is to be applied. The approach is no longer first to identify the scheme and then to pursue, possibly in relentless detail, a hypothetical reconstruction of what would have happened in the no scheme world. The role of the principle, as we understand the decision of the House of Lords, is to supplement the provisions of the Land Compensation 1961 to the extent that is necessary to ensure that the claimant receives no more or less than fair compensation. Whether the application of the principle is needed for this purpose, how the scheme is to be identified in order to achieve it and the extent to which features of the no scheme world need to be constructed will depend on the facts of the particular case.

It is noted that both parties defined the scheme as something significantly larger than the area of the CPO itself. While the CPO extended to some 19 acres, the Millennium Dome (the claimant's scheme) took up some 181 acres while the local authority's scheme extended to 294 acres. The Tribunal noted the House of Lords view that the *Pointe Gourde* principle had come to be interpreted too widely, and had regard to Lord Nicholls's six rules to be considered when applying the principle. It took the view that the starting point should be a narrow view that the scheme is only what takes place on the CPO land itself.

The Tribunal noted that based on their own definition of a scheme the acquiring authority and claimant had each "constructed the requisite features of the no scheme world, postulating, for example, what would have happened in terms of planning and the grant of permissions, the rate at which remediation would have occurred, including what grant aid would have been available and when, where on the peninsula development would have occurred, what it would have been and when it would have been carried out." In the view of the Tribunal this reflected the conventional application of the *Pointe Gourde* rule prior to *Waters* v *WDA*. Following *Waters*, however:

> The question to be asked for the purpose of applying the *Pointe Gourde* principle is not "what is the scheme underlying the acquisition?" But "how should the scheme underlying the acquisition be defined as an adjunct to the statutory disregards in order to give the claimant compensation that is neither more or less than what is fair?"

It concluded that the *Pointe Gourde* rule should be applied simply so as to leave out of account any increase in value of the reference lands resulting from the changes to the locality caused by the Dome development. It formed this conclusion having regard to the six principles set out by Lord Nicholls in the *Waters* case. The Tribunal decided that it should seek to confine the scheme to the area of the CPO unless an extended definition of the scheme is necessary in order to establish fair compensation. It was in no doubt that it is not appropriate to treat the scheme as extending to the whole of the Greenwich peninsula, as contended by the acquiring authority. This would breach the second of Lord Nicholls's principles, as it would be impractical to construct such a no scheme world. To treat the scheme as simply the CPO land itself would, however, allow the claimant to benefit from the urgent infrastructure works undertaken in respect of the nationally important Dome development and benefiting from the availability of lottery funds. This would result in a compensation payment in excess of that which is fair, and therefore the scheme to be left out of account should be extended beyond the CPO land to include the Dome but not so far as to include the complete redevelopment of the Greenwich peninsula.

It seems likely that when applying the *Pointe Gourde* rule in the future the scheme will be defined in whichever way is necessary to enable the courts to arrive at a compensation figure they consider to be fair.

(c) Arbitrary adjustments to valuations

If the valuer is successful in identifying the scheme underlying the acquisition, he must then assess market value under rule 2 and make an adjustment to the figure which reflects the effect of that scheme. This will pose great problems, particularly if the scheme itself is vague, and because general inflationary or deflationary trends in market values will be difficult, if not impossible, to distinguish from the effects of the scheme. Drawing guidance from reported cases will be of little use, as the adjustments made often appear to be arbitrary as *Pointe Gourde* rule gives carte blanche to the Tribunal to alter the compensation award (see *MacDonald* v *Midlothian County Council* (1975)).

7.7.4 The Planning Assumptions

(a) Existing use value and development value

The basis of valuation for land compulsorily taken is market value, with certain qualifications. The market value of land will largely be determined by the use to which it may be put. In a perfect market, and in the absence of planning controls, land will move to its most profitable use. But an imperfect market, with planning controls, is our inheritance. Land is not always put to its most profitable use.

For this reason, land put to a certain use may have a development value in excess of its existing use value. Whether this development value can be exploited depends upon planning permission (see (b) below), but the valuer must always enquire as to the possibility that development value, and hence two alternative bases of compensation, exists. It is even possible that two or more valuations will be required, as a combination of existing use value and a range of developments may produce the highest valuation, which is the market value and the measure of compensation.

Example 7.1

Take a smallholding of 6 ha including a detached house. Planning permission exists for residential development, but it is thought that the land will not become ripe for development, that is demand will not be forthcoming at the full residential development value, for a period of 10 years. Residential building land is worth £1,850,000 per ha; agricultural land is worth £25,000 per hectare and lets (with a farmhouse) at £500 per ha pa and (without a farmhouse) at £250 per ha pa. The house in 0.4 ha would be worth £700,000.

Valuation A (existing use value)

House and 0.4 ha	£700,000	
5.4 ha. @ £25,000 per ha	£135,000	
Total		£835,000

Valuation B (development value)

6 ha @ £1,850,000 per ha	£11,101,000		
PV 10 years @ 8%	0.4632		
		£5,141,520	
Plus rent @ £500 per ha pa	£3,000 pa		
YP 10 years @ 6%3	7.3601		
		£22,080	
		£5,163,000	

Valuation C (existing use value/development value)

House and 0.4 ha		£700,000	
5.4 ha. @ £1,850,000 per ha	£9,990,000		
PV 10 years @ 8%	0.4632		
		£4,627,390	
Plus rent @ £250 per ha	£1,350 pa		
YP 10 years @ 6%	7.3601		
		£9,936	
		£5,337,326	

In the open market the property would sell for its highest value and therefore, on the facts of this case, valuation (c) will prevail and a claim for say, £5,337,000 should be made.

(b) Planning permission and demand

In the *Camrose* case, Lord Denning said: "Even though the 233 acres had planning permission, this by itself does not create value. It is planning permission plus demand".

This crystallisation of the preliminary valuation process is of great aid to valuers when assessing compulsory purchase compensation. Lord Denning must have been distinguishing development value, as the existing use value of property does not depend upon the grant of planning permission (barring illegality of use).

In order to estimate existing use value, the valuer must consider the market value of property in its current use. His valuation is an estimate of the likely selling price. Price is the product of the interrelationship of supply and demand; property, being heterogeneous, has an arguably fixed supply, so great emphasis must be placed upon the likely demand for the property.

Establishing development value is less straightforward. Again, the demand for property in its proposed use must be considered but so must planning controls. The valuer must ascertain whether planning permission for the proposed development is necessary and, if so, whether it exists or may be assumed.

The analysis and measurement of the demand for a particular property at a given time is a task for which the valuer is uniquely qualified. His knowledge of the market provides the essential expertise in establishing what is essentially a matter of fact. The ascertainment of the existence of planning permission, on the other hand, is also a question of law. Owners of property which is about to be compulsorily acquired may be suspicious regarding the objectivity of the local planning authority considering the grant of planning permission for property which the same authority is in the process of buying. The planning assumptions (sections 14–17 of the 1961 Act) attempt to avoid such doubts, though the extent to which they succeed is open to question. In the absence of these provisions the valuer would be left with the task of assessing market value, which will include any realistic development value to the extent that it would be reflected in an arms length sale in the no scheme world. By specifically setting out a list of required assumptions, this rule at best forces the value to do what he should have done anyway: at worst, it requires him to make assumptions which would be unrealistic in the no scheme world and therefore force an unfair compensation award.

(i) Section 14 states that the assumptions to be made under sections 15 and 16 are additional to existing planning permissions which may be in effect. The valuer should obviously take existing planning permission into account, and presumably would do so even without the assistance of this statutory obligation.

Section 14(3) ensures that the lack of an actual or assumed planning permission does not carry with it the implication that such permission will never be obtained. This provision allows the valuer to take hope value, into account; that is the extra value attaching to land in the expectation that planning permission might one day be obtained in respect of it.

In *Corrin v Northampton Borough Council* (1980), compensation including hope value was awarded where land without an actual or assumed planning permission was found to have an increased market value in the expectation of future residential development being allowed.

Where two planning permissions had been attained in respect of the same land and the subsequent permission was inconsistent with the earlier one, it was held that the earlier permission must be treated as being abandoned and a valuation for compensation had to be carried out disregarded the earlier permission (*Thomas Langley Group Ltd v Borough of Royal Leamington Spa* (1973)).

A new subsection (5) introduced by section 64 of the 1991 Act relates solely to road schemes and requires the additional assumption that "no highway is to be constructed to meet the same or substantially the same need as the highway" which gives rise to the acquisition. Where a new road generates development value by creating access to land, the planning assumptions are intended to ensure that compensation is assessed based on the value in the "no scheme world", rather than on development value which would not have been realised if the road underlying the acquisition had not been constructed. An ingenious claimant may attempt to argue that if the actual road underlying the acquisition is ignored then there remains a need for a new road, perhaps on a slightly different alignment, and therefore it could be anticipated that access would be provided to his development site in the near future. The new subsection (5) makes it clear that such an argument will not succeed by requiring that compensation is assessed on the assumption that neither the road underlying the acquisition, nor any road designed to resolve the same traffic problem, will be constructed.

(ii) Section 15 deals with assumptions not directly derived from development plans and allows the assumption of two major heads of planning permission. First, planning permission may be assumed to permit development of the subject land in accordance with the proposals of the acquiring authority (section 15(1)). If those proposals are the construction of a road or the provision of open space, this assumption may be virtually worthless, but if the authority's proposed use is commercially valuable, then the impact on value could be substantial.

Second, under section 15(3), as amended by schedule 6 of the 1991 Act, planning permission may be assumed for development of any class specified in part I of schedule 3 of the 1990 Town and Country Planning Act. To explain this assumption requires an explanation of schedule 3; an historic provision which was originally schedule 3 of the 1947 Town and Country Planning Act. Schedule 3 was one of a number of concessions granted to landowners when the 1947 Act introduced the modern system of planning controls. This schedule defines what may be called marginal development including the rebuilding of existing property. It has no practical benefit: unlike the general development order it does not allow development falling within schedule 3 to actually take place. It merely requires an assumption that planning permission would be granted for this marginal development for compensation purposes only. The extent, and application, of schedule 3 has been greatly reduced by a number of repeals during the last 60 years, and the only real surprise is that it has not been removed in its entirety. All that remains of schedule 3 is part I which, simplified and summarised, defines marginal development as:

(a)   the rebuilding of any building which stood on the site in 1947 and his since been destroyed or demolished and

(c)   the use as two or more separate houses of a building which, at the valuation date, was a single house.

The second limb rarely applies. However, the first can cause some bizarre compensation awards. If for example you own a plot of land in the green belt, it is very likely that planning policies will mean that it will be virtually impossible to obtain planning permission to build a house, or anything else, on the plot. Your chances will not be significantly improved if you discover that there used to be a house on the site in 1947, which was demolished 25 or so years later. Your only hope of ever selling the land at a price which reflects development value is if you are fortunate enough to have the land compulsory purchased. In that event section 15 and schedule 3 require the assumption that planning permission would be granted to rebuild the house, and compensation will reflect this extremely valuable assumption! This assumption may be particularly valuable in purchase notice cases, where a landowner in this situation would be well advised to engineer a compulsory purchase situation (see section 7.9).

Similar planning assumptions are included in other enactments such as Part I of the Land Compensation Act 1973, which refers to compensation for injurious affection where no land is taken, and section 144 of the 1990 Act which refers to the definition of existing use value where a purchase notice is served.

Because the application of schedule 3 can generate such obviously unfair results, the courts have used a variety of measures to avoid paying compensation reflecting development value where none exists in the real world. It does not, for example, follow that just because planning permission has to be assumed, it necessarily increases value: it will only do so where there is a market demand for the development so permitted (see *Camrose* v *Basingstoke* below). In *Old England Properties* v *Telford and Wrekin Council* (1999) the Lands Tribunal decided that the planning permission to be assumed did not have to be for a house identical to the original, but did have to be for a reasonably similar house on broadly the same footprint. On the facts of that case planning permission had to be assumed for a house so awkwardly shaped and located, on the edge of the site, that no-one would want to build it and therefore the assumption added nothing to value. Reviewing previous decisions the Tribunal was able to define the following rules to be followed in such cases.

(1) The rebuilding envisaged under paragraph 1(a) of schedule 3 to the 1990 Act must constitute the rebuilding of the original building and not the erection of a new building; this is a question of fact and degree in each case.

(2) Rebuilding must take place on the site of the foundations of the original building, subject only to any minor deviations necessary to allow the permitted increase in size.

(3) There is no compulsion on rebuilding to create a slavish copy of dimensions, appearance and materials of the original building.

(4) A notional planning permission under section 15(3) of the 1961 Act may be incapable of implementation and therefore valueless.

In *Dutton & Black v Blaby District Council* (2005) the Tribunal found that even applying the above criteria it was unable to avoid the conclusion that schedule 3 required an assumption that planning consent would be granted to build a single house on land where there was no realistic development potential in the real world. Even an ingenious argument by the acquiring authority that any assumed planning consent had been abandoned had to fail. While accepting that "it was illogical and unreasonable for the acquiring authority do be required to pay compensation in 2005 in relation to a use which had been deliberately abandoned by the then landowner 43 years earlier, with no intention to resume that use" the tribunal confirmed that this is exactly what schedule 3 does require. However, just because there is planning permission, it does not necessarily follow that this adds significantly to value. The tribunal made deductions from the unrestricted value of the building plot (considered to be over £50,000) to reflect a number of major problems, including restricted access and services, and a lack of garden or parking space. As a result it was able to award compensation of just £500.

(iii) Section 16 deals with special assumptions in respect of certain land comprised in development plans, and falls into three major parts.

(i) If the relevant land consists, or forms part, of a site defined in the current development plan as the site of proposed development, it shall be assumed that planning permission would be granted for that development. For example, if the whole of a particular land holding is earmarked as the site of a sports hall and indicated as such on the structure plan and local plan (with any amendments), then planning permission may be assumed without question for that development upon acquisition of the site.

(ii) If the relevant land is not defined in the plan as the site of proposed development but is shown on the plan as an area allocated primarily for a specific use, then planning permission may be assumed for development for that use; but only if it is development for which planning permission might reasonably have been expected to be granted in respect of the relevant land.

In *Provincial Properties Ltd v Caterham and Warlingham UDC* (1972), the subject land formed part of an area allocated, or zoned, for residential use. However, because the land occupied a prominent position, its development would spoil the scenic advantages of the locality. Planning permission could not be reasonably expected and the planning assumption could not be made.

(iii) If the relevant land consists of, or forms part of, an area subject to comprehensive redevelopment, it shall be assumed that permission would be granted for any development falling within the planned range of uses, ignoring the fact that the area has been defined as an area of comprehensive redevelopment and subject to any conditions that might reasonably have been imposed and, again, the proviso that planning permission might reasonably have been expected to be granted for the proposed development.

Generally, section 16 is of more use to the claimant and his valuer, particularly where the acquisition is for some public work or use which has an ill-defined market value. Often sections 15 and 16 will coincide: for example, where the acquiring authority purchase development land shown on the development plan to form part of an area allocated for residential development for council housing purposes.

Section 16(7) ensures that the effect of the acquiring authority's proposals does not cloud the objective assessment of what planning consent might reasonably have been obtained if no part of the land were to be acquired (see *Richardsons Developments* v *Stoke-on-Trent City Council* (1968)).

A major difficulty with the application of section 16 in general is that planning system has changed considerably since 1961, and it is often difficult to relate the the current system of local plans to the documents referred to in this section (see *Essex County Showground Group Ltd* v *Essex County Council* (2006) below).

(iv) Section 17 as amended by section 65 of the 1991 Act deals with certificates of appropriate alternative development. These are effectively hypothetical planning applications under which the local planning authority can be requested to state what, if any, planning use would have been allowed on land if it had not been required for the scheme underlying the acquisition. This system is necessary to resolve disputes as to whether land has development value, as any actual planning application would be rejected on the grounds that the land is required for the scheme underlying the acquisition.

As a result of the 1991 Act amendment, an application may be made even where land is defined in a development plan as an area of comprehensive development or is zoned residential, commercial or industrial. This is useful where land is in an area zoned for a particular use but there is a dispute as to whether it would have been suitable for development in isolation.

The certificate issued by the local planning authority will state that if the subject land were not to be acquired by an authority possessing compulsory purchase powers, then planning permission would have been granted for any development for which the land is to be acquired and either:

(a) that planning permission would also have been granted for development of one or more classes specified in the certificate (whether or not these classes were set out in the application), but would not have been granted for any other development or

(b) that planning permission would not have been granted for any other development.

Unlike a planning application, therefore, the planning authority must consider all possible classes of development, not just those specified in the application.

If the applicant is aggrieved by the contents of a section 17 certificate he has a similar right of appeal to the Secretary of State to that enjoyed by an applicant for planning permission.

The expenses reasonably incurred in connection with obtaining a section 17 certificate can form part of the compensation claim, as can expenses incurred in appealing, but only where the appeal is wholly or partly successful (section 17(9A) inserted by section 65(3) of the 1991 Act).

In *Ryefields* v *Staffordshire County Council* (1972), land to be acquired and allocated as a school site had been given the benefit of a section 17 certificate specifying residential development and compensation was awarded on that basis. A similar certificate had been awarded when land was acquired for a road in *Portsmouth Roman Catholic Diocesan Trustees (Registered)* v *Hampshire County Council* (1979).

In *Sutton* v *Secretary of State for the Environment* (1984) the Secretary of State was held to have been in error on three counts when confirming a section 17 certificate:

(i)   the possibility of development on a site comprising the appellant's land and other land not owned should not have been excluded from consideration

(ii) the fact that the possible development was of an exceptional nature and not development of a particular class was not good reason for leaving it out of consideration and

(iii) the likelihood, or otherwise, of the development actually taking place was not a matter to be considered.

In applying the planning assumptions as a whole, the valuer should follow the procedure outlined below. He should establish:

(a)   the existence of any current planning permissions

(b)   the acquiring authority's proposed use of the land

(c)   the extent and effects of schedule 3 rights under the 1990 Town and Country Planning Act

(d)   whether the site is defined as the site of a particular development on the development plan

(e)   whether the site forms part of an area allocated primarily for a particular use on the development plan and, if so, whether planning permission would reasonably be expected to be granted for such development

(f)   whether the site forms part of an area subject to comprehensive redevelopment, what the planned range of uses under the plan is, and what planning permission would reasonably be expected to be granted for such development

(g)   whether an appropriate alternative use for the land exists and whether a valuable section 17 certificate has been, or might be, issued.

The completion of these enquiries will enable a full battery of alternative valuation bases to be evaluated, including existing use value, hope value (if it exists), and development value (if any of the above enquiries are fruitful).

The existence of planning permission is, as Lord Denning confirmed, valueless if there is no demand to put the land to its permitted alternative use. The valuer must establish whether demand exists: this is largely a matter of sound comparable market evidence coupled with an element of projection.

Demand may be easily identified. A vacant plot in the middle of a residential urban area will sell under most circumstances and good comparable evidence will be the sole weapon necessary for the valuer in such a case. But virgin land outside a city boundary with planning permission for residential development may not be an attractive proposition for developers, particularly in the absence of infrastructure on, or adjacent to, the site.

In *Viscount Camrose* v *Basingstoke Development Corporation* (1966), the claimant owned land including 94 ha which had the benefit of an assumed planning permission for residential development in accordance with the acquiring authority's scheme under section 15 of the 1961 Act. Ignoring the effects of development under the Basingstoke town expansion scheme (section 6 and the *Pointe Gourde* rule), the Court of Appeal was not convinced that sufficient demand would exist for the purchase of such a large expanse of virgin land given natural expansion of the town and the compensation award reflected agricultural land value plus an allowance for the future prospects of demand materialising on at least part of the land.

A more correct approach, which has found judicial favour, would be to defer the full development value of the land in chunks as the valuer estimates the time delay before each chunk becomes ripe for development.

Two particular problems concerning the application of the planning assumptions were considered by the Lands Tribunal in *Essex County Showground Group Ltd* v *Essex County Council* (2006). The case concerned the acquisition of a small part of a site which, by the date of the hearing but not at the valuation date, had planning permission for the first new racecourse to be built in England for 80 years. The question was whether at the valuation date planning permission for a racecourse could be assumed, given that the land was designated in the local plan to be in an area where there was a strong presumption against development. The Tribunal first had difficulty in interpreting section 16 which is very specific in setting out the planning assumptions which can be drawn from structure and local plans. The problem is that the whole planning policy system has changed since since 1961 and the plans referred to the Act no longer exist. On the facts of the case the Tribunal decided that the current local plan for the area was reasonably equivalent to the structure and local plans referred to in the 1961 Act, but the potential for confusion remains unless section 16 is updated. The Tribunal decided that there was nothing in the local plan which assisted the claimant in his argument that racecourse consent should be assumed. However, section 14 requires that no realistic development potential should be left out of account solely in the grounds that it does not fall within the planning assumptions. The second question facing the Tribunal was the extent to which hindsight can be taken into account in applying this principle, as in this case planning permission was actually granted for the racecourse two years after the valuation date. At the time of the consent the local planning authority was positively enthusiastic about the racecourse proposal, in spite of the fact that it required them to make an exception to the local plan. The Planning and Compensation Act 2004 attempts to clarify just this situation by inserting a new section 5A into the 1961 Act. This states among other things that: "No adjustment is to be made to the valuation in respect of anything which happens after the relevant valuation date". In the Tribunal's view, however, this subsection does anything but provide clarification:

> We do not find subsection (2) an easy provision to understand. The prohibition is against an "adjustment" to the valuation and this certainly suggests that it is not intended that events happening after the valuation date should be made inadmissible in all respects for the purpose of a rule (2) valuation. On the other hand we find hard to understand the concept, but forward by counsel, of post valuation-date events being taken into account solely to the purpose of checking some previous conclusion. It is difficult to see the utility of such a check if there is a prohibition on adjusting the valuation in the light of it. What has to be determined in the present case is whether a reasonable planning authority could have been expected to grant planning permission for a racecourse of the valuation date. In determining this question it would seem to us to be unfair and unreal to exclude from consideration the actual grant of planning permission for this purpose some months after the valuation date.

Having made this statement, it may be somewhat surprising that the tribunal then concluded that planning permission for a racecourse could not reasonably have been expected at the valuation date, because the strong policy objections would not have been shown to be outweighed by the other considerations.

Having established that planning permission exists or can be assumed, the valuer should not proceed with the quantification of development value before assessing the likely demand for the land at a price reflecting that permission. In the majority of cases the likely demand will speak for itself; but in those cases where that is not true, the valuer has a particularly difficult and responsible task.

7.8 A valuation model

7.8.1 Pointe Gourde, *planning permission, and demand*

In *Myers* v *Milton Keynes Development Corporation* (1974), the acquiring authority employed a devious argument in a case concerning similar facts to those which begat the *Camrose* case. The argument, which was accepted by the Lands Tribunal was, simply stated, that the assumption of planning permission under section 15 creates an increase in value which is entirely due to the scheme underlying the acquisition, and should be disregarded under *Pointe Gourde*. The application of this decision would result in any section 15 assumption becoming worthless and a 1961 statutory measure would have been rendered impotent by a 1947 common law rule of dubious origin. The Court of Appeal thankfully reversed the Lands Tribunal decision and concluded that although the section 15 planning assumption may be said to help create an increase in value which is due to the scheme underlying the acquisition, the *Pointe Gourde* rule could not defeat the planning assumptions.

In *City of Birmingham District Council* v *Morris and Jacombs* (1976), outline planning permission had been granted for residential development of the subject 8 ha site. When detailed permission was granted, it was conditional upon 0.95 ha being reserved partly as a rear access to existing houses and partly as open space. Of this land, 0.3 ha were specifically reserved as an access road and it is this land which formed the basis of the compensation dispute.

Development in accordance with the permission was completed, whereupon the claimants submitted a planning application for residential development of the 0.3 ha strip, knowing that this would be refused and hoping to found a purchase notice (see 7.9). The purchase notice was confirmed and the compulsory acquisition of the strip commenced. The value of the land as an access road was agreed at £4,000; if capable of being developed residentially it was worth £15,000.

The Lands Tribunal was influenced by the *Pointe Gourde* rule, concluding that *Salop County Council* v *Craddock* (1969) had confirmed that the rule applied to decreases as well as increases in value, and awarded £15,000 compensation on the basis that the land had suffered a decrease in value due to the scheme underlying the acquisition and that this decrease should be disregarded.

The result was that the claimants had developed their land in accordance with a planning permission. They did not appeal against the condition, yet they obtained compensation as if the condition had never existed.

This could not stand. The Court of Appeal reversed the decision and substituted an award of £4,000. The *Pointe Gourde* rule did not apply because there was no scheme and because, after the *Myers* case, it cannot affect the planning position of the subject property.

Similarly, section 6 affects the demand factor only. It is arguable that the words "other than the relevant land" in schedule 1 are included to ensure that the section 15 planning assumption remains unaffected. Section 9 is specifically concerned with a lack of demand due to blight, while section 5 rule 3, the final valuation proviso to the willing seller rule, deals with the demand of the acquiring authority as a special purchaser which should be disregarded in specific circumstances. The planning factor, once established, is not affected by these rules; it is the demand factor that should be checked against them.

It is now possible to build a valuation model describing the path which the valuer should follow when assessing compulsory purchase compensation (Figure 7.3).

Compensation may be based upon existing use value, development value, or a combination of these. While existing use value will be established by evidence of demand for the property at that price in its existing use, development value will only be applicable if, as well as demand for the property in its proposed use, planning permission has been obtained or may be assumed.

Figure 7.3 Compensation for land taken: a valuation model

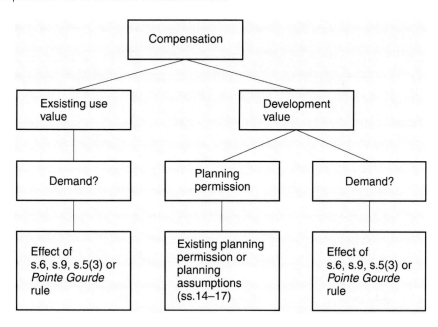

Section 5 rule 3, section 6, section 9, and the *Pointe Gourde* rule leave the planning factor unchallenged but all have regard to the demand factor. Their effect is not confined to development value but should also be considered when assessing existing use value.

It is this valuation model, drawn from Court of Appeal decisions and self-evident valuation principles, which suggests that the decision in *Jelson v Blaby District Council* (1977) (see p 238) was incorrect. The dispute concerned residential development value. There was no argument over demand and two alternative values were agreed between the parties, one reflecting planning permission for residential development, the other ignoring it.

There was no existing planning permission. The acquiring authority had no proposals for a use of the land beyond its retention as open space, so no valuable section 15 assumption could be made. On the development plan there was no allocation of the land for residential purposes. A nil section 17 certificate had been awarded and had been confirmed on appeal.

There was no planning assumption to be made other than that the land was to remain in its existing use. Yet residential development value was awarded.

It is suggested that valuers should draw little guidance from this decision which refutes Lord Denning's eminently sensible valuation hypothesis: it is planning permission plus demand which creates value. The correct approach was adopted by the Court of Appeal in *Abbey Homesteads Group Ltd v Secretary of State for Transport* (1982). The scheme should be treated as if it was abandoned immediately prior to the valuation date; not as if it had never existed. The basis of compensation is open market value, in its actual physical condition at the valuation date, as if offered for sale by a hypothetical willing seller. The *Pointe Gourde* rule, and the planning assumptions, should then be applied within that framework; not used to overrule it.

7.9 Purchase notices

It is possible in certain circumstances, using one of two procedures, for the owners of land to force the local authority to purchase their interest. This is sometimes called inverse compulsory purchase, but it is a curious fact that this reverse procedure is the only example of compulsory purchase which strictly earns that appellation, the usual case being more deserving of the term compulsory sale. As most landowners consider compulsory purchase an unwelcome imposition, it may by questioned when anyone would want to force it upon themselves. The first procedure involves service of a purchase notice, which is useful where a planning decision has rendered land virtually worthless, and therefore the owner may prefer to be rid of it. We will later consider a blight notice, which applies where land is blighted by planning proposals, and the owner may wish to sell it now, rather than waiting for the lengthy and uncertain compulsory purchase procedure to run its course.

This section on the purchase notice could equally be placed in the planning compensation chapter of this book, as it is actually a remedy for a planning decision. However, it is considered here because the result of its application is compulsory purchase, and because it is useful to compare and contrast it directly with the blight notice.

In order to have a purchase notice accepted, the owner should comply with the requirements of Part VI of the 1990 Town and Country Planning Act and, in particular, section 137 of that Act:

(1) Where, on an application for planning permission to develop any land, permission is refused or is granted subject to conditions, then if any owner of the land claims —

    (a) that the land has become incapable of reasonably beneficial use in its existing state; and

    (b) in a case where planning permission was granted subject to conditions, that the land cannot be rendered capable of reasonably beneficial use by the carrying out of the permitted development in accordance with those conditions; and

    (c) in any case, that the land cannot be rendered capable of reasonably beneficial use by the carrying out of any other development for which planning permission has been granted or for which the local planning authority or the Secretary of State has undertaken to grant planning permission.

he may ... serve on the council of the district or London borough in which the land is situated a notice requiring that council to purchase his interest in the land in accordance with the following provisions of this Part of the Act.

The local authority may then confirm the notice and proceed with the acquisition, or refuse to confirm and send the application to the Secretary of State. He, in turn, will confirm or refuse to confirm the notice, or even grant an alternative planning permission which will render the land capable of a reasonably beneficial use.

In summary, the test is whether, as a result of an adverse planning decision, land is incapable of a reasonably beneficial use. This is one of those vague terms which are loved by lawyers, and often interpreted in different ways by the courts, in order to engineer a fair result in a particular case. It is, therefore, difficult to find guidance in precedents, for how the term will be defined in the future. It is clear that reasonably beneficial use does not simply mean that the land is less useful or valuable (see *R v Minister of Housing and Local Government, ex parte Chichester RDC* (1960)): in fact, in *General Estates Co Ltd v Minister of Housing and Local Government* (1965) land used as a sports field and refused planning permission for residential development was not found to be incapable of a reasonably beneficial use in its existing state, although in *Wain v Secretary of State for the Environment* (1981) the temporary use of part of the subject land for agricultural operations was held to be insufficient to frustrate the owner. It appears that the use of land in its existing state has to be so insubstantial as to render the existing value of the property nil, or nominal. This was taken to the extreme in *Colley* v *Secretary of State for the Environment* (1988) when it was held by the Court of Appeal that a small plot of land with a few trees was capable of reasonably beneficial use as commercial woodland, even though it was subject to Tree Preservation Orders. However the facts of this case were somewhat unusual, as there were also revocation order proceedings in respect of the same plot of land (see p 330). The Lands Tribunal had already bent the law to award compensation under the revocation proceedings (though their decision was subsequently reversed by the Court of Appeal), and were in danger of having to compensate again for the same loss if a purchase notice was accepted. It should not, therefore, be assumed that the strong line taken in this case will be always followed in the future.

As the acquisition is deemed to be compulsory, the valuation model presented at 7.8 above should be followed: it is apparent that compensation under a purchase notice may be based on the development value of the land.

This is confusing as it appears that it is a refusal of planning permission, and hence a prohibition of development, which has given rise to the purchase notice. A value may exist, however, because in such a case the hypothetical planning assumptions under sections 15–17 may not coincide with the actual attitude of the acquiring authority and the land may be found to have development value by employing these assumptions.

Section 15(1) (the acquiring authority's own proposed use of the land) may be of benefit where the original planning application is refused because the land is allocated for a local authority scheme which involves a valuable use, such as town centre redevelopment.

Section 15(3) (schedule 3 rights) is likely to be of positive benefit in certain cases (see example 6.2 below). This anachronistic provision can be used to force planning authorities to pay compensation for development value which no longer exists in the real world. This was the case in the *Colley* case referred to above. The land concerned was, in 1947, the site of a house which was demolished in the 1960s. By the valuation date it was in an area where planning policies indicated a strong presumption against residential development. However, if a purchase notice had been accepted, schedule 3 would have required the assumption of planning consent to rebuild the house which had stood on the site in 1947.

Section 16 (assumptions derived from the development plan) may be of use but, in those cases where it is necessary to consider whether planning permission might reasonably be expected to be granted, the actual refusal of planning permission might create an unarguable precedent. In certain cases, however, section 16(7) may be of use. If permission is refused because the land is to be required for the acquiring authority's purposes in the future and these proposals do not equate with the zoning of the land on the development plan, then the section 16(7) requirement that the expectation of planning permission is to be considered in the absence of these proposals will be useful.

In *Devotwill Investments* v *Margate Corporation* (1967), the claimants had applied for permission for residential development of 0.55 ha of land. This was refused because part of the land was required for road-widening purposes. The claimants served a purchase notice, which was accepted. Despite the actual planning refusal, the combined effects of the development plan and section 16(7) enabled the assumption of planning permission for residential development to be made, and the land was to be valued on that basis.

Section 17 is likely to be of benefit in many purchase notice cases. Where appropriate alternative development exists in potential, a planning application will still be refused on the grounds that the land is required for the local authority's scheme. Serving a purchase notice, and then applying for a section 17 certificate, enables the applicant to establish whether, and which, development value would have been applicable in the no-scheme world.

An example of the application of the planning assumptions in a purchase notice case is given below.

Example 7.2

A light industrial unit of 400 m^2, situated on the edge of a small southern town, has been destroyed by fire. It was originally built in 1938. It occupied a site which is now situated in the Green Belt and planning permission for its replacement by a modern warehouse unit has been refused. No valuable alternative planning permission would be granted. A purchase notice has been served and accepted.

Compensation will be assessed as follows: the existing use value of the now vacant site is nominal, being a bare site in the Green Belt. Development value must be assessed in accordance with the requirement that both planning permission and demand must be established. Employing the planning assumptions, the acquiring authority have no proposals for use of the land beyond its retention as undeveloped open space. The development plan indicates the area of Green Belt land as land which is not to be developed and a section 17 certificate would be issued as a "nil" certificate due to the Green Belt.

Section 15(3) allows the assumption of planning permission for any development specified in Part 1 of schedule 3 of the 1990 Town and Country Planning Act. Included is the rebuilding of a construction in existence before 1 July 1948 so long as the cubic content of the original building is not exceeded by one-tenth.

Would there be demand for the site for such a use? This is easily established as there is a shortage of industrial space in the area and the site is easily accessible.

Compensation should therefore be assessed on the basis that the site may be developed by the construction of a modern factory unit of (400 m^2 + 10%) = 440 m^2 by reference to good comparable evidence or, failing that, a residual valuation/development appraisal.

Further specific points concerning compensation under purchase notices are considered in Chapter 10.

7.10 Blight notices

While a purchase notice is served as the result of an adverse planning decision, its corollary, the blight notice, is the result of an adverse planning proposal. The aim of serving a blight notice is to force the authority to acquire land which it has earmarked for eventual acquisition in order to prevent hardship being suffered by affected owners whose interest will at some time be compulsorily taken from them. This option is not available to everyone: affected owners are specifically limited to the following groups:

(a) residential owner-occupiers of a private dwelling
(b) owner–occupiers of agricultural units
(c) owner–occupiers of any unit, the annual value of which does not exceed £29,200 (this limit being subject to periodic variation).

An owner–occupier is a freeholder or a leaseholder with at least three years' unexpired term, with six months occupation preceding the service of the blight notice (see section 168 Town and Country Planning Act 1990).

In the case of a non-domestic property annual value means the rateable value. In the case of domestic properties, which are not now subject to rating, it is the figure certified by the valuation officer as being 5% of the compensation which would have been payable under compulsory purchase. For non-domestic property which is exempt from rating, the valuation officer can be requested to state the rateable value which would have applied if it was a relevant non-domestic hereditament. For mixed hereditaments the hypothetical rateable value can be added to the actual rateable value of the commercial part to establish the "annual value" for blight notice purposes.

Blight situations are specifically listed in schedule 13 of the 1990 Act as amended by the 2004 Planning and Compulsory Purchase Act, and by schedule 15(14) of the 1991 Act which specify cases of prospective public development which would not be undertaken in the natural course of events by the private sector, examples being land indicated on the development plan as land proposed for a highway or highway improvement, and land which a local authority proposes to acquire in pursuance of their powers relating to general improvement areas. It should be noted that these definitions are quite strict, and it is possible for land to be quite thoroughly blighted in the real world, and yet not fall within these legal definitions of blight. For example, when a highways authority decides to build a by-pass it will often identify several alternative alignments and embark on a lengthy consultation process. During this process, properties affected by all of the alternative alignments will be quite thoroughly blighted, but only once there is a commitments to one firm line for the road will affected properties fall within the necessary statutory definitions for the purpose of a blight notice.

In any case, the owner-occupier will only establish his right to serve a blight notice and (if no counter-objection is made) force the authority to acquire the interest as if it were compulsory acquisition if:

(a) he has made reasonable endeavours to sell that interest and
(b) "he has been unable to sell it except at a price substantially lower than that for which it might reasonably have been expected to sell if no part of the hereditament or unit were comprised in land of any of the specified descriptions" (section 130(1)(c), 1990 Act).

As a result of an amendment contained in schedule 15(13) of the 1991 Act, it is not necessary to fulfil those requirements where a compulsory purchase order has been confirmed and notice to treat has not yet been served. This reflects criticism that the marketing exercise is often a waste of time where CPO proceedings are at an advanced stage, though it assists only the most extreme cases. There is a similar exemption in respect of land authorised to be acquired under a special enactment.

The burden of proof is on the owner–occupier in establishing as a matter of fact that these conditions have been complied with. The evidence of a valuer is essential in proving the second point, while the first point is established by proof of reasonable exposure of the property to the market.

Once served, the blight notice will be accepted unless the appropriate authority issues a counter-notice within two months, stating a specific ground of objection (sections 151 and 162(5) of the 1990 Act). These include:

(a) the land is not blighted within any of the legal definitions
(b) the authority does not propose to acquire the land (effectively lifting the blight)
(c) it proposes to acquire part only
(d) it does not propose to acquire the land within 15 years (applicable only in certain cases such as highways)
(e) the claimant does not have a qualifying interest or
(f) the claimant has not satisfied the requirement to make reasonable endeavours to sell.

Once a blight notice is accepted, compulsory acquisition is then set in motion and the valuation should again follow the recommended model with specific reference to section 9 of the 1961 Act. Blight notices are rarely the result of a diminution in the development value of the subject property. They are specifically intended as a redress to decreases in existing use value. The demand for, and market value of, the property are the only points at issue: section 9 of the 1961 Act, as well as the *Pointe Gourde* rule requires that that value shall take no account of any decrease in value attributable to an impending compulsory acquisition, roughly in accordance with the applicable statutory blight situation.

The valuer is left with the task of assessing the market value of the property as if it were not to be acquired but subject to the usual compulsory purchase rules, requiring good comparable evidence of similar unaffected property. Therefore, disturbance and even severance and injurious affection may be payable, but subject to the usual rules and limitations. In *Budgen* v *Secretary of State for Wales* (1985) the costs of abortive attempts to market the property were disallowed by the Lands Tribunal as being not "an expense naturally and reasonably attributable to the acquisition". Such expenses would have been incurred if there had been no blight and no restriction on disposal of the property in the open market.

A new blight provision is introduced by section 62 of the 1991 Act, which amends section 26 of the 1973 Act and section 246 of the Highways Act 1980. Acquiring authorities have the power to mitigate the adverse effect of public works by purchasing by agreement land which is severely affected by the construction and use of those works, providing the vendor is an owner–occupier. This power is now extended to allow the purchase of such land in advance of requirements, thereby assisting those whose properties are not actually required for the works, but are none the less blighted by their proximity to such works. This is, however, purely a discretionary power, not a legal entitlement.

There are minor compensation provisions relating solely to blight notice situations in section 157 of the 1990 Act but these do not affect the general principle that once a blight notice is accepted, then the usual compensation provisions apply.

7.11 Development value: access and site assembly

When land is to be compulsorily acquired for the purposes of development by the acquiring authority, the demand for the land with the benefit of an assumed planning permission for development must be considered. The acquiring authority itself may be considered as a potential purchaser (see the *Indian* case, p 229) but the prospects of neighbouring land in different ownership being developed under the acquiring authority's scheme must be disregarded under section 6 of the 1961 Act as long as such development would not have occurred but for the scheme in question.

This presents the valuer with the concept of the no-scheme world in which he must envisage the market for the land in the absence of the authority's proposals. The general effect of such a presumption has already been considered (see p 234), but two further specific ramifications require detailed consideration.

7.11.1 Access and ransom value

Where the potential of a piece of land depends upon access to a public right of way, which constitutes the vast majority of cases apart from helicopter landing pads, and where access to that right of way is only possible by traversing land in different ownership, the problem (or bonanza) of the ransom strip emerges.

In some cases, land with enormous development value may be prevented from realising its potential by the refusal of the owner of the access strip to allow the owner of the back land to purchase or use that strip.

In a free market, an acceptable solution to the problem will normally be arrived at by negotiation, with the owner of the back land prepared to give up some of his development gain. For example, the owner of half an acre of land with planning permission for residential development may be prevented from realising its £150,000 development value by its lack of access, which may only be obtained through the car park of an adjoining office building, the owners of which have refused to grant a right of way. Without an access, the land is worth £15,000 as a garden extension, or even as a car park extension, to adjoining owners.

Frustrated, the owner places the property on the market at an asking price of £150,000. No prudent purchaser would pay this price: he would make exhaustive enquiries concerning the access problem even to the extent of establishing the price that the office owners would accept for an access. If, for example, this price was revealed in negotiations as a minimum of £60,000, the maximum bid the potential purchaser would make for the back land is fixed at £90,000, to make his total outlay £150,000. It is arguable that £90,000 is thus the market value of the back land — and that £60,000 is the market value of the access strip — as long as there is a likelihood that this price would be accepted. Bearing in mind that the current owner has few alternatives, this is likely to be the case.

Upon compulsory acquisition of back land with no access, bearing in mind the market value basis of compensation, this should be the valuer's approach. It was confirmed by the Lands Tribunal in *Stokes* v *Cambridge Corporation* (1961). This case concerned 5.1 ha of land with an assumption of planning permission for industrial development. There was one possible access over adjoining land in different ownership. It was held that the value of the back land depended upon the cost of the access; that the cost of the access would be a proportion of the gain in value of the back land resulting from its potential for development; and that a reasonable proportion of the gain which should be paid to the owner of the access strip was one-third.

The application of the *Stokes* principle to the above example results in the following valuation of the back land:

Value of land with access		£150,000
Less cost of obtaining access @ one-third		
of increase in value of back land:		
Development value	£150,000	
Existing use value	£15,000	
Gain in value	£135,000	
Take one-third as ransom payment	0.333	
		£45,000
Value of land without access		£105,000

The *Stokes* rule should not be applied without consideration. The *Stokes* principle is an excellent precedent and follows market psychology, but the award of one-third of the gain to the access owner is not to be seen as an inviolable principle.

Exactly the same logic may be applied to the valuation of the access land itself. The value of the access strip is that which would be paid in negotiation, the existing use value of the land fixing a minimum price, the total gain in value of the back land due to the prospects of development fixing a maximum, and the relative negotiating strength of the parties determining the settlement price.

For example, in *Ozanne* v *Hertfordshire County Council* (1991), 50% was the decided proportion (see p 229). The argument in that case was that both parties had to co-operate to realise any development value. Both, were in an equally strong negotiating position, and the fact that the ransom strip was much smaller than the development site, did not mean that the owner of the larger area would be able to negotiate a higher proportion of the ransom value. It may well be that *Ozanne*, rather than *Stokes*, will be followed in the future, but only where there is one single key to unlock development value.

Where there is more than one possible access, the ransom value principle may still apply, but the negotiating strength of the owner of the ransom strip is much weaker, as the development land owner can negotiate with other parties if an agreement is not reached. It follows that the ransom payment will be much lower, eg 15% in *Batchelor v Kent* (1992) where the ransom strip was only the most convenient of five possible means of gaining access.

An extraordinary result of applying the ransom value principle was illustrated in *Reaper Ltd* v *Merseyside Waste Disposal Authority* (1998). The claimant owned an area of land which had planning consent for out of town retail development, and had a perfectly satisfactory access. A compulsory purchase order was made on a large part of his land, but not including the access. A claim was submitted based on full development value. However, the acquiring authority argued, correctly, that the actual ownership should be ignored and the correct basis is open market value is if offered for sale by a hypothetical willing seller. That hypothetical seller has no access, and would have to negotiate the purchase of an access from one of two adjacent landowners (one of whom was of course, in the real world, Reaper Ltd). The Tribunal accepted this argument and made a deduction of 25% from full development value to reflect a ransom payment to acquire an access. As a result the compensation paid to the claimant was only 75% of its value in the no-scheme world.

It should be remembered that the owner of the access strip would not sell at a price below the existing use value of his land as he could obtain that price from anyone in the market.

7.11.2 Site assembly

The problem of assembling a site in order to secure a parcel of development land of optimum size has also been considered by the Lands Tribunal. Half a hectare of land may have the benefit of an assumed planning permission for industrial development for which it might be worth £200,000. The land may be fragmented into ten separate equal parcels of ownership, for example freehold allotments, each of which would be too small to develop individually for industrial purposes. The value of each plot is not, in this case, £20,000. Considerable time and expense on behalf of a prospective developer will be necessary to merge some or all of these plots to attain a site of economic proportions. Any one of the 10 owners may hold a key to the development and hold out for a greater price, and this should also be taken into account unless no one plot has any particular advantage over the others in this respect, and any extra value for a single plot would simply depend upon the sequence of acquisition.

In *Nash* v *Basildon Development Corporation* (1969), a deduction of 40% of the industrial development value of the subject site of 0.17 ha (which was too small to develop individually) was made to account for the marriage of sites and other factors. The percentage which should be applied will depend upon the number of sites to be merged, the likely attitude of vendors, the risk of the merger being frustrated, and the amount of fees likely to be involved.

In these circumstances ransom value will apply only in exceptional cases. In *County and District Properties v Harrow* (1992) land comprising almost the whole of a site for a town centre redevelopment scheme was being acquired from a single claimant. The acquiring authority argued for deducting the cost of buying the rest on a ransom value basis at 50%. However, the lands tribunal did not accept that approach. Some office development could take place on the subject land in isolation and the other plots were undevelopable so their owners were in a poor bargaining position. Rather than looking at a share of the development value in respect of the main site, the claimant's valuer deducted only development value of the additional land, plus an uplift of 30% to reflect any enhanced value to the development as a whole . The Tribunal felt that this was generous. Ransom value will only be applied where the land is a genuine key to unlocking development value which would otherwise remain untapped.

The method of valuation to be adopted in a ransom value situation is illustrated by the following example.

Example 7.3

600 m² of land has the benefit of planning permission for the erection of a single detached house. Similar development sites sell for £100,000 with unrestricted access. The land has no existing access. There are 3 possible routes on to a distributor road through adjoining plots. All 3 plots are currently used as garden land for which purpose they are worth £4000 each.

Value of 600 m² single house plot		£100,000
Less cost of access, equal to a proportion of the gain in value of whole due to access:		
Value with access	£100,000	
Value without access, say	£8,000	
Gain	£92,000	
Two alternative accesses, say 20%	£18,400	
Check existing use value of access land	£4,000	
Take higher		£18,400
Value of subject land		£72,600

7.11.3 *General*

The valuation of development land is usually attempted either by the use of direct capital comparison, employing evidence of transactions involving land with similar potential, or by the residual method of valuation.

The Lands Tribunal has made it abundantly clear that it does not favour the use of the latter method; its concern being that a relatively modest change in one of the underlying assumptions can have a large impact on the residual valuation: *County and District Properties* v *Harrow* (1992). Valuers can, therefore, manipulate the figures to produce a valuation which favours their client. This reticence appears outdated as, firstly, the residual valuation is almost exclusively used in the market place and, secondly, similar assumptions and adjustments are applied in the direct capital comparison approach. They are, however applied implicitly in an overall adjustment to the comparable evidence, rather than being explicitly set out in a valuation where they are open to analysis and criticism. In *South Coast Furnishing Co Ltd* v *Fareham Borough Council* (1977), the claimants relied upon a residual valuation to value six plots totalling 0.3 ha while the acquiring authority relied upon comparable evidence of land sales. The residual method was rejected as being not acceptable.

Landau v *Tyne & Wear Development Corporation* (1996) is one of a number of cases in recent years which has indicated a slight relaxation in the Tribunal's attitude to the residual method of valuation. In this case the claimant's valuer produced several development appraisals, but the acquiring authority argued that little reliance should be placed on them as residual valuations are fraught with danger as many elements were exceptionally sensitive. However there was little alternative evidence available and in its decision the Tribunal commented that:

> The passage of time has not made the criticisms of the residual approach any less valid. However, in this case significant parts of the appraisal are agreed between the parties and there is substantial common ground in

others, together with extensive evidence relating to the disputed elements. As I have found that there is inadequate evidence of comparable transactions, I have given consideration to a residual valuation based upon the evidence of the development appraisals in respect of the claimant's scheme.

The Tribunal member then borrowed the computer programme used by the claimant's valuer, and produced his own appraisal!

7.12 Merger of interests and marriage value

There are two broad types of transaction which may result in a merger of interest and the possible release of marriage value. These are:

(a) the merger of two or more legal interests in the same property
(b) the merger of two or more contiguous pieces of land in separate ownership.

A marriage value will be created in either case if the value of the single, merged interest exceeds the total of the values of the separate interests before the merger. Where an authority with compulsory powers acquires one of these interests, it may be argued that some of the marriage value which may be created upon merger attaches to the market value of the interest, particularly if it can be shown that an adjoining owner would be in the market for the subject land.

Prior to its amendment by the 1991 Act, it could be argued that section 5(3) of the 1961 Act allows the acquiring authority to purchase at prices ignoring any value to a particular purchaser (such as an adjoining owner). However, even before the 1991 Act removed the special purchaser limb to rule 3, it was well settled, after *Lambe* v *Secretary of State for War* (1955) (see p 230) that the rule cannot apply to transaction type (a) above. In that case the acquiring authority, the owner of the leasehold interest in the subject property, acquired the freehold reversion in that property. It was found that the merger of interests would create a marriage value and that the sitting tenant could offer more for the reversionary interest as a result. Compensation was awarded on that basis and rule 3 was held not to apply to the merger of legal interests in the same property.

This approach was confirmed in *Mountview Estates Ltd* v *London Borough of Enfield* (1968). The claimants in this case owned the leasehold interest in two adjacent terraced houses which were to be acquired at site value under the Housing Act 1957. The tribunal remarked, following the *Lambe* case, that any special value to a leaseholder due to a merger of his interest with that of the freeholder may be taken into account in assessing compensation. John Watson, member of the tribunal, went further to preclude the possible argument that the actual freeholder may have had no interest in merging the interests by acquiring the leasehold interest:

> Whether the actual freeholder was rich or poor, wise or foolish, easy-going or obstinate — those personal characteristics are as irrelevant as whether he was black or white. For the purpose of the hypothetical exercise the parties are to be conceived as two passengers on the Clapham omnibus, devoid of prejudice or eccentricity, who in any given circumstances might have been expected to behave reasonably.

This is hard to reconcile with the majority decision of the Court of Appeal in *Trocette Property Co Ltd* v *Greater London Council* (1974). The subject property constituted a leasehold interest with 11.5 years to run. The site had development potential which could only be realised by the merger of the freehold and leasehold interests. It was adequately demonstrated that the freeholder (the GLC, which was being forced to purchase under purchase notice procedure) had no interest in redeveloping the site by merging the interests and this fact should be taken into account.

Megaw LJ said (and this was the majority opinion):

> I see nothing in the legislation ... which compels or permits one to ignore, in assessing compensation for the leasehold interest, evidence of a fact which would be known to the buyers in the market.

The latter decision, being of higher authority, should be regarded as correct and the actual attitude of interested parties may well be relevant.

Two later cases illustrate possible valuation approaches. *Honisett and Kennedy v Rother District Council* (1977) concerned the acquisition of a leasehold interest in half an acre of land with permission for residential development. It was agreed that the value of the freehold interest in possession of the subject land was £15,000 and that a prospective buyer would be interested in merging freehold and leasehold interests. What was to be valued was the 27.5 years' remainder of a ground lease, the rent paid being 5 pence pa, but no attempt was made to value this interest using the investment method or income approach. Instead, the valuation approach was to deduct the likely cost of acquiring the freehold reversion plus fees from the value of the freehold in possession, as follows:

Net value of freehold in possession (road costs deducted)			£14,250
(a) Cost of acquisition of freehold reversion	£14,250		
PV 27 years @ 6%	0.207		
		£2,950	
(b) Fees on acquisition		£340	
			£3,290
Value of leasehold interest			£10,960
Say			£10,900

This valuation approach awards all marriage value (if it exists) to the leaseholder, suggesting that the freeholder in this particular example is in so weak a position that he could insist upon none of it.

This may be contrasted with the attitude of the parties and the tribunal in *Hearts of Oak Benefit Society v Lewisham Borough Council* (1977). Here compensation was to be assessed in respect of the freehold interest in eight terraced houses subject to leases which expired 14 years after the valuation date. The dispute between the parties concerned the extent to which the special value of the freehold interest to the head lessees should be taken into account. It was agreed that a large marriage value existed; that, following the *Mountview* case, with no reference to *Trocette*, the hypothetical leaseholders might be interested in purchasing the freehold interest; and that any marriage value would be distributed between the parties.

To conclude, where legal merger of interests including the subject interest is possible and an element of marriage value might be released, the actual bargaining position of the parties (although not their real-life financial situation) should be taken into account in assessing the likely distribution of that marriage value, which should be included in the market value of the interest to be acquired.

The physical merger of sites follows similar rules since the amendment of section 5(3) of the 1961 Act to allow the bid of a special purchaser to be taken into account. Even before this change, however,

Rathgar v *Haringey London Borough* (see p 229) avoided rule 3 when the value of access land was under consideration. And in *Dicconson Holdings Ltd* v *St Helens MB* (1979) it was held that the value of access land might be up to one-third of the value of the back land released for development (following *Stokes* v *Cambridge Corporation* (1961) by implication), again avoiding rule 3.

It appears that these valuations are best carried out by an examination of the likely selling price of the subject land in the market, uncluttered by legal abstractions.

7.13 Leaseholds and freehold reversions

Where an authority with compulsory powers wishes to acquire a property in order to demolish it or to put it to a particular use, it will require absolute possession of that property. Hence, where that property is subject to a lease it will be necessary to ensure that the rights of both the freeholder and the leaseholder(s) are extinguished before proceeding with the scheme for development of the property.

The freehold reversion will be acquired subject to the normal rules of market value and sections 47 and 50 of the 1973 Land Compensation Act (see below, p 265). The leasehold interest, however, may be dealt with in either of two ways: a notice to treat may be served on the leaseholder and the leasehold interest acquired subject to the rules concerning the date of valuation discussed at 7.4 and 7.5, or, alternatively, the acquiring authority as the new freehold reversioner may serve notice to quit upon the leaseholder in accordance with the terms of the tenancy and any relevant act such as the 1954 Landlord and Tenant Act (see Chapter 1) or the Housing Acts (see Chapter 2). Section 30 of the 1954 Act enables the landlord to obtain possession if he requires the premises for his own occupation or if he intends to demolish or substantially reconstruct the premises. In these cases compensation will be payable in respect of improvements (in the former case only) and disturbance equal to three or six times the rateable value of the holding, or the amount of a disturbance payment under section 37 (1) and (4) 1973 Land Compensation Act (in both cases). In the case of residential property, the duty to rehouse tenants imposed by section 39 of the 1973 Land Compensation Act complements the right to possession given by section 90(1)(a) of the 1977 Rent Act and Part III (i) of schedule 2 of the 1988 Housing Act, where suitable alternative accommodation is provided.

The choice between acquisition and eviction may depend upon the date when possession is likely to be required, as this is liable to fix the date of valuation. The value of a wasting leasehold interest may be compared at this date with the likely burden of compensation (if any) under the 1954 Act.

The value of the unexpired term of a short tenancy must be assessed bearing in mind the requirements of sections 47 and 50 of the 1973 Land Compensation Act. The right of the business tenant to apply for a new tenancy under Part II of the 1954 Act must be taken into account under section 47, and the fact that a residential tenant is to be rehoused by the acquiring authority should not be used as a device to reduce the compensation payable in respect of his interest (section 50).

The valuation implications of section 20 of the 1965 Act and the 1973 provisions are not straightforward. In a conventional valuation, a business tenant's interest would carry the same value whether a new tenancy is to be granted or not, as that tenancy would usually be in return for a full market rent. However, if improvements which have added to the rental value of the premises have been carried out and are to be ignored in fixing the new rent, then the assumption of a new tenancy may add to the value of the lessee's interest.

Also, the right to a new tenancy even at a full market rent may often increase the value of an interest, due to the goodwill and extra key money which contribute greatly to the value of commercial leaseholds. Key money cannot always be analysed in terms of the profit rental it purchases: instead, it may represent capitalised trading profits which may be offered in order to gain a foothold in an area of a particular attraction. If market evidence exists to suggest that such key money would be paid, there is no reason to ignore that fact, and the requirements of section 20 suggest that a single premium encompassing profit rental (albeit extremely short lived), an allowance for fittings and goodwill, and key money should determine the quantum of compensation to be paid.

It is well settled (see Chapter 2) that residential tenancies of less than one year's duration have no market value: the effect of section 20 is therefore meaningless in respect of them, so far as it relates to compensation for land taken.

The valuation of leaseholds should follow market evidence rather than precedents set by the Lands Tribunal. The use of dual rate valuations for commercial leaseholds is questionable to say the least and an excellent case may be made for single rate leasehold valuations both because of the failings of a dual rate approach (see Baum, Mackmin and Nunnington 2006), and because such an approach more logically reflects the purchaser's attitude. The choice between dual rate or single rate was considered in *Wilrow Engineering Ltd and Stapleton* v *Letchworth UDC* (1974), when conventional wisdom prevailed, but there is no guarantee that the same attitude would be reasserted in a new case.

The valuation of freeholds subject to a lease is also affected by sections 47 and 50 of the 1973 Act. In the case of business premises, the right of the tenant to apply for a new tenancy shall be taken into account when valuing the freehold. This will be of importance if improvements are to be disregarded in fixing the new rent, or if for any reason the value of the freehold is less as a result of the characteristics of the tenant, who might represent a poor covenant.

When valuing the freehold interest in residential property, there shall be left out of account any increase in value resulting from the tenant having been rehoused by the acquiring authority, so the freehold is to be valued taking into account the fact that it is encumbered by a tenancy, even though this might not be the case at the date of valuation (section 50, 1973 Act). This enactment is the result of the calamitous (for acquiring authorities) decision in *Banham* v *London Borough of Hackney* (1970) where the acquiring authority, having rehoused the statutory tenant before the date of valuation, was forced to pay vacant possession value for the freehold interest.

The valuation of tenanted freehold commercial property presents no special difficulties. The valuation of tenanted freehold residential property has conventionally been executed employing the investment method or income approach, which was the approach adopted both in *Basted* v *Suffolk County Council* (1979) and *Fawcett* v *Newcastle-upon-Tyne MDC* (1980). Evidence now exists to suggest that such a valuation approach is the exception rather than the rule, and property let on regulated tenancies is often valued at a percentage of vacant possession value varying between 50% and 75%, and even more if there are good prospects of obtaining possession. The return to open market rents as a result of the 1988 Housing Act made the investment method realistic once again, though a proportion of vacant possession value remains the preferred approach in the open market. With most residential tenancies now being assured shortholds, vacant possession value less a small allowance for the delay and risk involved in obtaining possession, will often be appropriate. Whichever approach is adopted in the market should be followed in the assessment of compensation.

If part only of property subject to a lease is acquired, section 19 of the 1965 Compulsory Purchase Act provides that the rent for the whole premises shall be apportioned between the part retained and the part taken. Disputes over the apportionment shall be heard by the Lands Tribunal, which was the case in *Dixon* v *Allgood and Secretary of State for Transport* (1981), in which it was confirmed that the apportionment should not be pro rata according to the relative sizes of the parts taken and retained, but should be arrived at by considering their relative attractiveness and rental values as separate units.

Having determined the apportioned rents, the valuation of the part taken should be ascertained by capitalising its rental value, and not by subtracting the value of the retained holding from the value of the whole prior to the acquisition. It is suggested that the latter approach includes compensation for severance, calculated incorrectly (see Chapter 8).

Where it is not reasonable to acquire part only of any property, the acquiring authority may be forced to acquire the whole. This is considered immediately below.

7.14 Material detriment

Section 8 of the 1965 Compulsory Purchase Act gives the Lands Tribunal power to decide that a claimant should not be required to sell part of a house, building or manufactory, unless that part can be severed from the rest without causing material detriment to the remainder, or to sell part of a park or garden attaching to a house unless that part can be taken without seriously affecting the amenity or convenience of the house. Section 58 of the 1973 Land Compensation Act amplifies this provision by providing that, in deciding whether material detriment has been caused, the Lands Tribunal should take into account not only the effect of the severance but also the use to be made of the part proposed to be acquired. Section 8 is often described as the all or nothing provision but it should be remembered that it is not a right, as it is subject to the discretion of the Lands Tribunal, and that it is not universal, as only certain categories of claimant may benefit, particularly where land only is acquired.

Ravenseft Properties Ltd v *Hillingdon LBC* (1969) established that even though a right to severance compensation might exist, and the receipt of such compensation would place the claimant in the same position as if no land had been taken from him, the right to utilise section 8 of the 1965 Act is not affected. This right is the ability to require the acquiring authority to purchase the whole, and not simply part, of the property. If the right is established, the claimant may have the choice between compensation for the part taken plus severance compensation, and compensation for the whole land taken.

The *Ravenseft* decision clarified section 8 to a certain extent. It appears that virtually all types of building will qualify. In order to test whether material detriment is suffered, it is necessary to establish whether the remainder would be less useful or less valuable in some significant degree. In the case of garden land, one is left to test whether the amenity or convenience of the house is seriously affected.

The latter case is classically illustrated by a road-widening acquisition which may take a small strip of front garden for the improvement of a bend in a country lane, where material detriment is unlikely to be suffered, and its contrast, where the whole of a front garden is taken for an urban motorway and material detriment is the unavoidable result. The provisions of section 58 of the 1973 Act should be remembered in differentiating between the two cases.

When part of a building is taken, the valuer's evidence and opinion becomes of the essence in determining whether the remainder is less useful or less valuable in a significant degree. It may be necessary to produce valuations before and after the partial acquisition in order to ascertain this.

Example 7.4

Imagine that 1,500 m² of a 2,000 m² supermarket and its complete adjoining car park are to be acquired.

The authority have agreed to make good the building after its partial demolition for a town centre redevelopment.

Valuation before:
2,000 m² supermarket with adequate parking facilities

Rental value @, say, £300 m² pa	£600,000 pa	
YP perp. @ 7%	14.2857	
		£8,571,420
Say		£8,600,000

Valuation after:
500 m² shop with no parking facilities.

Rental value @, say, £160 m² pa	£80,000 pa	
YP perp. @ 8%	12.50	
		£1,000,000

In such a case it should easily be established that the remainder is less useful or valuable in a significant degree and the authority should be required to purchase the whole.

There is, however, evidence that the Lands Tribunal will be generous in its application of the provision.

In *London Transport Executive* v *Congregational Union of England and Wales (Incorporated)* (1979) there was a dispute concerning the acquisition of the back land of a church including a church hall, but including no part of the church itself. Even in these wide circumstances it was decided that the whole had to be acquired, including the church itself.

In *McMillan* v *Strathclyde Regional Council* (1982) the Tribunal decided not to require the acquisition of the whole property where road widening brought a new footpath within 15 ft of the front wall of a house. The Tribunal decided that following the Ravenseft decision, the test was whether the remainder after the part is taken is less useful or less valuable in some significant degree compared with the property as existing before the acquisition. This was not the case because of an unusual obligation on the owners to create a road and footpath in the same location if the work had not been done by the local authority.

In the event that the Tribunal should decide not to exercise its discretion to require acquisition of the whole, the effort may not have been completely wasted as section 8(i) requires that the compensation will then be assessed including compensation for severance in respect of land taken as well as land retained. This appears not to be the case otherwise (see p 277).

7.15 Housing Act acquisitions

Until 1990, a major exception to the general market value provision which forms the basis for the assessment of compulsory purchase compensation was provided by the Housing Act 1985, which replaced similar provisions in the Housing Act of 1957. The basic principle was that where a house declared to be unfit for human habitation was subject to compulsory purchase, then compensation should be based on site value, ignoring any value attributable to the house itself.

The obvious inequity of the system was alleviated to some extent by a system of special payments, principally owner-occupier supplements and well-maintained payments, each of which had rigid qualification rules.

Under Part IV of schedule 9 of the 1989 Local Government and Housing Act, the special compensation provisions relating to unfit properties were repealed and site value compensation is no longer applicable. Compensation for the compulsory purchase of unfit houses is now assessed on the same basis as for a fit property, and the redundant owner-occupier and well-maintained payments are repealed. The new provisions will apply where the compulsory purchase order was made after 1 April 1990, though where the date is earlier, there are transitional provisions contained in the Local Government and Housing Act 1989 (Commencement Order No 5 and Transitional Provisions) Order 1990 (SI 1990, No 431).

Special compensation provisions do, however, remain in relation to closing and demolition orders (see section 58(iv)(a) of the 1985 Act which was inserted by para 75 of schedule 9 of the 1989 Local Government and Housing Act). Any depreciation in the value of a house which is caused by serving a demolition or closing order is compensatable. It should be noted that the compensation must arise from the order itself, not from the disrepair, and that there are circumstances where such an order may enhance the value of the house, for example where it results in re-housing of the tenants.

Apart from these limited provisions, compensation for the compulsory purchase of unfit property is now assessed in accordance with the general compensation code, including entitlements to home, basic and occupier's loss payments, as well as disturbance payments, where appropriate.

7.16 Examples

The following examples are intended to illustrate the use of the valuation model in establishing a logical valuation format. Certain information must be assumed, and the answers are intentionally skeletal.

Example 7.5

Mr A owns 1,000 m² of land in an area zoned on the development plan for industrial use. The local authority is to acquire the land under compulsory purchase powers, for the development of an industrial estate comprising individual units of 1,000 m² at a ground coverage of 50%.

The land is currently used for car parking, for which use it is worth £120,000 per hectare. Industrial land in a similar area fetches £500,000 per ha.

Advise Mr A.

Base 1: Existing use value (car park site). Establish demand by analysis of comparable
evidence. Consider effect of ss.5(3), 6, and 9 of the 1961 Land Compensation Act.
1,000 m² of car parking land @ £120,000 per ha.   £12,000

Base 2: Development value. Establish:
    (a) Planning permission:
        (i)   Section14: Existing planning permissions? Hope value?
        (ii)   Section15: Acquiring authority's proposed use: as part of an industrial estate of 1,000 m² units at 50% coverage. Subject land too small for individual development. Schedule 3 rights?
        (iii)   Section 16: Development plan zoning as part of industrial estate. Reasonable expectation of planning permission? Only as part of a larger scheme at 50% coverage.
        (iv)   Section 17: can a certificate of appropriate alternative development be obtained. Assume that a certificate for industrial use is issued.
    Sections 15, 16 and 17 coincide: planning permission may be assumed for development of the subject land as part of an industrial estate of 1,000 m² units at 50% site coverage.

    (b) Demand:
    Establish demand by analysis of comparable evidence of industrial land sales. Effect of sections 5(3), 6, and 9?

1,000 m² of industrial land @ £500,000 per ha	£50,000	
Less deduction of 30% for site assembly	£15,000	
		£35,000

The claim should be based on development value.

Example 7.6

Mrs B owns a house in a 1.2 ha plot. She has planning permission to erect a second house on a 2,000 m² site within this plot. The value of the existing house is £450,000 in its 1.2 ha plot.

On the development plan the land is zoned for residential purposes but is to be acquired as part of the site of a school.

The 2,000 m² plot is worth £100,000 with the benefit of the existing planning permission.

Advise Mrs B.

Base 1: Existing use value (house in 1.2 ha). Establish demand by analysis of comparable evidence. Effect of sections 5(3), 6, and 9?
Value of house in 1.2 ha plot   £450,000

Base 2: Development value. Establish:
(a) Planning permission:
(i)   Section 14: Existing planning permission for 2,000 m² plot. Hope value?
(ii)   Section 15: Acquiring authority's proposed use: as part of school site. Schedule 3 rights?
(iii) Section 16: Development plan zoning for residential purposes. What permission can reasonably be assumed? Say 20 units to the ha.
(iv)   Section 17: not required as residential use is established by sections 14–16.

Planning permission therefore exists or may be assumed for the following: a single house on a 2,000 m² site; residential development at 20 units to the ha; development as part of a school.

(b) Demand:
Establish demand for land for all above purposes by analysis of comparable evidence. Effect of sections 5(3), 6, and 9?

(i)	Single 2,000 m² plot	£100,000	
	1 ha at a density of 20 units to the ha @, say, £50,000 per plot	£1,000,000	
			£1,100,000
(ii)	1.2 ha. at a density of 20 units to the ha @, say, £50,000 per plot		£1,200,000
(iii)	Value as part of school site say £500,000 per ha		£600,000

Base 3: Combination of existing use value and development value.

Existing house in (say) 2,000 m² plot	£375,000	
1 ha at a density of 20 units to the ha, @, say, £50,000 per plot	£1,000,000	
		£1,375,000

Claim £1,375,000, combining existing use value and development value.

Example 7.7

Mr C owns 60 ha of agricultural land allocated on the development plan as being part of a town extension scheme. The land lies 1 mile from the town boundary at its nearest point and is required for residential use (40 ha) and open space (20 ha).

Advise Mr C.

Base 1: Existing use value (agricultural land). Establish demand by analysis of comparable evidence. Effect of sections 5(3), 6, and 9?

60 ha @ £10,000 per ha   £600,000

Base 2: Development value. Establish:
  (a) Planning permission:
    (i)   Section 14: Existing planning permissions? Hope value?
    (ii)   Section15: Acquiring authority's proposed use: as part of town extension scheme, for residential land at low density, and as open space. Schedule 3 rights?
    (iii)   Section 16: Development plan zoning as above.
    (iv)   Section 17: not required

Planning permission may therefore be assumed for the development of 40 ha as low-density housing, with 20 ha to be used as open space land.

  (b) Demand:
Establish demand by analysis of comparable evidence. Would all 40 ha be ripe for immediate development? Demand at full residential values may only materialise over a period: assume 20 ha would be ripe for development in 10 years, the remainder in 20. Effect of sections 5(3), 6, and 9? Any development of adjoining land under the same scheme must be ignored in so far as it affects the value of the subject land (section 6). What would have happened in the absence of the scheme? Natural expansion of the town may have been spread rather than concentrated: ignore development between town boundary and subject site.

20 ha of open space land @ £20,000 per ha		£400,000
20 ha of residential development land @		
£1,000,000 per ha	£20,000,000	
PV 10 yrs @ 8%	0.4632	
		£9,264,000
20 ha of residential development		
land @ £1,000,000 per ha	£20,000,000	
PV 20 yrs @ 8%	0.2145	
		£4,290,000
Rent from 20 ha		
@ £500 per ha pa	£10,000 pa	
YP 10 yrs @ 6%	7.3601	
		£73,600
Rent from 20 ha		
@ £500 per ha pa	£10,000 pa	
YP 20 yrs @ 6%	11,4699	
		£114,700
		£14,142,300

Claim should be based on development value.

Example 7.8

Mr D owns the freehold interest in a shop of 5 m width and 25 m depth. It is currently let at a rent of £12,000 pa net, with a review in three years, while comparable evidence suggests that zone A rents of £300 per m^2 are currently being achieved.

The local authority is to acquire part of the shop, constituting the front 7 m for road-widening purposes.

Advise Mr D and Mr E, the lessee.

Base 1: Existing use value (as shop). Establish demand by analysis of comparable evidence. Effect of sections 5(3), 6, and 9? Consider possible blight (sections 9). In addition, is material detriment caused to the remainder so that the acquiring authority must acquire the whole? Assume not, after considering before and after valuations.

Before: Full rental value (@ 150/m^2 zone A pa). arrived at as follows:

Zone A:	5m × 7m × £300	=	£10,500 pa
Zone B:	5m × 7m × £150	=	£5,250 pa
Zone C:	5m × 11m × £76	=	£4,180 pa
Full rental value			£19,930 pa

After:

Zone A:	5m × 7m × £300	=	£10,500 pa
Zone B:	5m × 7m × £150	=	£5,250 pa
Zone C:	5m × 4m × £76	=	£1,520 pa
Full rental value			£17,270 pa

New rent is 87% of old: rent will continue to be paid at 87% of current level: £10,440 pa.

Compensation will be payable to both freeholder and leaseholder on the basis of the difference in values of their interests before and after the acquisition.

Freehold, before:			
Rent	£12,000 pa		
YP 3 yrs @ 8%	2.5771		
		£30,926	
Reversion to rent	£19,930 pa		
YP rev. perp. 3 yrs @ 8%	9.9229		
		£197.764	
			£228,690
Freehold, after:			
Rent	£10,440 pa		
YP 3 yrs @ 8%	2.5771		
		£26,904	
Reversion to rent	£17,270 pa		
YP rev. perp. 3 yrs @ 8%	9.9229		
		£171,368	
			£198,272
Compensation			£30,418

Leasehold, before:

Rent received	£19,930 pa		
Rent paid	£12,000 pa		
Profit rent		£7,930 pa	
YP 3 yrs @ 11%		2.4437	
			£19,379

Leasehold, after:

Rent received	£17,270 pa		
Rent paid	£10,440 pa		
Profit rent		£6,830 pa	
YP 3 yrs @ 11%		2.4437	
			£16,690
Compensation			£2,689
Say			£2,700

Base 2: Development value. Establish:
    (a) Planning permission:
        (i)   Section 14: Existing planning permissions? Hope value?
        (ii)   Section 15: Acquiring authority's proposed use: for road widening. Schedule 3 rights?
        (iii)   Section 16: Development plan zoning coincides with present use, as shop property.
        (iv)   Section 17: Assume certificate can be obtained in line with commercial zoning.

There are no valuable planning assumptions, so that no alternative claim base exists.

For further examples, including compensation claims in respect of land taken, refer to the ends of Chapters 8 and 9.

Severance, Injurious Affection, and Betterment

8.1 Introduction

Ricket v *Metropolitan Rail Co* (1867) established a fundamental principle in compulsory purchase compensation: compensation is the amount required so far as money can do so, to put the owner in the same position as if his property had not been acquired.

Compensating the owner at market value for the land taken from him goes only part of the way to attaining this end. So, in addition, monetary compensation will be payable to reimburse the claimant for "disturbance or any other matter not directly based on the value of land". This covers such easily identifiable losses as removal expenses and loss of trade, and is discussed in chapter 9.

Compensation for land taken plus disturbance will not always fulfil the requirements of the principle expressed in the *Ricket* case, however. We are left with an area which is not covered by compensation for land taken, but which is directly based on the value of land. This area is described by the generic terms severance and injurious affection. Strictly speaking, severance is merely one form of injurious affection but it is more usual and convenient to consider the terms as separate but complementary claims.

(a) Injurious affection may best be likened to a private nuisance, which is not actionable because it is carried out with statutory authority.

   The land remaining after the acquisition of part of a holding might suffer depreciation due to the acquiring authority's proposed use of the acquired part. For example, land acquired from the back garden of a dwelling house for the extension of a sewage works may result in depreciation of the house due to the malodorous nature of the proposed adjacent use.

   Even where no land is acquired from the claimant he may suffer a loss in land value due to the activities of a statutory authority which, apart from the statutory immunity of that body, may or may not have given rise to an actionable claim for nuisance. The construction of an urban motorway close to a rest home would provide the ideal example where considerable loss of value might be suffered although no land is taken.

(b) Severance may be suffered where part of a claimant's holding it acquired, and the value of the remainder plus the compensation paid for land taken do not equate to the original value for either of two reasons.

(i) The value of a holding when divided into two parts may be less than the value of the single entity. For example, development land will have a value which can be assessed by calculating the difference between the value of the completed development., and its construction cost. Costs per acre may reach a minimum at a certain size of development due to economies of scale. The reduction of the unit may increase the costs per acre and reduce the value of the remaining land. Alternatively consider a plot of land with planning consent for a house, half of which is acquired. Neither the half taken, nor the half retained may be sufficiently large to accommodate a house on its own. Both the land retained, and the land taken will, therefore, suffer considerable loss of value, which should be compensated under any fair system.

(ii) The requirements of the acquiring authority might split the land into three parts, the combined values of which do not summate to the value of the whole. Land required for a motorway may be driven through development land. After construction of the motorway, the claimant will be left with two parcels of land on opposite sides of the motorway, plus a right to compensation for the land taken. While each parcel may suffer from the depreciation exemplified at (i) above, one or both of the two parcels might suffer extreme devaluation due to its small size, lack of access or its transformation to uselessness by means of its new awkward shape: all the result of "severance".

For any of these reasons, severance and injurious affection compensation may be payable in order to enable the acquiring authority to compensate in accordance with the *Ricket* principle. It is not always able to do so: the reasons for this where part of holding is acquired are considered at 8.2.8 below.

The tortuous qualification rules for injurious affection where no land is taken are considered at 8.3.

(c) The statutory authority engaged in a programme of public works may take measures to mitigate the losses of potential claimants and others. Such indirect compensation is dealt with at 8.4.

(d) There is no provision which requires a landowner to pay a statutory authority an amount equating to the increase in the value of his land where (for example) a road scheme reduces the traffic flow past his dwelling. The difficulties of quantification of such a betterment tax together with political expediency ensure the absence of such a parallel measure from our compensation/ betterment code. However, the owner of land of which part is acquired is required to suffer a deduction to be made from his compensation if the value of the retained land increases in value. The set-off provisions are considered at 8.5.

(e) Finally, public works may cause physical damage to land. Compensation for such damage is discussed at 8.6.

8.2 Injurious affection to retained land where some land is taken

8.2.1 *General basis*

Section 7 of the 1965 Compulsory Purchase Act reads:

> In assessing the compensation to be paid by the acquiring authority under this Act regard shall be had not only to the value of the land to be purchased by the acquiring authority but also to the damage, if any, to be sustained by the owner of the land by reason of the severing of the land purchased from the other land of the owner, or otherwise injuriously affecting that other land by the exercise of the powers conferred by this or the special Act.

This enactment gives statutory authority for the payment of compensation for all three categories of severance and injurious affection referred to at 8.1(a) above. It should be noted that the section specifically refers to cases where land is purchased from the claimant (hence the use of the phrase "acquiring authority"). The wording is broad and comprehensive, and except for the problem discussed at para 8.2.8 below, will normally ensure that a qualifying claimant is fully compensated for any loss suffered.

Where part only of land is taken it is advisable for the valuer to ask himself two questions:

(a) Has material detriment been suffered by the remainder? (see Chapter 7, p 266, and section 8, 1965 Act.)
(b) Do before and after valuations indicate that injurious affection has been suffered?

If the acquiring authority is not to be forced into acquisition of the whole under section 8 of the 1965 Act, a before and after valuation of the land retained should quantify the injurious affection claim. This sum will be added to the market value of the land taken, and disturbance if appropriate, to produce the total compensation claim. It is advisable not to compare the value of the original holding with the value of the retained land in a before and after valuation for the reasons discussed at 8.2.8 below. However, this strict separation of the heads of claim may not always be practical where the land taken represents only a small part of the whole (see example 8.1 below).

8.2.2 Separation of claim items

When reference is made to the Lands Tribunal, it will be necessary to give "details for the compensation claimed, distinguishing the amounts under separate heads and showing how the amount claimed under each head is calculated" (1961 Act, section 4).

Hence, having established that a claim for injurious affection compensation exists, it will be necessary to split the total into claims for land taken and for injurious affection.

Example 8.1

For example, assume that a strip of front garden land 5m × 10m, is to be acquired from the plot of a large house worth £420,000 before the acquisition. The acquisition is for road-widening purposes; the value of the house after the acquisition is £360,000.

Total compensation claim:

Value of house before acquisition		£420,000	
Value of house after acquisition		£360,000	
Total depreciation			£60,000

Check: value of land taken:			
50 m² of garden land @ £90 m²		£4,500	
Injurious affection therefore	£60,000		
	£4,500		
		£55,500	
Total depreciation			£60,000

The injurious affection may be caused by at least three separate factors: depreciation due to reduced garden size which considerably reduces the number of potential purchasers and is not completely reflected in the rule 2 value of land taken; visual intrusion due to the fact that the road is now closer to the front window of the house; and noise, smell etc. from the increased traffic volume due to the road widening. It is not necessary to split the injurious affection claim into constituent factors of causation, however.

8.2.3 1973 Act, section 44

Section 44 of the 1973 Land Compensation Act extended what was formerly a restricted item of claim of which the depreciation referred to above is an example. It reads:

> Where land is acquired or taken from any person for the purpose of works which are to be situated partly on that land and partly elsewhere, compensation for injurious affection of land retained by that person shall be assessed by reference to the whole of the works and not only the part situated on the land acquired or taken from him.

It is apparent that in example 8.1 the effects of the whole road-widening scheme as it reduces the value of the claimant's house should be taken into account, even if part of the road is to be constructed on land not acquired from the claimant. This saves the valuer from the considerable problems which had previously arisen when the law required that only works carried out on land taken from the claimant could be considered.

8.2.4 Disturbance

A claim for injurious affection will always be based on a loss in land value. The resultant compensation will not include any loss of profit suffered by the occupier of the land: this is a matter for disturbance (see *Cooke* v *Secretary of State for the Environment* (1973)). It is generally, though not universally, accepted that loss of profits and other disturbance losses are allowable as part of a claim under section 7. Where, however, no land is taken and section 7 does not apply, compensation is limited to depreciation in the value of land and buildings only.

8.2.5 Retrospective claims

Although the claim for injurious affection must be made in the light of values subsisting at the valuation date, the real damage suffered by the claimant may not become apparent until several years afterwards. In *Waterworth* v *Bolton Metropolitan Borough Council* (1979), the acquisition of part of the claimant's land resulted in the deferment of the receipt of planning permission for development of part of the land. By the time of the tribunal hearing the actual length of deferment was known, and the Tribunal felt it could take this fact into account, and was not bound to base the compensation on the deferment period which might have been anticipated at the valuation date. This decision was confirmed by the Court of Appeal in 1981.

In *Chamberlain* v *West End of London and Crystal Palace Railways Co* (1863), it was concluded that "a person seeking to obtain compensation under these Acts must once for all make one claim for all damages which can be reasonably foreseen". The *Waterworth* decision perhaps questions this ruling and the suggestion of Corfield and Carnwath (1978: 326) may have gained authority. In their opinion:

Logic would seem to point to a further claim being permissible where, for example as a result of a change of plan by the authority concerned damage occurs which could not have been foreseen at the date of the initial claim.

However, acceptance of the authority's offer in "full and final settlement" will inhibit the prospects of re-opening the dispute.

8.2.6 Planning assumptions

A question will often arise as follows: in determining the extent of depreciation to retained land, what assumptions concerning demand and planning permission should be made concerning that land? It is apparent from section 14 of the 1961 Land Compensation Act that the planning assumptions do not apply to land other than that which is to be acquired. Yet an Australian case, *Melwood Units Pty Ltd* v *Commissioner of Main Roads* (1979), established a more vague set of assumptions which should be made by the valuer. Land of 37 acres was owned by the claimants who intended to employ it in the construction of a drive-in shopping centre. A road scheme prevented the use of the whole land for this purpose.

It was held that in assessing compensation for severance the valuer should envisage what use could have been made of the land if the acquisition had not proceeded.

Care should be taken, however, to ensure that this assumption is consistent with the actual proposals of the acquiring authority: the *Melwood* decision was questionable because it was assumed that the whole 37 acres could have been used as a shopping centre even though the land was purchased in the knowledge of road proposals and the shopping centre was anyway enhanced in value by the future existence of the road.

8.2.7 Rights over land

Where a right over land is acquired, the owner of the right is entitled to compensation for injurious affection to retained land and section 7 of the 1965 Act applies as if a parcel of land had been acquired: see *Turris Investments Ltd* v *Central Electricity Generating Board* (1981).

8.2.8 The Hoveringham problem

In order that justice is done, the following equation should describe the quantification of compensation for injurious affection:

Value of holding before acquisition – Value of retained land plus compensation for land taken
= Injurious affection compensation

Unfortunately, this is not always the case and it appears that acquiring authorities are prevented from engineering an equitable solution by the courts' interpretation of the law.

Consider the following example. The local authority wishes to acquire a strip of land of 10m × 100m (one-tenth of a hectare) for a road-widening scheme. The strip forms part of a 1 ha plot of land which is zoned on the development plan for industrial purposes.

Industrial land sells on the open market at £400,000 per ha; amenity land sells at £10,000 per ha.

Pro rata, the development value of the land acquired would be £40,000: but the demand to purchase a 10 m wide strip for industrial development would be extremely limited, as it could not be developed in isolation. Its existing use value is £1,000; the claimant has an offer of compensation in that sum and retains land worth £360,000, leaving him with assets worth £361,000 after the scheme, compared to £400,000 before. He has lost £39,000 in respect of severance suffered on the land taken.

Will injurious affection compensation cover such a loss? It is arguable that a reasonable interpretation of section 7 of the 1965 Act would justify this equitable result as the claimant has suffered damage "by reason of the severing of the land purchased from the other land of the owner".

It appears, however, from the decision of the Court of Appeal in *Hoveringham Gravels Ltd* v *Chiltern District Council* (1977) that section 7 cannot be employed to attain this result. The court said:

> It seems to us clear that the section on its true construction is envisaged as an additional head of compensation for the owner of the land taken by reason of other retained land of his being less valuable to him through that retained land being severed from or otherwise injuriously affected by the compulsory acquisition of the land taken.

In short, injurious affection compensation is payable in respect of a depreciation in value only of the claimant's retained land, and not for depreciation in the value of the land acquired from him. This may result in injustice in cases such as that considered above. (For a fuller discussion of this problem see Sams 1982.) This point was amply illustrated in *Abbey Homesteads Group Ltd* v *Secretary of State for Transport* (1982) which concerned a long, narrow strip of development land acquired for a new road. The land taken was too narrow to be developable on its own, but would have been developed in conjunction with the adjoining land if it were not for the road scheme. In spite of the obvious inequity the tribunal was obliged to award compensation on the basis of the market value of an undevelopable strip of land.

The before and after valuation should consequently be applied to the retained land only in order to calculate injurious affection compensation. While the before and after valuation of the whole plot prior to and after the acquisition may more fairly indicate the measure of the claimant's loss, it may not produce the correct compensation figure. In *ADP & E Farmers* v *Department of Transport* (1988) this method of valuing land taken and land retained separately was preferred by the Lands Tribunal.

Example 8.2

The following example will serve to illustrate the quantification of compensation in what is usually called a "severance" case and a more general injurious affection claim (see 8.1 above).

The claimant owns a house and garden, the total plot size being 1 ha, its value before the acquisition being £360,000. The local authority wishes to construct an access road to its nearby new housing development to the rear of the property (see Figure 8.1). It has powers to acquire extra land in the area if necessary for the scheme.

Advise the local authority if:

(a) it acquires a strip of land total size 200 m2 for the access road severing the claimant's house from the greater part of his garden

(b) it acquires the garden land also (total land taken 0.75 ha) to avoid an element of the severance claim

(a) This is a classic severance problem. In addition to a loss of land and injurious affection to the house, the claimant will suffer a depreciation in the value of his holding caused by the severance of the rear garden from the house.

The valuer must envisage the new use of the rear garden land. Can the claimant gain easy access to it and use if (for example) as an allotment? Can he let it to an adjoining farmer? Can he sell it as garden land to an adjoining householder?

Figure 8.1 The property involved in Example 8.2.

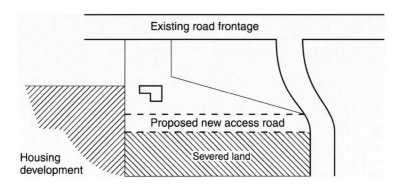

Assume that there is evidence in the locality that garden or amenity land sells in small parcels at a price representing around £16,800 per ha, but the existence of a single possible purchaser for this particular land reduced the price that would be paid for the severed land to a nominal £3,000.

The following claim should be made:

(i) Land taken:
200 m² @£16,800 per ha   £336

(ii) Injurious affection:

(1) To house, as retained:

Value of house with 0.25 ha of garden as part of the original holding, say £360,000 — (0.75 ha @ £16,800 per ha) = £12,600	£347,400	
Less value of house with 0.25 ha of garden land after acquisition, say	£300,000	
Depreciation in value of house		£47,400

(2) To severed land:

Value of severed land as part of the original holding; 0.73 ha @ £16,800 per ha	£12,264	
Value of severed land after acquisition	£3,000	
Depreciation in value of severed land		£9,264
Compensation for injurious affection		£56,664
Total compensation		£57,000

Check:

Value of holding prior to acquisition		£360,000
Value of house after acquisition	£300,000	
Value of severed land after acquisition	£3,000	
Compensation	£57,000	
		£360,000

The claimant is placed in the same position (as far as money can do so) after the acquisition, as there is no depreciation in value of the acquired land. Note, however, that following the *Hoveringham* decision the land taken should be valued as a narrow strip of land with only one likely purchaser. Such a valuation may produce a figure for land taken lower than £336 in which case the claimant would not receive sufficient compensation to put him in the same financial position after the acquisition. While the difference in figures is insignificant in this example, it can be very large where the land taken has development potential.

(b) In practice, this is the more common severance and injurious affection situation. If the local authority takes the garden land in addition to the strip, there may still be injurious affection to the retained land caused by nuisance from the new road and severance from the loss of part of its garden.

(i)	Land taken:		
	0.75 ha @ £16,800 per ha		£12,600
(ii)	Injurious affection:		
	Value of house with 0.25 ha of garden land		
	as part of the original holding, say	£347,400	
	Less value of house with 0.25 ha of		
	garden land after acquisition, say	£300,000	
	Compensation for injurious affection		£47,400
	Total compensation		£60,000

Check:		
Value of holding prior to acquisition		£360,000
Value of house after acquisition	£300,000	
Compensation	£60,000	
		£360,000

Again the claimant is placed in the same position (as far as money can do so) after the acquisition as there is no depreciation in value of the acquired land.

8.3 Injurious affection where no land is taken

8.3.1 General basis

While section 7 of the 1965 Act gives statutory authority for the payment of injurious affection compensation to landowners who are partially expropriated, much more restrictive rights to compensation may exist for other landowners from whom no land is actually acquired. Examples of cases where injurious affection may be suffered without land being taken include road-widening schemes which convert the quiet road outside the front garden into a dual carriageway, thereby reducing the value of properties on the road without necessarily taking land.

This category of injurious affection case may lead to a compensation claim by two alternative routes. The original of these routes is the result of judicial interpretation of section 68 of the 1845 Lands Clauses Consolidation Act, re-enacted with modifications in section 10 of the 1965 Act. For ease, it is useful to refer to this category as section 10 compensation as a contrast to the right afforded by section 7 of the same Act. The more modern alternative is the right afforded by Part I of the 1973 Land Compensation Act, and known as a Part I claim.

8.3.2 Section 10, 1965 Compulsory Purchase Act

The relevant wording in section 10 is:

> This section shall be construed as affording in all cases a right to compensation for injurious affection to land which is the same as the right which section 68 of the Lands Clauses Consolidation Act 1845 has been construed as affording ...

It is only possible to speculate on what persuaded the parliamentary draftsman to produce such a contrived and unhelpful provision. This was a consolidation act and his role was simply to set out the law as it already stood. He clearly felt this to be too difficult a task, and therefore decided simply to repeat the original provision. However, to find this he had to go all the way back to 1845, and even then, he found that section 68 actually says little which is specific to the subject. He therefore had to re-enact the interpretation of section 68 ie the case law arising from it, rather than section 68 itself.

Section 10 compensation is dependent upon the fulfilment of four requirements set out in the leading case arising from section 68. These are known as the *McCarthy* rules, from the House of Lords' decision in *Metropolitan Board of Works* v *McCarthy* (1874). This case summed up the decisions in earlier cases including *Hammersmith City Rly Co* v *Brand* (1867) and *Ricket* v *Metropolitan Rail Co* (1867).

The rules are as follows:

(a) The injurious affection must result from the execution of the works and not their use

This rule by itself restricts the usefulness of section 10 compensation to an enormous degree. Section 7 compensation is not, of course, subject to any such rule, and under that provision injurious affection compensation may be payable for losses resulting from the execution and use of the public works necessitating the acquisition of part of the claimant's land. Interpretation of the word "execution" is, as would be expected the nub of the problem. Judicial interpretation has been generally conservative, and it appears that compensation will only be paid (for example) where the loss results from the actual building of a road and not its subsequent use.

The valuer may be posed with insurmountable problems of interpretation of this rule. Generally, a house affected by a road-widening scheme is depreciated in value by the use of the road and not the actual construction process, which is a temporary inconvenience. Building a road or railway so as to block an easement of way will give rise to compensation while a loss of peace and quiet resulting from its construction will not.

(b) Compensation can only be claimed in respect of depreciation in value of rights in land

Consequently, a loss of trade is not compensatable unless it can be shown to have a direct effect upon the capital value of the premises.

Infringement of an easement such as a right to light may also be suffered as a result of the execution of public works. In such a case, following *Eagle* v *Charing Cross Rail Co* (1967), compensation should be payable.

A summary of this rule is that the test is simply to show that the market value of the claimant's interest in his land has been depreciated by the execution of the works.

(c) The cause of injurious affection would be actionable as a tort was it not for the existence of statutory powers

It is this rule which most severely limits the application of section 10. Statutory powers protect an authority such as a district council from an action in (to take the most likely tort) nuisance, when public works interfere with the individual's enjoyment of his use of land. It is often said that a claim for injurious affection is the wounded party's means of redress where statutory authority prevents such an action.

The main difficulty is that to have a valid action for nuisance, it would be necessary for the claimant to demonstrate that he has suffered injury to a personal legal right, over and above that suffered to the public at large. Thus a claim for a loss of privacy would fail, because this is not actionable as a tort (see *Re Penny and South Eastern Rail Co* (1857)). For most practical purposes this rule is unlikely to be met except where a claimant has suffered interference to an easement or wayleaf, such as a right of access.

(d) The action giving rise to the depreciation of the claimant's land must be authorised by statute

If the action were not authorised by statute then a claim in tort would be possible. It is only where such a claim is not possible, due to statutory authority, that section 10 is required. This may be a question of degree: the authorising Act may give certain powers which the authority might exceed by an *ultra vires* act, as in *Clowes* v *Staffordshire Potteries Waterworks Co* (1872). Compulsory purchase requires statutory authority, so this rule will normally be satisfied if a compulsory purchase order is in place.

Section 10 compensation is rarely attainable, due in particular to the third of these rules. Many cases of depreciation in land values, and all cases of other types of loss, as a result of public works (sometimes called worsenment) therefore remain uncovered by this compensation route.

In the rare cases where all four rules are satisfied, it is necessary to consider the basis on which compensation will be assessed. Clearly compensation is limited to depreciation in land values, caused by the execution of the works. Does this mean any such depreciation, or only depreciation arising from the interference with the private right which would have given rise to a claim in tort, but for the statutory authority. The former would be neater, as section 10 would then supplement the Part I claim, which covers losses due to the use of, but not the execution of, the scheme. However, the latter appears more logical given that section 10 appears designed to replace a right which would otherwise exist in tort.

In *Wildtree Hotels Ltd* v *Harrow London Borough Council* (2000) the House of Lords was asked to consider just that question. It unanimously confirmed that compensation is limited to that arising from the claim which would have arisen in tort, though it did note that such losses could be of a temporary nature, and did not necessarily have to be permanent.

8.3.3 Part I Claim

An additional right to compensation was introduced by the enactment in 1973 of the Land Compensation Act. Part I of that Act gives a route to compensation for injurious affection where no land is taken. The result is by no means a comprehensive right to compensation, but the right given by Part I is considerably less restrictive than section 10 compensation.

Part I of the 1973 Act deals with "compensation for depreciation caused by use of public works", presenting an adjunct to the *McCarthy* rules restriction to execution and not use. As it is more easily envisaged that the use of (say) an airport is detrimental to the capital value of a dwelling than its construction, Part I is more easily interpreted by the valuer. Again, payment of compensation is dependent upon several rules to be found in Part I. The assessment of compensation in theory is straightforward and is aided by consistent judicial interpretation.

(a) The depreciation must be suffered by an interest in land as a result of physical factors. The factors are listed in section 1(2) as noise, vibration, smell, fumes, smoke, artificial lighting, and the discharge on to land of any solid or liquid substance.

(b) The depreciation must be caused by the use of public works, listed in section 1(3) as any highway, any aerodrome, and any works or land provided or used in the exercise of statutory powers. The latter blanket term appears to cover all cases and to render the former cases redundant.

(c) The interest must be that of an owner-occupier or (where the subject property is a dwelling) an owner. If the land is not a dwelling, then the annual value for rating purposes should not exceed a prescribed amount as laid down for the same purposes as a blight notice under section 149(3)(a) of the 1990 Town and Country Planning Act — currently £29,200 but subject to regular review.

The assessment of compensation is governed by section 4(2).

In assessing depreciation due to the physical factors caused by the use of any public works, account shall be taken of the use of those works as it exists on the first day of the claim period, and of any intensification that may then be reasonably expected of the use of those works in the state in which they are on that date.

The compensation payable shall be assessed by reference to prices current on the first day of the claim period, that is upon expiry of one year after commencement of the use of the public works. Thus, the valuation exercise is as follows: assess the capital value of the land and buildings one year after commencement of the use of the public works ignoring the effect of those works; assess the actual capital value of the land and buildings at the same date; and calculate the difference.

Calculation of the market value of the property leaving the scheme out of account may be relatively easily proven by reference to comparable evidence of sales in the market of similar properties unaffected by any scheme. This is not usually true of the task of actual calculating the actual market value. The valuer is forced to seek comparables otherwise identical and, in the case of a road scheme, a similar distance from a road generating similar levels of noise, fumes etc. The sensitivity of location as it affects value is a factor which often confounds this exercise. He must demonstrate that any difference between the two figures is attributable to the physical factors referred to in the Act, rather than to the construction and existence of the works. This evidence is likely to be impossible to find and consequently, many of the cases heard by the Lands Tribunal have been established by employing arguments that the value of the property must logically be, rather than proving that it has been, affected.

In *Marchant* v *Secretary of State or Transport* (1979), for example, a bungalow in a small rural village was affected by noise from the M20, sited 600 m away. The claimant made no attempt to establish a loss in value by means of comparable evidence, but instead employed an acoustics expert to report upon the noise levels created by the road comparing those levels with the recommended maximum noise levels to be found in the Final Report of the Committee on the Problem of Noise (HMSO, Cmnd 2056). He claimed a depreciation of 10% in the value of the house as a result and was awarded £1,000 or around 5%. It is interesting to note that percentage deductions such as this appear to be derived from rating practice. Prior to the 1973 Act there was no direct right to compensation, but houses were subject to the rating system and it was common for the lands tribunal to award percentage reductions in rateable value for houses affected by road schemes and other public works.

In *Arkell* v *Department of Transport* (1983) compensation was arrived at by valuing the property assuming the presence and use of the public works but without the physical factors. The actual market value of the property was then deducted to leave a figure representing the effect of the physical factors. However, the same member of the Tribunal, WH Rees, preferred an alternative approach in *Maile and Brock* v *West Sussex County Council* (1984). In this case there was sound evidence that the actual value of the property at the valuation date was £32,000. Adopting the *Arkell* approach the claimant arrived at a valuation of £37,250, assuming the presence and use of the road but without the physical factors. He therefore claimed compensation of £5,250. The authority's valuer first valued the property in the no-scheme world at £35,000 and then assessed injurious affection caused by "physical factors", taking 5% of this value to produce a figure of £1,750.

Mr Rees reminded the claimant that in the earlier case both valuers had adopted the same valuation basis and given the choice he had in this case, he preferred the local authority's method. Making a small adjustment to the percentage he awarded compensation at £2,000. It should be noted that the no-scheme value of the property (£35,000) less the compensation (£2,000) does not leave the actual value of the property (£32,000). The missing of £1,000 represents depreciation caused by the presence of the road which is not compensatable under the 1973 Act, as it is not attributable to the physical factors.

It should be noted that the loss of a view or general ambience of a location are factors which are not compensatable under Part I. Thus, in *Shepherd and Shepherd* v *Lancs County Council* (1977) the use of nearby land as a tip resulted in no compensation being payable as the claimants failed to prove that the depreciation suffered by their property (which was not doubted by the Tribunal) was the result of the physical factors listed in the Act.

A Part I claim is more likely to produce compensation than the route provided by section 10 of the 1965 Act. Nevertheless, the scope for compensation where no land is taken is very restricted and as Davies suggests, the landowner "should move heaven and earth to convey a square metre (or less, if necessary) to an acquiring body rather than have it acquire adjoining land yet none from him".

8.4 Mitigation

Part II of the 1973 Act includes provisions enabling the Secretary of State to make regulations concerning the mitigation of the injurious effect of public works. He may impose a duty on statutory authorities to insulate, or pay grants towards insulation of, property or alternatively confer a power on such authorities to do so. Noise Insulation Regulations set out the detailed obligations imposed on acquiring authorities. In respect of road schemes this normally means that local authorities will take noise readings to each elevation of a house before the scheme is built. It will then either estimate, or on completion, calculate noise levels following completion of the scheme (the former will allow works to be of benefit during construction as well as later). If the difference in noise levels exceeds the limits set out in the Noise Insulation Regulations, sound insulation will be offered. This will take the form of secondary glazing to windows, and often the replacement of doors. Ventilation ducts will also be installed, as it would defeat the objective if windows had to be opened to provide ventilation.

In addition, responsible authorities may acquire land for the purpose of mitigating any adverse effect which the existence, or use, of any public works has on the surroundings. This work may include tree planting as a means of soundproofing or visual protection.

While Part II gives no right to compensation, the works offered by the authority in mitigation of injurious affection should be taken into account in the quantification of any injurious affection claim under Part I. This will be the case whether or not the offer is accepted by the claimant (1973 Act section 4(3)).

In severance cases, other works towards mitigation of injurious affection may be undertaken by acquiring authorities. For example, where a motorway is driven through farmland the depreciation of the severed land normally compensatable may be reduced by the construction of connecting bridges between the two parcels.

The general principles of compulsory purchase compensation require that the expropriated owner is placed in the same position as if no land had been taken from him. Consequently, the replacement of a boundary hedge or fence should be undertaken by the acquiring authority where land is severed although the statutory authority for such work to be undertaken is less than comprehensive. For a fuller discussion of fencing, see *Cuthbert* v *Secretary of State for the Environment* (1979): see 9.6.

8.5 Set-off, or betterment

It might be argued that governmental bodies and other statutory authorities are especially generous concerning what is generally known as the "compensation-betterment problem". Activity in the property market by such a powerful force as a county council can result in a major redistribution of land values; increasing development value here, reducing existing use value there. In theory. the ideal solution would be for all adverse effects to be fully compensated, with a betterment levy being imposed on all beneficiaries. As already discussed the compensation obligations are less than comprehensive, but betterment provisions are even more limited; in fact there is no route by which increases in value can be extracted from the fortunate landowner who retains all of his land. There are, however, provisions enabling a statutory authority to reduce the quantum of compensation payable for land taken or injurious affection where other lands in the same ownership are benefited by the same scheme. No landowner is required to pay negative compensation where the betterment exceeds the compensation payable: the minimum quantum of compensation is nil.

The general set-off provision is to be found in section 7 of the 1961 Land Compensation Act. Figure 7.2 (p 235) was employed to illustrate section 6 of the same Act. That diagram may be modified to illustrate section 7 (see Figure 8.2).

Figure 8.2 Diagram to illustrate section 7 of the Land Compensation Act 1961

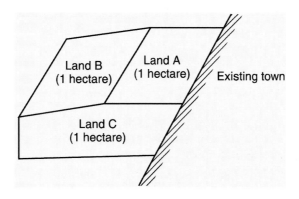

Lands A and B, in separate ownership, are to be acquired by the local authority. Land A is ripe for development and worth £80,000; land B is worth only £40,000 as it is separated from existing roads and services by land A.

Land C, in the same ownership as Land B, is worth £60,000, being partially ripe for development.

As land A is acquired and developed the values of both B and C will rise to £80,000 as they too become ripe for development. The subsequent acquisition of land B is at a price of £40,000 due to effect of section 6. But the claimant has received a benefit in the form of an increase in value of land C to the tune of £20,000.

In order that the landowner is in the same position prior to and after the scheme, the compensation award should be £40,000 – (the increase in value of C) £20,000 = £20,000. Prior to the scheme he held land worth £100,000; subsequent to it, he holds land worth £80,000 and should therefore receive £20,000 compensation.

This is effected by section 7, which provides that where:

> the person entitled to the relevant interest is entitled in the same capacity to an interest in other land contiguous or adjacent to the relevant land there shall be deducted from the amount of the compensation which would be payable apart from this section the amount (if any) of such an increase in the value of the interest in that other land as is mentioned in subsection (2) of this section.

Subsection 2 refers to schedule 1, Part I of the Act, which has already been discussed in relation to section 6 (see p 234).

There is a similar provision to section 7 with specific reference to highways acquisitions in the 1980 Highways Act (section 261). In *Grosvenor Motor Co Ltd* v *Chester Corporation* (1963), the claimant company claimed compensation for the compulsory purchase of part of a strip of land acquired for road widening. The company owned other land behind the frontage of the new road where it ran a garage. It was held that the value of the land acquired was £6,899, but that the creation of a new frontage to the garage increased the value of the retained land by £8745. Thus no compensation was payable.

In *Portsmouth Roman Catholic Diocesan Trustees (Registered)* v *Hampshire CC* (1979), land was acquired for a road scheme. There was no doubt that planning permission for residential development of contiguous lands would have been deferred until the road was built. The acquiring authority attempted to rely upon section 261 in order that betterment resulting from the grant of planning permission should be set off against the value of land taken. However, the Lands Tribunal considered that the grant of planning permission is not the kind of benefit which the set-off provisions are designed to collect and that the type of benefit qualifying is "one which is directly referable to the purpose for which the land is authorised to be acquired such as where the coming of the road will provide access to the retained land of a new or improved kind". It should be noted that the requirements of section 7 are less specific: it is not certain that the same result would have been obtained had the authority been able to rely upon it and the Portsmouth decision may be questioned as being unnecessarily restrictive concerning the betterment principle.

An interesting problem arises if the benefited contiguous land in the same ownership such as land C in Figure 8.2 is later acquired under the same scheme. The landowner, having received £20,000 compensation when land B was acquired would be faced with an offer of only £60,000 for land C due to the provisions of section 6, were it not for section 8 of the same Act. Where section 7 has already been applied to retained land the same increase in value should not be disregarded by section 6 upon subsequent acquisition of the land. Consequently, £80,000 compensation should be paid leaving the landowner with £100,000 in cash rather than in land value.

There is also a set-off provision in the 1973 Land Compensation Act, section 6. Where a Part I claim is payable as a result of injurious affection caused by physical factors, the compensation shall be reduced by an amount equal to any increase in the value of the subject land, or any other land contiguous or adjacent to it, attributable to the use of the same public works.

The fairness of these set-off provisions continues to be disputed by both acquiring authority and claimant, though from different points of view. It is clearly inequitable for a landowner to be allowed to treat as a windfall an increase in land value caused by public works, except where he is unfortunate enough to have other land taken from him for those works.

8.6 Damage to land

The right to injurious affection compensation given by section 7 of the 1965 Compulsory Purchase Act is designed to compensate for damage sustained by the owner of that land. To date, such damage has been seen in terms of a loss in value caused by economic factors, but there are cases where the act of a statutory authority may create physical damage, giving cause to a claim for compensation or, to place the subject nicely in context, damages.

In the majority of cases, the physical damage caused by (say) the cutting of a canal through development land will be put right by accommodation works (considered below), while the monetary loss in land value will be compensated in the normal way. There are several particular examples, however, where special comment is merited.

Injurious affection may be suffered by a property as a result of work done to deepen, widen, or straighten brooks, streams, or rivers by local drainage boards. In such cases the Land Drainage Act 1991 provides for full compensation to be paid. In *Strutt's Kingston Estate Settlement Trustees* v *Kingston Brook Internal Drainage Board* (1979), where a bridge collapsed in floods as a result of river-straightening work, this was held to be the cost of reinstatement of the bridge. In *Weeks* v *Thames Water Authority* (1979), on the other hand where a loss in land value was suffered by a riverside cottage, the usual basis of injurious affection compensation was employed.

It appears that the Lands Tribunal, and indirectly the valuer, is left with a long rein in his interpretation of full compensation which, it is suggested should be an attempt to reflect the claimant's loss.

The construction of a public sewer gives rise to a similar right to compensation under The Water Industry Act 1991. In *George Whitehouse Ltd* v *Anglian Water Authority* (1978), a sewer laid in a public highway caused both a loss of custom and the necessity to clean forecourt cars regularly due to dust and mud thrown up by excavation. It was held that the claimants had a right to compensation under both heads, again to cover their loss.

There are further specialised compensation provisions in respect of other statutory undertakings, such as gas pipelines and telecommunications.

8.7 Accommodation works

As an alternative to the payment of compensation to a claimant who retains land, the acquiring authority will often carry out works (known as accommodation works) in order to mitigate the effects of the scheme. Fencing along the boundary of a new road would be a typical example of such works, which normally belong to, and are maintained by, the claimant.

There is no statutory basis for such works, and therefore no obligation either on the authority to provide them or on the claimant to accept them, except by agreement between both parties. It will, however, often be considerably more convenient and economical for the authority to provide accommodation works in connection with a scheme of public works, than for each individual claimant to arrange for works to be carried out and to include the expense in his own compensation claim. From the claimants point of view, this saves him the trouble of organising his own works, and, as cost does not necessarily equal value, the cost of doing the work may be more than the increase in his claim for severance and injurious affection. Accommodation works can, therefore, be considered as an alternative to the payment of compensation and will only be provided in respect of works, the cost of which could form part of a compensation claim.

In considering this question, the usual rules of mitigation of loss, betterment, and value for money must be taken into account (see p 296).

Other examples of accommodation works have included the provision of a bridge to reunite severed parcels of land, the making good of gable walls when adjoining terraced houses are acquired, and even the provision of a new outside WC where part of a rear yard was taken.

Surveyors' fees will normally be paid for negotiating accommodation works based upon the cost of those works, again following the principle that that sum could otherwise form part of the compensation claim. For a detailed consideration of surveyors' fees in accommodation works: see *Mahood* v *Department of the Environment for Northern Ireland* (1985).

Example 8.3

The following example illustrates many of these measures as a means of compensating injurious affection. A full answer to the question involves a consideration of disturbance and other payments which have yet to be dealt with. The outline answer given below concentrates, in note form, on claims for land taken and injurious affection only.

Mr Dale is the freehold owner of 2 ha of vacant land at the rear of 1–15 (odd) Pickwick Road on the outskirts of a small seaside town. Included within his freehold ownership is a strip of land some 2m wide between nos 9 and 11. This strip of land is subject to a restrictive covenant in favour of the owner of no 11 restricting the use to pedestrian access only.

All the houses in Pickwick Road have frontages of some 7 m and would sell for about £150,000 each, freehold and with vacant possession. All are owner-occupied except no 9 which is held by a tenant who is paying £90 per week (exclusive) for the whole on a weekly regulated tenancy.

A compulsory purchase order is to be made for housing purposes in respect of Mr Dale's land (including the access) together with no 9 which will be demolished so that the existing access can be widened to provide vehicular access to the new estate.

Residential building land in the area with planning permission is fetching about £1,000,000 per ha.

The existing use value of the 2 ha of vacant land is assumed to be £200,000.

Prepare claims for Mr Dale, the freeholder of no 9, and any other compensatable interests quoting relevant statutory provisions and case law.

It is assumed that a reasonable vehicular access may only be formed by relaxing the restrictive covenant over the 2 m strip and purchasing and demolishing either no 9 or no 11 Pickwick Road.

Under section 84 of the Law of Property Act 1925 the owner of land subject to a restrictive covenant may apply for relaxation or discharge of that covenant to the Lands Tribunal which may award such compensation to the owner of the dominant tenement as it deems appropriate.

In *SJC Construction Ltd* v *Sutton London Borough Council* (1974), the *Stokes* principle was adopted (see p 264) and 50% of the increase in value of the servient tenement (in this example the access strip and back land) was awarded as compensation for relaxation of the covenant. In this case, 50% would not be appropriate as access depends not only upon relaxation of the restrictive covenant but also upon the demolition of no 9 (or no 11). A lower percentage deduction — say 15% to 20% — might instead be appropriate.

Figure 8.3 Diagram illustrating property

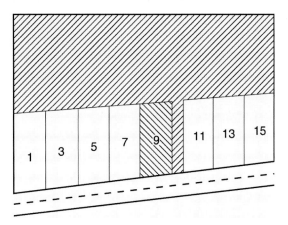

The *Stokes* principle is again applicable in determining the market value of nos 9 and 11 to reflect their use as alternatives to a developer wishing to purchase both the back land and the access strips. A market analogy should be followed. What percentage of the gain in value of the back land resulting from its accessibility would be given up to the access owner by a bidder in the market?

In the *Stokes* case, this was settled at 33%; in this case, a lower percentage of (say) 15% should be used to reflect the substitutability of nos 9 and 11 as alternatives.

(a) Mr Dale

The land taken amounts to 2 ha of vacant land with an assumed planning permission for residential development (1961 Act, section 15), plus 2 m × (say) 20 m access strip. Assuming that no more valuable planning assumptions may be made, this has a value of £2,000,000. However, in order to exploit this value, a vehicular access must be acquired and the restrictive covenant must be relaxed. Assuming that it would not be economical to purchase and demolish any of nos 1–7 and 13–15 in preference to extending the existing strip, then any purchaser's bid for the back land would be reduced by the cost of purchasing no 9 or no 11 and relaxing the restrictive covenant. If he were to buy no 11, the two parties would negotiate an agreement whereby the owner of no 11 transferred his interest including the benefit of the restrictive covenant. The minimum price he would possibly accept would be £150,000, while the maximum bid the intending purchaser could possibly make would be £1,800,000 and a settlement would be reached between these figures reflecting a split of the gain of £1,800,000 (2 ha. @ £1,000,000 per ha less existing use value of £200,000) between the parties. A split of 65/35 in favour of the purchaser results in a price of £180,000 + [0.35 × £1,800,000] = £630,000 being paid for no 11. This is made up of the following:

Allowance for relaxation of covenant: 20% of gain	£360,000
Allowance for ransom value of land: 15% of gain	£270,000
	£630,000

On the other hand no 9 might be purchased. After allowing for the relaxation of the restrictive covenant as agreed with the owner of no 11 for an inducement of £400,000, the purchase of no 9 is likely to be a more logical course of action for the potential owner of the back land as it is tenanted and intrinsically less valuable. An inducement would have to be paid to the tenant to move (see chapter 2), say £20,000. The market value of the freehold interest in no 9 (as tenanted) is assumed to be around two-thirds of vacant possession value, say £100,000. As an alternative to this a *Stokes* addition of 15% of the gain in value (£270,000) would be payable.

Relaxation of restrictive covenant	£360,000
Inducement to tenant	£20,000
Stokes payment for No 9	£270,000
Total outlay if No 9 is used as access	£650,000

Assume, therefore, that No 11 is purchased and the restrictive covenant relaxed for a total outlay of £820,000.

Value of Mr Dale's interest with access	£2,000,000	
Cost of access	£630,000	
Value of interest without access		£1,370,000

Severance and injurious affection — nil. The whole of Mr. Dale's land is taken.
Disturbance — nil. Mr. Dale is not an occupier and does not qualify.

(b) Owner–occupier of no 11

No 9 is actually to be acquired and demolished but it is assumed that the restrictive covenant benefiting no 11 must also be relaxed.

Following the assumptions outlined above, compensation of £360,000 (20% of the gain) should be payable for this.

The relaxation of a restrictive covenant qualifies as the taking of land so that the normal rules of injurious affection apply. No 11 may well be reduced in value as a result of the increased volume of traffic and the building of a small housing development to the rear.

Assume the value of the house after the acquisition is £120,000. Compensation of £30,000 is therefore payable for injurious affection.

Total compensation: £360,000 + £30,000	£390,000

It is unlikely that the owner-occupier of no 11 could rely upon section 8 of the 1965 Act and force the acquisition of the whole of his property (see p 266).

Disturbance will be payable, but is likely to be minimal. It could include costs such as extra cleaning of windows and driveways during the construction period.

(c) Freeholder of No 9

The whole of the interest of the freeholder is to be acquired so no right to injurious affection compensation arises.

Value of interest, based on its ransom value (see above):	£270,000

(d) Tenant of no 9

This tenant, who might have received an inducement to vacate in the normal course of events, is entitled to no compensation under section 20 of the 1965 Act for his interest as the value of his unexpired term is negligible (see p 264). This loss is balanced by the right to rehousing which he gains upon compulsory acquisition of the property (see p 317). Disturbance will also be payable under section 37 of the 1973 Act and will include removal costs, redirection of post, etc.

(e) Owner–occupier of No 7

While the freeholder of No 7 will have no land acquired from him he will be injuriously affected in several ways. He may lose a right of support; traffic will flow directly past his new flank wall; a new housing estate will be built to his rear.

Accommodation works should reduce the burden of compensation in respect of the first of these. If not, the cost of supporting and making weatherproof the new flank wall (say £20,000) will be awarded. The second and third factors may give rise to a claim under section 10 of the 1965 Act as governed by the *McCarthy* rules, or a Part I claim under the 1973 Act.

Under the *McCarthy* rules the depreciation in value which might be compensatable must arise from the execution and not the use of the public works. It can be difficult enough to assess the total depreciation suffered but the need to distinguish between these two causes creates even greater problems. While the other rules may be fulfilled it will be advisable to pursue, in addition, a Part I claim. It should be noted that the 1973 Act, section 8(7), precludes the payment of compensation under both Part I and "under any other enactment" for the same depreciation: it is suggested however, that the execution and use of the public works will produce two separate causes for depreciation and two claims under section 10 and Part I.

Disturbance is not payable as there is no expropriation of an interest in land.

Pursuing a Part I claim in isolation, three rules were identified on p 283. All three are fulfilled as the interest is that of an owner-occupier and the causation is a highway coupled with other "works or land provided or used in the exercise of statutory powers". Compensation is limited to the loss caused by "physical factors", which in this case includes noise, vibration, fumes, etc. The claim should be established by the use of comparable evidence before and after the works are complete or empirical argument employing noise level readings, etc. The claim is based on depreciation, say £20,000.

This may be reduced by mitigation works (such as soundproofing) carried out by the responsible authority under Part II of the 1973 Act.

(f) Other owner–occupiers

It may be possible to establish claims under section 10 or, more likely, Part I for all other owner-occupiers in respect of an increased traffic flow but not, it should be stressed in respect of a loss of privacy or view to the rear of their property.

Again, work carried out in mitigation may reduce or abrogate the responsibility of the authority to pay compensation.

Disturbance

9.1 Introduction and general principles

Section 5 of the 1961 Land Compensation Act lays down the six rules which might almost be termed a code for the assessment of compulsory purchase compensation. The last of these is the rather flimsy justification for calling disturbance a statutory matter:

> The provisions of rule (2) shall not affect the assessment of compensation for disturbance or any other matter not directly based on the value of land.

While rule 6 is often quoted as the legal basis of the disturbance claim, it is clear from the wording that such a right is presumed to exist already. The rule simply gives instruction on how it is to be assessed. In fact, the right to disturbance compensation has no clear statutory basis at all; it has developed from case law and common practice arising from the 1845 Act.

This instruction is necessary to remove any conflict between the well-established common law right to compensation for disturbance and the requirements of rule 2, which establishes market value as the basis of compensation for land taken. In the open market sale of a house, no allowance is apparently made in the purchase price to cover the vendor's removal costs: yet the *Ricket* principle, considered at the outset of the previous chapter, requires that compensation should put the owner in the same position as if no property had been acquired from him, so far as money can do so. In addition to compensation for the market value of the land taken and for injurious affection, disturbance is the third balancing factor which covers all the other losses and expenses forced upon the owner of land which is compulsorily acquired. It is clear that the market value principle would, in any event, have little application to many aspects of the disturbance claim, such as redundancy payments and the cost of disconnecting services on the closure of a business. There are, however, circumstances where market value could be relevant, for example the value of the goodwill of business when it is forced to close. Although there may be sound comparable evidence of how much would be paid for the business in the market place, rule 6 makes it clear that this is not the correct basis. Instead, compensation should be based on value to the owner. This is a much more subjective calculation; the only certainty being that value to the owner must be equal to or higher than market value. If the claimant valued his business at less than market value, he would presumably have sold it already.

Rule 6 refers in addition to "any other matter not directly based on the value of land". It is commonly accepted that this refers to the costs of the negotiation of the sale and the costs of conveyance of the vendor's title to the acquiring authority, usually charged by the vendor's surveyor and solicitor respectively.

An equation may therefore be established:

Compensation (that sum of money required to place the owner in the same position as if no land had been acquired from him) = Market value of land taken + Severance and Injurious affection to land retained + Disturbance + Professional fees

Unlike the first two elements of the claim, the rules for establishing whether a claimant qualifies for disturbance are few and relatively straightforward. The same can be said of the rules relating to whether any item claimed is allowable. It tends to be in the quantification of the loss that problems arise. First, however it is necessary to consider a number of common law guidelines which have become established over the last century and a half, in respect of qualification for, and quantification of, the disturbance element of the compensation claim.

9.1.1 Qualification rules

There are two rules to be considered when deciding whether a claimant qualifies to claim disturbance under rule 6. A right to disturbance compensation only arises where a claimant is (i) expropriated and (ii) dispossessed (evicted from occupation).

Disturbance compensation is payable only when a claim for land taken is established: consequently expropriation must be suffered. A valuable interest in land must be acquired from the claimant, and this has been interpreted to mean a freehold, or a leasehold with more than one year unexpired. Disturbance compensation under rule 6 is not, for example, payable to a statutory tenant of residential property. The Lands Tribunal affirmed in *Kovacs* v *City of Birmingham* (1984) that a claimant who is not in occupation of the land being acquired cannot be entitled to disturbance compensation under section 5(6) of the Land Compensation Act 1961.

However, enactments subsequent to the 1961 Act have rendered this rule virtually impotent. Section 20, 1965 Act gives a similar right to claimants with at least an annual tenancy. Section 37 Land Compensation Act gives a right to a "disturbance payment" to anyone displaced from lawful occupation, which will include the weekly tenant referred to above (see p 315).

Dispossession must also be established. In this case possession actually means occupation — "disturbance must, in my judgment, refer to the fact of having to vacate the premises" (Davies LJ in *Lee* v *Minister of Transport* (1966)). Disturbance is, therefore, not payable to a claimant who owns land, but is not in phyisical occupation of it, such as a developer, an investor or the owner of a vacant property.

This could create a situation where the compensation paid to the owner of an investment property is insufficient to purchase a replacement of similar value once professional fees and other incidental costs are taken into account. The obvious inequity of this general principle has been recently tempered to a limited extent by an amendment to section 10 of the 1961 Land Compensation Act ,which was introduced in Schedule 15 of the 1991 Planning and Compensation Act. Where a person who is not in occupation of property is subject to compulsory purchase, and he acquires a replacement property within one year of entry, the charges and expenses incurred in the purchase can be claimed as if he were in occupation and entitled to disturbance compensation.

Disturbance compensation is payable as part of the purchase price of the land acquired. No claim for disturbance compensation arises where the expropriated business has no interest in the land. This disqualification may be in question where the individual who has an interest in the property is a director of the company carrying on the business or where the occupant company is the subsidiary of a larger company, which has an interest in the premises. Although for all practical purposes the owner and the occupier are one and the same, it is possible for the acquiring authority to argue that neither qualifies to claim disturbance compensation. The owner does not qualify as he has not been dispossessed; the occupier because he has not been expropriated of an interest in land. In some cases the courts have chosen to lift the corporate veil and treat the owner and occupier as one and the same.

In *Smith, Stone and Knight* v *Birmingham Corporation* (1959), the occupation of the subject premises by a company subsidiary to the freeholder company did not disqualify the occupants from disturbance compensation as the subsidiary company was held to be operating strictly as the servant of the parent company. In *Taylor* v *Greater London Council* (1973), however, these were regarded as rare exceptional circumstances, and the claim of the freeholder of the subject property for trade disturbance in respect of a travel agency, of which he was a director, was dismissed. The travel agency was seen to be a separate legal persona with no interest in the subject premises.

In *DHN Food Distributors Ltd* v *London Borough of Tower Hamlets* (1976), the subject land was owned by a subsidiary company, while the parent company owned the occupying business. The parent company held all the shares in the subsidiary. The court of Appeal decided, first, that the parent company held an irrevocable licence to occupy the land which gave a right to disturbance compensation and, second, that the corporate veil may be lifted and the two companies could be treated as one and the same, particularly because all shares in the subsidiary company were held by the parent.

However, in two other subsequent cases it was held that the corporate veil should not, in the particular circumstances of each case, be pierced. The first of these is *Woolfson* v *Strathclyde Regional Council* (1978), a Court of Appeal case which questioned the *DHN* decision and distinguished it from the subject case because the landowner had less than complete control of the tenant company. This reflected the decision in *Taylor* v *Greater London Council* (1973).

In *Rakusen Properties Ltd* v *Leeds City Council* (1979), similar circumstances to those precipitating the *DHN* case resulted in a second questioning of that earlier award. Consequently, the *DHN* decision should not be taken as an inviolable guide: rather, one should apply the attitude of the Court of Appeal in the *Woolfson* case. Given total control of the subsidiary by the parent, is the corporate veil "a mere facade concealing the true facts"? If so, a right to disturbance compensation may be established even where the occupant company has no interest in the subject property.

The principle of lifting the corporate veil was taken to extremes by the Lands Tribunal in *John Edward Roberts and John Edwards (Bexley) Ltd* v *Ashford Borough Council* (2005), when it awarded disturbance compensation to a claimant who was neither expropriated nor dispossessed. Mr Roberts owned the land, and also the company which operated a plant hire business from it. However, by the valuation date the plant hire company had ceased trading and the land was let to a third party. As Mr Roberts owned almost all the shares in John Edwards (Bexley) Ltd, the tribunal had little difficulty in deciding that the corporate veil should be lifted, and therefore there had been expropriation of the owner-occupier's interest. It was, however, pointed out by the counsel for the local authority, that this argument failed to address the question of dispossession and that neither the company nor the individual was in occupation at the relevant date. The Tribunal, however, was satisfied that the plant hire business had only been forced to close as a result of the compulsory purchase and the member could not "see any reason why a claim for disturbance for losses suffered by the company is to be excluded for the reason that the company was no longer in occupation at the date of vesting. The purpose of lifting the veil is to reflect the reality of the situation — that the company's losses are Mr Roberts's losses — and I can see no reason why he should be deprived of compensation for losses suffered by the company in consequence of the acquisition in circumstances where, if the business had been in his name, he could have obtained compensation". Compensation for disturbance was, therefore, awarded even though there was clearly no dispossession at the valuation date.

9.1.2 *Quantification rules*

There are also only two main rules to consider when deciding whether any item of loss claimed is allowable, though there are a number of subordinate rules. A loss is compensatable as disturbance "provided, first, that it is not too remote, and, second, that it is the natural and reasonable consequence of the dispossession of the owner" (Romer LJ in *Harvey* v *Crawley Development Corporation* (1957)).

The same principles were expressed slightly differently by the Privy Council in *Director of Buildings and Lands* v *Shun Fung Ironworks Ltd* (1995), one of the subordinate rules being promoted to the status of a third main rule. The conditions for the payment of disturbance compensation were in that case stated to be:

(a)   causation: the loss must have been caused by the compulsory acquisition
(b)   remoteness: the loss must not be too remote
(c)   duty to mitigate: the claimant must act reasonably in seeking to mitigate his loss.

Therefore the rules of disturbance are very simple. A claimant who is expropriated and dispossessed can claim any loss which is not too remote, and which is a natural and reasonable consequence of dispossession, providing he does his best to minimise his loss. The lack of detail in these rules does, however, leave plenty of room for uncertainty, and it is necessary to consider some subordinate rules which have been developed to assist in their interpretation.

9.1.3 *Other rules of disturbance*

(a) *Value for money*

Where an expropriated and dispossessed owner replaces lost property, no compensation will be payable for disturbance in respect of any extra expenditure, over and above that necessary to put him in the same position as before.

In *Harvey* v *Crawley Development Corporation* (1957) Denning LJ considered a dispossessed house owner and said: "If he pays a higher price for the new house, he would not get compensation on that account, because he would be presumed to have got value for money."

This principle, which has become known as the "value for money" rule, does not relate purely to the extra cost involved in replacing the lost house, shop, or factory: it can relate to any item of claim where it is apparent that the claimant places himself in a better position after the acquisition. The *Harvey* case does, however, illustrate the most important ramification of the principle. If a houseowner is dispossessed of a small bungalow in a poor area of town for £200,000 and is only able to buy a similar house in a better part of town for £300,000, the extra £100,000 cannot be claimed as disturbance compensation. Nor can the additional surveyors fees, and other relocation costs, applicable to the purchase of a £300,000 house, compared to one worth around £200,000. In certain circumstances this may have unfortunate effects, but in the majority of cases the rule serves to prevent abuses of the compensation code, where market value is the measure.

Standard items in the disturbance claim, which will be discussed later, are also subject to the rule. For example, take a manufacturing company whose premises are acquired as part of land required for a road scheme. A timber sign is used to advertise the firm's name on the premises. Upon removal, a new, larger, illuminated plastic sign is used on the new premises. The cost of this sign should not be allowed, as the firm is presumed to have got value for money: the allowable amount should be the notional cost of removing and re-erecting the old sign.

In *Smith v Birmingham Corporation* (1974), an engineer's workshop was acquired. The claimant found alternative premises and spent money extending and improving them. The cost of these improvements was reflected in the market value of the new premises and was therefore not compensatable (see also *Goss v Paddington Churches Housing Association Ltd* (1982), where the cost of a new carpet was held not to be compensatable).

In some circumstances, on the other hand, such expenditure may be allowed if it is necessary to maintain the previous standard and level of output without being reflected in the value of the premises in the market (due to its peculiar relevance to a particular operator) or in increased profitability. In one negotiation, a metal processing factory needed a 150mm gas main to fire its furnace. The alternative premises it found had no gas supply so a new supply was installed. It was agreed that the provision of gas would enhance the value of the new property, but that most users would need no more than a 50mm supply. The additional cost of a 150mm main supply, compared to 50mm, was agreed to be allowable as it was not reflected in value for money.

In *J Bibby & Sons Ltd v Merseyside County Council* (1979), the claimants incurred increased operating costs at their new premises. Brandon LJ said:

> It seems to me that it would be right to award compensation in respect of such items if it were shown, firstly, that the applicant, as a result of the compulsory purchase, had no alternative but to incur the increased operating costs concerned and, secondly, that he had no benefit as a result of the extra operating costs which would make the incurring of them worth while.

In the *Bibby* case, the Lands Tribunal had held that the claimants' extra operating costs were worthwhile, presumably due to increased profitability, and the claim was therefore dismissed.

The value for money principle is the necessary progression from the overriding principle of compensation: the claimant "gains the right to receive a money payment not less than the loss imposed upon him in the public interest, but on the other hand no greater" (Scott LJ in *Horn v Sunderland Corporation* (1941)). In other words, an item of claim must provide something no better than was present at the old premises and the claimant must not receive value for money for that item. The principle was confirmed and expanded upon in *Director of Buildings and Lands v Shun Fung Ironworks Ltd* (1995) in which Lord Nicholls stated that "a claimant is entitled to be compensated fairly and fully for his loss", but " is not entiled to receive more than fair compensation".

(b) The claim must be consistent

The *Horn* principle is useful in establishing a more specific limitation to disturbance compensation. In that case, agricultural land which was ripe for building was acquired at a price reflecting development value. The dispossessed owner claimed disturbance in addition to development value.

This was clearly inconsistent with the basis of the claim for land taken. In order to realise the development value of the land in the absence of compulsory acquisition, relocation would in any case have been necessary and the incentive for this relocation is given by the extra development value of the land. Consequently, no disturbance compensation should be paid where the claim for land taken is based wholly on development value and the valuer should always consider the following two bases:

(a) existing use value + disturbance or
(b) development value.

The highest of these will form the basis of the claim.

Difficulties may arise where part of the claimant's holding is valued on the basis of development value and part at its existing use value. The example shown in chapter 7 on p 268 is a suitable illustration; in such a case disturbance compensation may only be awarded in respect of the land valued at existing use value, in that case the dwelling, and not for the land valued at development value.

(c) *Consequence means causation*

Until fairly recently it was regarded as well settled that expenses incurred before notice to treat will not earn disturbance compensation as, it was argued, they cannot be said to be "the natural and reasonable consequence of the dispossession", if they precede the dispossession order.

This principle could work unfairly as it did in *Bostock, Chater & Sons Ltd* v *Chelmsford Borough Council* (1973), where the affected company expended time and money in seeking alternative accommodation after the compulsory purchase order was confirmed, but before notice to treat was served, and was disallowed such expenditure.

Doubt was first thrown upon this dubious principle in two Scottish cases, *Smith* v *Strathclyde Regional Council* and *Sim* v *Aberdeen City Council* (both 1981). In the former, steps were taken before notice to treat to ensure that any temporary closure of the business was a short as possible. Because such steps were taken in mitigation of the claimant's loss, compensation was awarded for these expenses. This logical and reasonable approach was followed in the latter case and was then given the approval of the Court of Appeal in *Prasad* v *Wolverhampton Borough Council* (1983).

It is now clear that the test of whether losses are a "natural and reasonable consequence of dispossession" is one of cause and effect rather than of timing. This was affirmed again by the Lands Tribunal for Scotland in *Campbell Douglas & Co Ltd* v *Hamilton District Council* (1983) which sets out the principles involved in some detail. It was given the authroity of the Privy Council in *Director of Buildings and Lands* v *Shun Fung Ironworks Ltd* (1995).

(d) *The claimant must mitigate his loss*

The mitigation principle is that a claimant must take all reasonable steps to minimise his claim: see *Quartons (Gardens) Ltd* v *Scarborough RDC* (1955)). In general this simply means acting prudently as if the claimant was personally bearing the cost: for example, obtaining several quotes from removal firms and accepting the best, rather than the most expensive. Consequently, as relocation is usually cheaper than total extinguishment of a business, a dispossessed and expropriated business must move to alternative premises and continue in business, if this is reasonably practicable, rather than cease trading and claim compensation for total loss of profits from the acquiring authority.

In *Linden Print Ltd* v *West Midlands County Council* (1987) the Lands Tribunal made it clear that the onus is on the acquiring authority to establish that the claimant has not mitigated his loss. It was considered unreasonable to expect the claimant to relocate in the two months available following notice to treat. In spite of the principle established in the *Prasad* case, it was also considered that the claimant had no obligation to mitigate his loss prior to notice to treat.

The principle can cause difficulties where the claim arises from the service of a blight notice. In one actual case a shopkeeper decided to sell his corner shop and business on the open market, but was unable to do so as the property was blighted by a road junction improvement scheme. A blight notice was duly served and accepted, but at the time negotiations for compensation commenced, an identical shop next door was vacant and available on the open market, and was unaffected by the scheme. The acquiring authority quite legitimately argued that the claimant had a duty to mitigate his loss by relocating next door, rather than opting for total extinguishment of the business. Thus, a claimant who would have been able to sell his business in the open market in the absence of a scheme was prevented from doing so. See p 305 for a more detailed consideration of this point.

Disturbance compensation is a capital sum payable as part of the total compensation for land taken. Nonetheless, it may represent (at least in part) compensation for lost income or loss of profits. Such income might have been taxed by income tax or corporation tax but for the compulsory acquisition: disturbance compensation is taxed as a capital gain in many circumstances. Where tax rates differ, a discrepancy in the amounts of tax that would have been, and actually are, charged might arise.

Previously, a rather tortuous method of adjustment of compensation had become established practice (see *West Suffolk County Council* v *W Rought Ltd* (1957)). Since *Wood Mitchell & Co Ltd* v *Stoke-on-Trent City Council* (1978), however, the valuation world was made aware that the Finance Act of 1969, which had passed almost unnoticed, had enabled the Inland Revenue to apportion compulsory purchase compensation in such a way that the "income element" is distinguished from the "capital element", so that the claimant is left in the same tax position despite the compulsory acquisition. Apparently, the Inland Revenue had not been fully aware of this either, and issued a Statement of Practice, SP 8/79, which ensures the use of the 1969 Act provision and leaves valuers in the fortunate position of having to make no adjustment to the compensation figure for the incidence of tax.

These general principles lay down a framework within which the detail of a disturbance claim should be constructed. There is no finite list of admissible heads of claim: any head of claim will be justly compensatable, provided the requirements of the foregoing principles are adhered to.

Although there is no limit to the types of loss which may be incurred by a dispossessed owner and compensated as disturbance, it is useful to consider the most common types if only to illustrate the application of the general principles. These will be considered under two heads: the items of claim which may result from any compulsory purchase acquisition, and those items which specifically arise from the disturbance of a business occupier.

9.2 Heads of claim: general

9.2.1 Costs of finding alternative accommodation

The fees charged by the surveyor and solicitor in negotiating the compensation sum and conveying the title of the interest to the acquiring authority are other matters and are paid in most cases as of right (see p 298), regardless of the qualification rules for disturbance. They are, therefore, best considered as an additional payment over and above the compensation claim, rather than as a part of it.

Other fees may be incurred in the process of removal, however, in addition to advice on the notice to treat and these other matters. From *Harvey* v *Crawley Development Corporation* (1957) (see p 316), a loss may be compensatable provided it is the natural and reasonable consequence of the dispossession and it is not too remote. In the *Harvey* case it was held that this principle allowed the inclusion of fees and other costs incurred in finding and purchasing alternative accommodation for the household and business, including surveyors' and solicitors' fees, and this amount is not limited to those fees relating to the premises finally taken. Consequently, provided the claimant is attempting to mitigate his loss, the cost of more than one structural survey, building society fee, and search exercise may be awarded (see, for example, *Cole* v *Southwark London Borough* (1979)).

The value for money principle ensures that these fees must relate to premises of a comparable, and therfore not necessarily identical, market value to those being taken by the acquiring authority. In *Succamore* v *London Borough of Newham* (1978) a disturbance claim had been made in respect of the acquisition of a terraced house whose value was agreed at £7,500. The claimant tried to purchase a house 10 miles away, but the transaction did not materialise due to an adverse survey report. Eventually the claimant purchased a house 31 miles away for £13,250 and it was agreed that removal to such a distance was a reasonable consequence of the acquisition. The fees charged for the structural survey and local search in respect of the first house were awarded in addition to surveyors' and solicitors' costs in respect of the second house.

Solicitors' costs should arguably only be awarded in respect of the second house subject to a limit of that fee which would have been charged on the purchase of a house costing £7,500. On the other hand, the survey fee was allowed in full because "the work involved in the structural survey that was carried out may well have been no more than would have been involved in a survey of some older and less valuable property".

Other expenses may be incurred in finding alternative premises. For example, in *Smith* v *Lewisham London Borough* (1979) the cost of abortive journeys made by the claimant to view alternative premises was allowed in full, and the labour and time of an employee or director of a business spent in finding alternative premises and planning removals, and consequently mitigating the loss, should also be compensated (see *Drake and Underwood* v *London County Council* (1960)).

9.2.2 Removal expenses

The cost of removing from the acquired premises to the new accommodation is perhaps the most obvious item in a disturbance claim. As in all items of claim, it is the duty of the claimant to mitigate his loss, so more than one quotation should be obtained from removal firms, and the best evidence of a claim is the receipted invoice for the work carried out.

Removal costs may include the cost of temporary storage (see *Harvey* v *Crawley Development Corporation* (1957)) and damage to goods during transit. However, where insurance is available the cost of a suitable policy premium should be claimed in preference to the cost of repair of uninsured damaged goods, following the general duty to mitigate such a loss and reduce the claim accordingly (see *Coulson* v *Bury St Edmunds BC* (1969)).

Although it is incumbent upon the claimant to show that removal is a natural and reasonable consequence of the acquisition, the cost of removal from Deptford, London, to Spalding, Lincolnshire, has been allowed (*Smith* v *Lewisham London Borough* (1979)) due to the difficulty of finding similar accommodation in the vicinity of the acquired property at a similar price.

In *Roberts* v *Greater London Council* (1975) the cost of new school uniforms for the claimant's children was also allowed as a removal expense.

Where a regional development grant was available as an incentive to relocate to a particular area, the Court of Appeal held that the benefit should not be deducted from any compensation payable (*Palatine Graphic Arts Co Ltd* v *Liverpool CC* (1986)).

9.2.3 Publicising the move

Removal may necessitate the alteration of headed notepaper, business cards, signs, advertisements, and so on. For a business it may be reasonable to place an advertisement in the local newspaper notifying the change, while householders may wish to post "change of address" cards. Such reasonable expenses are to be allowed, provided obviously that no extra publicity for the company is engineered by the use of excessive advertisements.

9.2.4 Adaptation of fixtures, fittings, chattels and premises

These heads of disturbance may arise as separate items in the same claim, but care must be taken by the valuer to ensure that the heads are consistent with one another. For example, a family may have recently acquired a new gas cooker at a cost of £400 when notice to treat is served. Suitable alternative accommodation is found but gas is not connected to the new house. Three alternatives may present themselves: to convert the cooker for an alternative fuel; to supply gas to the new house; or to sell the cooker and buy an electric cooker. In the third case the allowable cost will be the loss on forced sale of the cooker, which is unlikely to command a market value anywhere near its cost of £400, and not the cost of purchase of a new cooker, as this is reflected in value for money. The lowest of the three costs should form the basis of the claim.

The choice between adaptation of the fixtures or the premises was the point under consideration in *Tamplins Brewery Ltd* v *County Borough of Brighton* (1970) and compensation was awarded on the basis that adaptation of the existing fixtures was uneconomic due to their age. Compensation was therefore awarded on the basis of the cost of adaptation of the premises, specifically the cost of replacing the equipment, less a saving in operating costs which resulted from the greater efficiency of the new equipment. This follows from the "value for money" principle, which must always be considered where the new premises are to be adapted.

The application of this rule will not disallow the cost of adaptations which are not reflected in the value of the premises, such as concrete plinths necessary to support heavy machinery of a specialised nature, laboratories for special processes, and so on.

Where a dwelling has been constructed or substantially modified to meet the special needs of a disabled person then, instead of basing the claim upon the market value of land taken plus a controversial list of disturbance items, the claimant may elect to claim compensation for land taken on a rule 5 (equivalent reinstatement) basis, thereby obviating many of the disturbance items (1973 Land Compensation Act, section 45).

The adaptation of business premises was considered in *J Bresgall & Sons Ltd* v *Hackney London Borough* (1976). The claimants removed to new leasehold premises where, as a condition of the lease, they were required to put the premises into repair. A claim for the cost of this work was disallowed, as the claimants were getting value for money in the nature of a low rent which reflected the lack of repair. Similarly, the cost of installing a toilet was disallowed as it constituted an improvement which would be compensatable under the Landlord and Tenant Act of 1927, and value for money would again be the result.

On the other hand, the installation of power points in order to upgrade the electrical system to that pertaining in the previous premises was an admissible item of claim, as was the cost of installing partitioning, because neither enhanced the value of the lease, and both were a special requirement of the particular occupier.

In *D Newton & Son Ltd* v *Lincoln County Council* (1985) items of claim which were allowed in full included the cost of fencing required to comply with a planning condition, and a estate agent's fee in connection with negotiations for a new lease. However, the Lands Tribunal did not allow the cost of service connections to the new property nor the cost of landscaping or architects' fees. The deciding factor was whether the costs incurred enhanced the value of the new premises — the value of money principle.

Smith v *Sheffield Metropolitan District Council* (1980) is a useful case concerning both the adaptation of fixtures and chattels and the loss on the forced sale of certain items, and illustrated the overriding principles bearing on these types of claim. The claimant must mitigate his loss, and the claim must be consistent.

9.2.5 *Additional transport costs*

It is conceivable that all sorts of claims might arise where a displaced occupier incurs extra travelling expenses by moving further away from his place of work, or even his mistress's boudoir. The obvious counter-argument of the acquiring authority is that the duty to mitigate would require removal to a nearby property, but such a move may not be possible.

In *Rutter* v *City of Manchester* (1974) such a situation arose, and the circumstances of the case resulted in the first reported award of compensation for the increased cost of travel to work, based on an all-in cost per mile, discounted over the working life of the claimant, less a deduction for the risk that he may accept or select early redundancy.

It was apparent that compensation was awarded in this case:

(a) because no alternative houses nearby were available, as he had lived in a clearance area
(b) because the claimant had worked for the same employer for 15 years and
(c) because he accepted rehousing by the local authority at the nearest of the three houses he was offered to his place of work.

Similarly overwhelming circumstances are almost certainly needed to establish a claim. For example in *Barker* v *Hull City Council* (1985) which concerned a similar domestic removal, a sum of £30 was awarded for increased travelling expenses, as against £1,000 claimed. While the principle of the *Rutter* decision was followed, the claimant had refused three offers of alternative accommodation nearer the acquired premises and had neither discounted his annual losses back to the valuation date, nor reflected the risk that the losses might not be incurred for the whole of his working life.

9.2.6 *Increased overheads*

Upon removal of a household or a business to new premises, the cost of occupying those new premises may be higher than those incurred by occupation of the compulsorily acquired property. For example, heating bills may be higher due to reduced thermal qualities of the new premises; rates may be higher due to a change of local rating authority; and, in the case of leasehold premises, the rent payable may be higher.

While increased overheads may reflect larger or more valuable premises, thereby attracting no compensation due to the value for money rule, it is possible that extra running costs may be incurred without any benefit to the occupant. For example, in *Roberts* v *Greater London Council* (1975) compensation was awarded for the extra cost of heating caused by the demolition of adjacent premises and the resultant loss of insulation (the claimant continuing to occupy the acquired property after the acquisition).

Normally, the increased rent of business premises is reflected in value for money as the extra size and/or better location of the new premises should enable increased turnover or profit. However, in *Metropolitan and District Railway* v *Burrow* (1884) the increase in rent was held to be compensatable. It appears that the following combination of circumstances may give rise to such a claim:

(a) the removal of the business to other premises is in mitigation of the claimant's loss
(b) there are no more suitable alternative premises
(c) the increased overheads do not benefit turnover or profit
(d) there is no prospect of sale or expansion of the business in order to take advantage of the extra size of the premises.

In *J Bibby & Sons Ltd* v *Merseyside County Council* (1979), offices were acquired and the claimants moved to newer, larger premises, incurring increased operating costs which did not result in greater efficiency or any commercial advantage. The Court of Appeal decided that no compensation should be awarded on this account because other premises had been available.

It may be that the higher costs of operating the new premises do contribute to greater efficiency, in which case the claim for increased overheads should be reduced accordingly (*Greenberg* v *Grimsby Corporation* (1961)), or even that overheads are reduced at the new premises, in which case the total claim should be reduced to reflect that saving: see, for example, *Tamplins Brewery Ltd* v *County Borough of Brighton* (1970).

The valuer should discount the increase or decrease in overheads at a rate of interest which reflects the growth potential of the costs and should attempt to calculate such a loss or saving logically without recourse to rough percentage increases or decreases (such as the 20% accepted in the *Tamplins* case as being a likely cost saving). This approach easily disguises the real position where, for example, operating costs in the new premises will not rise as sharply as those in the old.

9.2.7 Double overheads/unproductive overheads

A residential claimant, or one running a small business, is likely to be able to move out of his old premises and into his new on the same day. This may not be the case with a larger business which, particularly if it needs to trade continuously, may move gradually over a period of weeks or months. Such an occupier moving premises may have to pay two sets of rent, rates, and power bills during the period of removal. Such double overheads may arise in any case where the sale to the acquiring authority is not completed before removal to alternative premises commences.

From *B&T (Essex) Ltd* v *Shoreditch Corporation* (1958), it appears that the extra overheads claim should be in respect of the property which is substantially out of use i.e. the new premises initially, and then the old premises once the move is more than 50% complete. Where this rule is difficult to apply, the mitigation requirement may result in the claim being related to the lower of the two sets of overheads incurred. Such a claim may often be termed unproductive overheads.

9.2.8 Section 17 certificates

In establishing a claim for compensation for land taken it may be advisable for the claimant to obtain a certificate of appropriate alternative development (see Chapter 7). Solicitors' and architects' fees may be incurred when submitting an application for a certificate and (possibly) appealing against the local authority's decision.

9.2.8 Bridging finance

Where new premises are being purchased it may be necessary for the claimant to arrange for a "bridging loan" to finance and acquisition of (and alterations to) the new property before the compensation for the old property is received.

Even in a simple case, where simultaneous completion of the sale to the acquiring authority and acquisition of the new property is arranged, bridging finance might still be necessary to fund the 10% deposit usually paid on exchange of contracts.

Example 9.1

Mr A buys alternative accommodation for £170,000. He pays a deposit of £17,000 on exchange of contracts, financed by a bank loan. The purchase is completed simultaneously with the sale of his old house to the acquiring authority, two months after exchange of contracts.

If interest is charged at 1% per month the amount of interest charged is:

$$£17,000 \, (1.01)^2 - £17,000 = £341.70$$

An example of this type of claim can be found in *Roberts* v *Greater London Council* (1975), while a more general authority for the payment of bridging finance compensation is given by *Coulson* v *Bury St Edmunds Borough Council* (1969). In *Cole* v *Southwark London Borough* (1979) rather more complicated circumstances featured the payment of compensation for bridging finance interest where the claimant purchased his new house 16.5 months prior to completion of the sale of his old house to the acquiring authority. He argued that this was made necessary by certain works which had to be completed at the new house before he moved in.

The tribunal considered that 16.5 months was an excessively long period in which to complete such works and awarded compensation on the basis of six months' work (and bridging loan interest) being necessary.

This approach must be tempered where a new property is constructed as a replacement for the acquired premises. In *Service Welding Ltd* v *Tyne and Wear City Council* (1977) the claimants acquired a new site and erected and equipped a new factory, borrowing capital and incurring interest of £4,285 in doing so. The Court of Appeal held that this interest charge was part of the normal cost of constructing new premises, because a developer constructing a similar building would charge a price reflecting that cost. Consequently, the claimants received value for money and no compensation was payable.

Where advance compensation may be claimed (see p 317) the assumption will be made that such a claim will reduce any necessity for bridging finance, and the claimant should mitigate his loss in this way. This point is illustrated by *Simpson* v *Stoke-on-Trent City Council* (1982).

9.3 Heads of claim: business

9.3.1 Introduction

In addition to the foregoing heads of claim for disturbance compensation, which may arise in any compulsory acquisition, certain other losses might be suffered by business occupiers who are displaced as a result of compulsory expropriation. In particular, there may be a partial or total loss of profit, usually referred to as a loss of goodwill.

Goodwill has been defined as "the probability of the continuance of a business connection". Such a probability will usually command a market value and goodwill is often acquired where a business is transferred at a sum additional to the value of the premises from which the business is operated. The compulsory acquisition of such premises may result in the loss of the probability of a business connection. Compensation is required to place the claimant in the same position as if his land had not been taken from him; goodwill is a matter not directly based on the value of the land; and consequently the loss of goodwill is an acceptable item in a disturbance claim.

Goodwill might, perhaps, be regarded as the probability that old customers will continue to resort to the same place of business or continue to deal with a person or company. It can therefore be seen as having two incarnations: locational goodwill and personal goodwill. The former is both more easily identifiable and more common, because the attractiveness of the personality of a proprietor is eventually translated into locational goodwill.

The distinction is fundamentally unhelpful, except in special circumstances (eg a Wayne Rooney sports shop would have enormous goodwill no matter how often it was transferred). While it may be argued that personal goodwill can have no market value anyway, this too is unimportant, as rule 6 exempts disturbance from the market value principle established by rule 2. The member of the Lands Tribunal in *RC Handley Ltd* v *Greenwich London Borough* (1970) said:

> The value of the business to the dispossessed owner may be measurable by the amount at which his goodwill would have been saleable in the market on the basis of proved profitability; but this is a coincidental effect only, and the test must always be that of "value to the claimant", not that of "value in the market".

This was demonstrated in *Afzal* v *Rochdale Metropolitan Borough* (1980), where the compensation award was more than double the open market value of the goodwill.

What is of importance is the degree to which the profit-making potential of the business is affected by the compulsory acquisition of its premises. In the extreme case, it may be totally destroyed where it is not possible to move the business to alternative premises. Alternatively, removal of the business may result in a reduced profitability and a partial loss of goodwill. In either case, the amount of compensation to be paid should be the capitalised value to the proprietor, or business, of the lost future profits.

Total extinguishment of the business is almost certain to lead to a higher compensation claim than relocation, and because the claimant has a duty to mitigate his loss the business must attempt to find suitable alternative accommodation.

A dispute might arise between the authority and the claimant over the suitability of such alternative accommodation. For example, in *Knott Mill Carpets Ltd* v *Stretford Borough Council* (1973), the issue was whether the claimants were entitled to be paid £2,500 on the basis of a notional removal to alternative premises (which they had declined) or £9,000 for the total extinguishment of their business. The local authority had offered the claimant a unit in a new pedestrian precinct, but the claimant argued that the alternative premises were unsuitable. Because the alternative premises were available at a higher rent, because the claimants were unlikely to increase their turnover sufficiently to justify that rent, and because there was no evidence that they would have abandoned the business were it not for the acquisition, compensation was awarded on the basis of total extinguishment of the business. The poor viability of the company and the high rents charged for alternative premises were again of the essence in the award of total extinguishment compensation in *Bede Distributors Ltd* v *Newcastle-upon-Tyne City Council* (1973).

Following *Bede*, the Lands Tribunal determined in *Hall* v *Horsham District Council* (1976) that the substantive issue was whether "the claimant, in incurring his loss and failing to take mitigating steps, can be said to have acted 'unreasonably', ie 'in a way that no reasonable person would have acted'". The claimant in the *Horsham* case failed to accept the offer of alternative accommodation provided by the acquiring authority and his general conduct was unreasonable. Consequently, his claim for total extinguishment compensation failed.

The onus is on the local authority to show that the claimant has made inadequate attempts to mitigate his loss by relocating, and this will usually be a difficult task. In the *Knott Mill Carpets* case, the lands tribunal asked itself who was best qualified to decide whether the business could trade successfully at the alternative premises. Was it the claimant, who had run the business for many years, or the acquiring authority which knew nothing about selling carpets? In *Linden Print Ltd* v *West Midlands County Council* (1987) the Lands Tribunal awarded total extinguishment compensation for a printing business because a two-month period just before Christmas was quite inadequate for the company to relocate; because the company was reasonable in wanting to replace its freehold property with other freehold property; and because the local authority had failed to identify the steps which ought to have been taken by Linden Print in mitigation.

From the above decisions is evident that it is very difficult for an authority to argue successfully that a claimant has acted unreasonably in closing down rather than relocating. However, there have been one or two such cases in recent years. *Lamba Trading Company Ltd* v *City of Salford* (1999) concerned a large cash and carry warehouse which was acquired for road widening purposes by the Trafford Park Development Corporation. The principal issue between the parties was whether compensation should be assessed on the basis of total extinguishment of the claimant's business, or on the basis of a hypothetical relocation. For the claimant, who had ceased trading, it was argued that the closure of the business was entirely due to the compulsory purchase. The acquiring authority claimed, however, that suitable alternative accommodation was available, and that by not relocating into that accommodation the claimant had failed in its duty to mitigate its loss. The decision on the lands tribunal began by restating the principle that "the best judge of what is suitable and what is not must be the operator of the business". However, in this case there was evidence that the claimant had actually purchased alternative premises with a view to relocation, and then sold them arguing that they had proved to be unsuitable and pointing out that "nobody could know the cash and carry business as well as the person who was actually carrying on the business. The council's experts and witnesses were...conspicuously lacking in knowledge of cash and carry requirements. The suitability or otherwise of alternative premises would be ...best judged by the claimant, and it was not for others to second-guess this".

The Tribunal was unconvinced that there was a reason for deciding not to relocate, noting that the claimant may well have seen "an opportunity to diversify into other less cash intensive but more profitable areas, at the same time as picking up a significant payment for the extinguishment of the business". Accordingly the claimant did not take reasonable steps to mitigate the additional losses which would be caused by extinguishment, and compensation should be assessed based on the notional costs of relocation.

In some circumstances, no such dispute arises. Section 46 of the 1973 Land Compensation Act provides that, subject to certain rateable value limits and undertakings preventing disposal of the goodwill on the open market, a sole proprietor aged 60 or over on the date of dispossession may claim total extinguishment compensation as of right. Certain partnerships and companies have a similar right subject to age requirements made by section 46.

The principle that the claimant is the best judge of whether relocation is viable was confirmed by the Lands Tribunal in *Crowley and Jarvis (T/A Contraband Stores) v Liverpool PSDA and Liverpool City Council* (2007). In that case two potential alternative premises were identified for a large discount retailer. The claimant argued that neither was large enough for their purposes, and both would require remote storage which would cause an unacceptable increase in overheads. The Tribunal agreed, and awarded compensation on a total extinguishment basis. The claimant's were, however, less fortunate in their argument that the closure of their Birkenhead store was also a direct result of the CPO. In their view they relied on bulk buying, and the closure of the main Liverpool store rendered the Birkenhead branch uneconomic. The Tribunal disagreed, and decided that Birkenhead had been making losses for some time and would have been likely to close regardless of the scheme.

In *Bailey v Derby Corporation* (1965), total extinguishment compensation was not awarded where the ill health of the claimant prevented removal to alternative premises, but the ill health in that case arose after service of notice to treat and it is generally accepted that the duty to mitigate loss is a personal duty and must depend on the ability of the claimant to perform it at the material time. Consequently, in certain circumstances the ill health of the business occupier may prevent removal and justify the award of total extinguishment compensation.

Subject to qualification for disturbance compensation, therefore, a company may suffer a loss of profit as a result of its forced expropriation. The quantum of such a loss will depend upon the availability of suitable alternative accommodation and whether as a result the business removes or is extinguished.

A permanent loss of profit will often form the major part of a business disturbance claim. This may be total, where the business is extinguished, or partial where there is removal to a less profitable location or to smaller premises. In addition, there will almost certainly be a temporary loss of profits upon removal, and in any case there may be redundancy payments. These three heads of claim are considered below.

9.3.2 Permanent loss of profit

A business is totally extinguished as a result of the acquisition only where it is not reasonably possible for the claimant to find alternative premises within a sufficiently short distance of the acquired premises so as to retain at least part of the goodwill of that business.

In *Appleby and Ireland v Hampshire County Council* (1978), the sensitivity of profits to location was well illustrated. In that case, a precision engineering company's premises at Basingstoke were acquired as part of a road scheme. The company relocated the business 12 miles away at Alton, although it was agreed that alternative accommodation existed within Basingstoke. While a move to Alton was considered to be likely to incur an 85% loss of profit, the tribunal considered that a move within Basingstoke would have incurred a loss of only 50% and awarded compensation on that basis. In *Massie v Liverpool Corporation* (1974), the removal of a scrap metal business to a site 3.5 miles away resulted in a 90% loss of profit due to special advantages of the original location.

The distance at which a 100% loss of goodwill is likely to be incurred will depend on many factors, not the least of which is the type of business involved. Industrial concerns might be able to move a considerable distance without losing the whole of their established business, as in the *Appleby* case, while a corner shop is likely to lose all its custom by moving only half a mile.

Consequently, the very fact of removal should not blind the valuer to the possibility of a total loss of profit being suffered. In the majority of cases, however, total extinguishment claims concern businesses that are wound up or those that have several branches and continue trading despite the loss of one outlet.

The basis of a claim for total extinguishment is the capitalised value to the claimant of the likely average future net profits to be made at the acquired premises, in the absence of the compulsory acquisition, and subject to certain adjustments.

In order to arrive at the compensation figure, the valuer, perhaps with the aid of an accountant, should follow a certain procedure.

(a) The likely average future net profits should be calculated. This can only be achieved by a consideration of the accounts of the company, adjusted wherever it may be suspected that those accounts are less than perfectly accurate. In order to prevent especially good or bad years from having undue weight, it is usual for three years' accounts to be taken into consideration, and an average taken after the adjustments referred to at (b) below have been made. In periods of high inflation it is arguable that an adjustment of these future profits should be made to allow for their real value. Although it is future profits that are to be estimated, such extrapolation should only be made on the basis of previous profit, and any expectation of increases or falls in the profit level should be reflected in the capitalisation of the average figure.

The practice of taking three years' accounts should not be followed slavishly where to do so would not achieve the aim of assessing likely future profitability. In *Hussain* v *Oldham MBC* (1981) only the last year's figures were used as the business was new and was only just approaching its true trading potential. On the other hand, in *Reed* v *North Tyneside MBC* (1982) compensation was assessed on the average of four years' figures including projected figures for the final year based upon only part of a year's trading figures. The member of the tribunal expressed doubts about the projected year's figures and also the first year's figures which, in a period of high inflation, were relatively low in money terms. However, he felt that the divergences cancelled each other out.

(b) Having derived net profit figures for three previous years' trading, certain adjustments must be made. The aim of these adjustments is to ensure that the net profit figure fairly reflects the profitability of the business. Adjustments to be considered are as follows.

(i) Rent or profit rent. Where premises are owned, one of the deductions from income which must be made in order to ascertain net profit, is rent. The business is the same whether the premises are owned or leased. With a leasehold business, rent has already been deducted in the accounts, and in order to ensure that a freehold business is treated in the same way, the full rental value should be deducted to give a comparable net profit. To look at it another way, a business operating from freehold premises will show a net profit without deduction for rent. This profit figure is not, however an accurate assessment of the return on running the business. Some of that profit could be earned by simply closing the business and renting out the building. A notional rent needs to be deducted in order to arrive at the true profit of the business, over and above any income which could be derived if the business were to close. A third good reason for deducting notional rent is that the loss of the freehold property will be compensated separately under the claim for land taken. To calculate disturbance compensation based on profits which make no allowance for rent would lead to enhanced figure as a result of the property being freehold. The claimant would benefit twice for the fact that the property is owned. This is known as double counting — a trap which regularly arises in compensation claims. Similar principles apply where the property is leasehold, but held at a low rent under an old lease. In such cases, the accounts will show an artificially high net profit, as an element of the profit apparently made by the business is made up of part of rental value of the premises or profit rent. This will again be compensated separately by the acquiring authority as compensation for land taken: the profit rent must therefore be deducted from profit shown in the accounts to prevent the authority paying for it twice.

The notional rent or profit rent figure which should be deducted from profits may vary from year to year in accordance with recent rental trends, rent reviews, and so on.

(ii) Interest on capital. Besides the possibility that the company has capital tied up in a valuable interest in land, it will certainly have other capital investment, perhaps in machinery, stock, raw materials, and cash used for day-to-day trading.

Again, where the company to cease trading, such assets could be translated into cash, which could be invested to produce an income. This income — interest on capital — should also be deducted from net profits in order to leave a true figure of profit made by the business. Again, the justification is the need to arrive at the true profit derived from the business (sometimes called a super profit) over and above the income which could be generated if the business were to close. The rate taken should be a reasonable rate of interest which might have been earned from capital invested in a reasonably secure way during the accounts year.

(iii) Director's remuneration and owner's wages. Trading accounts should include as a deduction the salaries and wages of all employees. Directors and proprietors customarily derive an income from the business: this might show in the accounts as a directors remuneration, or be included in the overall wages figure. However, it is often the case that no income is shown for the owners or directors, and they simply take the profits. Alternatively the figure shown in the accounts as directors remuneration or owners' salary may not accurately reflect a fair payment for their labour — it may be artificially high where a sleeping partner takes a wage, or artificially low where an owner takes part of his income as a wage, and part as profits. If the valuer suspects that the remuneration of the proprietor or director indicated in the accounts is not the level of reward likely to be required for such services, it may be that the profit level apparently indicated is artificially low or high, and a further adjustment should be made to reflect a reasonable income for the work done by owners or directors. In *Shulman (Tailors) Ltd* v *Greater London Council* (1965), the Tribunal said:

I accept evidence that in the great majority of small family businesses profitability is more reliably measured by ignoring the directors' fees actually charged and substituting instead a reasonable figure being the value of the directors' services had they been employees.

Such smaller businesses may include in the accounts a wage or salary for the proprietor's wife or other connected persons as a means of reducing income tax liability. If it is suspected that the apparent recipient of this wage does little or nothing to earn it, then the amount should be added back to the profit shown. Any new proprietor taking over the business would not be forced to employ anyone and would make higher revealed profits. For an example of such a case, see *Zarraga* v *Newcastle-upon-Tyne Corporation* (1968), where no deduction was made in respect of the wages of a fish and chip shop proprietor's wife, as she could not fairly be classed as a paid employee.

In any one year it may be that an unusual expense arises, for example a golden handshake may be awarded to a retiring director. In such a case, the amount awarded should be added back to the profits to reveal the true profit level of that year.

Less straightforward is the deduction of the proprietor's remuneration where a small business or partnership is involved. It may be argued that in the case of a one-man business it is correct to deduct from the profits of the business an amount to reflect the proprietor's own labour should the accounts show no such deduction as a wage or salary. If the business produces a profit of (say) £20,000 pa, it may be that no one would pay a capital sum for the goodwill of such a business because that amount of salary could be earned elsewhere without the necessary risks of being self-employed. However, it is clear that there is a market for such business in the open market and, since *Perezic* v *Bristol Corporation* (1955), the Lands Tribunal appears to be inclined towards the view that someone would part with capital in order to be self-employed or, more strictly, that the business has some value to the owner who prefers to be his own governor.

In *Donald McKellar & Sons Ltd* v *Glasgow Corporation* (1975) the tribunal summed up the position as follows:

The question of whether such a deduction should be made to arrive at net profits has been dealt with in a number of Lands Tribunal cases in England. It has been held that in the case of a limited company with branches a deduction is appropriate: see *Cliffords (Dover) Ltd* v *Dover Corporation* (1965) and *Freeman Hardy & Willis* v *Bradford Corporation* (1967). In the case of *GE Widden & Co* v *Kensington* (1970), however, it was agreed by both the claimants and the acquiring authority witnesses that no deduction should be made for directors' fees: but in that case the business was run by the director himself and his wife. The matter was also considered in the case of *Longbottom* v *Bingley* (1974), where no deduction was made for the directors' remuneration in the case of a partnership.

Consequently, valuers should apparently be advised as follows: where a large company is concerned, directors' remuneration should be checked as being realistic, and adjusted where appropriate. Where a partnership, husband and wife business, sole proprietorship or company is concerned, no deduction from profit should be made to cover the proprietor's/director's remuneration where the business is so small that the operators of the business would not distinguish between profit and their personal remuneration.

(iv) General adjustments, such as saving in head office expenses. In *Reed Employment Ltd* v *London Transport Executive* (1978), the branch office of an employment agency was acquired. It was held by the Lands Tribunal that the extinguishment of the branch resulted in a saving by the head office of the claimant business of £5,000 pa. Consequently, £5,000 was deducted from profits in the quantification of compensation.

This illustrates that the above list of adjustments is not exhaustive, and that other adjustments to the accounts may need to be made in order to identify true net profit depending upon the accuracy and honesty of those figures. The valuer should check that a proper amount has been included in the accounts for repairs to, and maintenance of, the buildings (the last year's figure is often low, as the owner knows the business is closing soon). Stock should be valued at a realistic price (normally cost price unless it is obsolete) and the amount kept must not be too much or too little for the efficient running of the business. It may be that bank charges shown in the accounts include repayments of loans taken out to acquire or develop the business. These are a function of the way the business was acquired, and are not relevant to its annual profitability: after all once the loan is repaid, the payments will cease and the profits will rise, but the business has not suddenly improved. Mortgage payments should, therefore be added back to the net profit shown in the accounts. Realistic bank charges and credit card charges necessary for running the business should, however, remain deducted and this may require an apportionment where a single figure is shown for bank charges and interest. Where the valuer has reason to doubt the accuracy of these items he should adjust the accounts accordingly.

(c) Having established net adjusted profit figures from around three recent years' accounts, an average should be taken. Because the valuer is attempting to reflect the value to the claimant of his likely future profits, the annual figure should be converted to a capital sum by means of a capitalisation factor (sometimes called a multiple or YP) which takes into account the growth potential (or otherwise) of those profits.

Imprecise capitalisation techniques are generally used. In almost all cases a capitalisation factor of between 1 and 5 is applied to the adjusted average annual profit figure, reflecting the practice of the market for certain businesses rather than academically sound valuation techniques. A higher factor should be used where profits are strong, where there is an improving trend and, in particular, where the business has an element of monopoly. An example of this would be a scrapyard, as planning permission for a replacement business would be difficult to obtain. A lower factor should be used where the profits are poor and declining, and where the business is of a type which is in general decline, such as a shoe repairer. These capitalisation factors may seem low to valuers used to valuing investment property; after all, a capitalisation figure of 3 equates to a years purchase in perpetuity at 33.3%. However, it should be remembered that profits are a much less secure return than rents, and need to be earned with considerable effort, rather than simply collected.

In *Wakerley* v *St Edmundsbury Borough Council* (1977), a capitalisation factor of five was applied in the case of a sole proprietor, based on his life expectancy and a risk rate of 20%. This case has not, however, been regarded as part of the mainstream of disturbance cases, being specifically concerned with agricultural property. The majority of Lands Tribunal decisions have approved factors of between 1.5 and 3, depending upon the trend of profits.

The Lands Tribunal threw doubt on this traditional approach in *Crowley* v *Liverpool* (2007) (see p307). They were presented by the acquiring authority's surveyor with a goodwill calculation following the traditional approach, whereas the claimant employed an accountant who used a Price/Earnings approach. This involved taking the average adjusted net profit, after the deduction of proprietor's remuneration and tax, and employing a multiple based on P/E multiples of listed companies in the same retail discount sector. This multiple was reduced by 35% to reflect the fact that this was a relatively small business compared to the listed companies which included Woolworths and Poundstretcher. On this basis the accountant considered a multiple of 12.3 to be appropriate, though the resultant figure was then increased by 34% to reflect the control and freedom (control premium) enjoyed by the subject business compared to publicly listed companies, producing a claim of £1,166,512. Somewhat surprisingly, the tribunal adopted this approach in preference to the traditional method, noting its earlier criticism in *Optical Express Southern Ltd* v *Birmingham City Council* that "There is a lack of market evidence and the figure of YP is usually fixed by reference to settlements and decisions of the tribunal, which become self-perpetuating". While the P/E approach required substantial adjustments, it was preferred in respect of its treatment of owner's wages and anticipated rent increases. However, the procedure of deducting 35% because the claimant was not a listed company, then adding 34% as a control premium' was considered inappropriate. The Tribunal adopted a multiple of 8 on the P/E basis, plus a "control premium" of 20%, producing an award of £600,000. This figure equates to a multiple of 4.4 on the traditional average adjusted net profit (though before deduction for owner's wages). This is an exceptionally high YP, even having regard to the fact that the company was highly profitable.

(d) Many Lands Tribunal decisions provide examples of the calculations of a claim for total extinguishment of goodwill: see, for example, *Easton* v *Newcastle-upon-Tyne Metropolitan District Council* (1979) and *Watson* v *Warrington Borough Council* (1979), and *Reed* v *North Tyneside MBC* (1982). A simplified example is shown below.

Example 9.2

Jones & Sons Ltd own leasehold factory premises which are to be acquired. The company has three directors whose wives receive salaries for a small amount of secretarial help. The company has stock worth £30,000, cash of £4,500, and raw materials worth £9,000 in addition to plant and machinery whose current market value is £72,000. The full rental value of the factory is around £52,500 but Jones & Sons Ltd occupy the factory under a 14-year lease granted 4 years ago at a fixed rent of £22,500 pa.

The company cannot find suitable alternative premises. Net profits over the last 3 years have been shown in the accounts at £132,000, £135,000, and £153,000. Directors' fees have totalled £39,000, £39,000 and £45,000 respectively. Salaries for their wives have totalled £33,000, £33,000, and £36,000.

	Year 1 (£)	Year 2 (£)	Year 3 (£)
Net profit	132,000	135,000	153,000
Add wives' salaries[1]	33,000	33,000	36,000
	165,000	168,000	189,000
Less shortfall in directors' fees[2]	15,000	18,000	24,000
	150,000	150,000	165,000
Less wage for secretary[3]	15,000	16,500	21,000
	135,000	133,500	144,000
Less interest on capital: total £115,500 @ 5%[4]	5,775	5,775	5,775
	129,225	127,725	138,225
Less profit rent: FRV rising from £45,000 to £48,000 to £53,500[5]	22,500	25,500	30,000
Adjusted net profits	106,725	102,225	108,225

Average adjusted net profit	$\dfrac{£317,175}{3}$ =	£105,725
YP, say		2.5[6]
Compensation		£264,313
Say		£264,000

---

1   Wive's salaries are added as they appear to be excessive in relation to the work provided.
2   Directors' fees appear to be inadequate for the work done.
3   A secretary would have to be employed to replace the services provided by the wives.
4   Based on realistic interest rates in each of these years.
5   As the rental value of the factory has risen, so has the profit rent enjoyed by the business and hidden in the accounts.
6   Slightly rising trend in profitability.

As has already been seen, a partial loss of goodwill might be suffered by a business which moves to new, less profitable premises. It may be that an easily estimated reduction in trade will reduce profit proportionately and that the prospects of increasing profits at the new premises equate with the prospects which have been lost. In such a case, a percentage of the adjusted average net profit might be accepted as being lost, and the capitalisation of this lost profit by a suitable factor will reflect the partial loss of profit.

For example, in Example 9.2 Jones & Sons Ltd might alternatively have found suitable new premises where profitability was likely to be reduced by 20% due to increased transportation costs, the prospects for bettering this profit being unaltered.

The valuation for a partial loss of goodwill becomes:

	Average adjusted net profit	£105,725
	Loss of profit @ 20%	£21,145
	YP	2.5
	Compensation	£52,863
or:	Compensation for total loss	£264,313
	Loss of profit @ 20%	£52,863

The above is, however, a fairly unlikely scenario as a fall in turnover is rarely accompanied by a similar fall in overheads. Some costs, such as heat and light, and maintenance, may be fixed, while others, such as wages, may be fall less than the fall in turnover. Care must be taken, therefore, where an increase in overheads has a disproportionately large effect on profitability, where a reduction in turnover has a similar effect, or where the future prospects for profitability are altered. In these cases, it is suggested that a valuation of the goodwill prior to the move is compared with a valuation of the goodwill after the move, the difference representing the goodwill payable. In Example 9.2 profitability might be reduced by 40% but the prospects for increasing that profitability might be enhanced by the proximity of a motorway extension planned to be built. Compensation should be assessed on a before and after bases, as follows:

Valuation of goodwill before move:	£264,313
Valuation of goodwill after move:	
Average adjusted net profit	£105,725
Net profit level @ 60%	£63,435
YP (increased to reflect growth potential)	3.5
	£222,023
Compensation: £264,313 — £222,023	£42,290

Estimation of the quantum of likely lost profit will often be a matter over which specialist advice from (for example) an accountant should be taken. For an in-depth consideration of the estimation of this loss, see *Appleby and Ireland* v *Hampshire County Council* (1978).

9.3.3 *Temporary loss of profit*

Where the goodwill of a business is totally extinguished there may, in addition to the permanent loss of profit, be suffered a temporary loss of profit due to the natural running down of the activities of that business prior to expropriation. Where the business removes to new premises a temporary loss of profit is even more probable. In addition to losses sustained by running down activities in the old premises, there may be a period of complete closure during removal, and a "build- up" period in the new premises, which results in lost production and lost profits.

Where a temporary loss of profit is suffered, the loss may in part be due to increased or double overheads and care should be taken by the valuer to ensure that the there is no "double counting", by paying both double overheads and temporary loss of profits, and that the compensation claim in toto is consistent.

In *Pearl Wallpaper Holdings* v *Cherwell District Council* (1977), a paint and wallpaper business removed to nearby premises and the tribunal awarded compensation on the basis of disruption of the business for a period of one year after removal. No claim for a permanent loss of profit was made due to the adjacency of the two shops, so the claim for a temporary loss of profit was based on a comparison of the actual profits earned after the move with a projection of profits made at the old premises. This approach can be contrasted with the *Appleby* case, where a claim for temporary loss of profits during a run-down period, removal, and a run-up period was based on non-productive overheads.

The claim for a temporary loss of profit may be complicated by a parallel claim for a permanent loss. The very process of establishing compensation for a permanent loss of profit can, however, aid the assessment of a temporary loss. For example, a company might suffer a permanent loss of 10% of its average adjusted net profit figure of £25,000 pa. There will be a run-down period of six months prior to removal and a run-up period of six months after removal during which time production will run at an average of 80% capacity. A total of one month's whole production will be lost during removal.

The calculation of a claim for permanent loss of profit will necessitate the production of the average adjusted net profit figure. Using a rule of thumb approach, this sum may be used as the basis of the claim for a temporary loss of profit, by allowing simply 20% of this figure for each month during which profitabilty is reduced. This will not be particularly accurate and it would be better to calculate the actual profitabilty in each month, and compare this with the antcipated profitability had the business been unaffected. Even then, the calculation is likely to be complicated by changes in other factors, such as the local competitive environment, seasonality and the general state of the economy.

Where the average profit fails to reflect fairly the profit levels which will actually be affected (eg where profits are steadily increasing), it may be justifiable to use a projected profit level rather than the average figure employed in the calculation of the permanent loss of profit.

9.3.4 *Redundancy payments*

An item of claim peculiar to business loss is the payment of cash sums to employees who have to be laid off as a result either of total extinguishment of the business, or removal to smaller premises. It must be ascertained that the payments are the reasonable consequence of the acquisition and that redundancies were not expected in the normal course of events: see *GE Widden & Co v Royal Borough of Kensington and Chelsea* (1970). Payments are made partly from government funds and only the employer's portion of the payment should, of course, be claimed.

9.3.5 *Loss on forced sale*

When a business closes it will often be left with stock, equipment and plant and machinery which had a substantial value to the trading concern, but for which there is little demand if it is offered for sale second hand. The plant and machinery division of the valuation profession actually has separate bases of valuation for items in their working place, and if offered for sale. This can be particularly true of some items, such as shop fittings, which are very expensive to buy and fit, but are almost valueless second hand. On the other hand, some items, such as motor vehicles, are readily saleable in the open market and are likely to sell at a price fairly reflecting their value to the business.

The usual practice is for the claimant to arrange for the sale of these items, normally by auction, and to claim the difference between their value to the business, and the net sale price. In complex cases it may be necessary for the claimant and the acquiring authority to each employ their own specialist valuers to agree the in situ value. This loss on forced sale item of claim will usually arise where a business is forced to close, but may also apply where there is a relocation and some items, such as a built in blast furnace, are not removable. It may also arise on a domestic claim where, for example, someone with a gas cooker moves to an all electric house. Similar losses may arise in respect of stock which remains unused or unsold at the time of closure (even following a closing down sale). In this case the loss on forced sale claim is likely to be based on the difference between cost price and sale price at auction.

9.4 Disturbance payments

As discussed on p 294 above, a right to disturbance compensation only arises where a claimant is:

(a) expropriated and
(b) dispossessed.

Dispossession is factual: expropriation depends upon the acquisition from the claimant of a valuable interest in land, though section 20 of the Compulsory Purchase Act 1965 gives a similar right to tenants with an annual interest or more. Consequently, tenants of residential property, short leaseholders, and weekly or monthly tenants have no right to disturbance compensation. Inequities could, therefore, arise where, for example, a long-standing tenant of residential property had no right to recover his or her removal costs.

Consequently, the Land Compensation Act of 1973, which was designed to extend generally the ambit of compensation upon compulsory acquisition, included a provision entitling all persons displaced in consequence of the compulsory acquisition of land to a disturbance payment, provided:

(a) he is in lawful possession of the land from which he is displaced and
(b) he is not entitled to receive disturbance compensation.

Section 37 of the 1973 Act lays down fuller requirements for qualification, while section 38 established the amount of a disturbance payment as being:

(a) the reasonable expenses of the person entitled to the payment in removing from the land from which he is displaced and

(b) if he was carrying on a trade or business on that land, the loss he will sustain by reason of the disturbance of that trade or business consequent upon his having to quit the land.

The wording of (a) above would appear to be considerably more restrictive than general disturbance compensation, which can be described as all losses which are "the natural and reasonable consequence of the dispossession of the owner" (*Harvey* v *Crawley Development Corporation* (1957)). Initially, a disturbance payment was considered by many compensating authorities to amount to little more than the physical cost of removal, but this interpretation was widened over the years (see *Anderson* v *Glasgow Corporation* (1976) and *Nolan* v *Sheffield Metropolitan Borough Council* (1979)). Finally, almost as an aside, Lord Justice Stephenson commented in the Court of Appeal decision in *Prasad* v *Wolverhampton Borough Council* (1983):

> The language of these provisions [sections 37 and 38] is, to my mind, like enough to the language in which judges have stated the principle of fully compensating owners dispossessed by compulsory acquisition of their property to indicate the intention of Parliament to give to those classes of persons not previously entitled to disturbance the same right as those previously entitled to it enjoyed, not a reduced and lesser right.

In other words, a disturbance payment under sections 37 and 38 of the 1973 Act is to be treated in exactly the same way as disturbance compensation to a dispossessed owner. The views of Stephenson LJ were followed by the Lands Tribunal in *Barker* v *Hull City Council* (1985) where a small sum was allowed in respect of increased travelling expenses (see p 312).

9.5 "Other matters not directly based on the value of land"

As suggested on p 293, this head of claim, derived from the wording of section 5(6) of the 1961 Act, covers the costs of negotiating compensation usually charged by a surveyor and the costs of conveying the title of the compulsorily acquired property to the acquiring authority usually charged by a solicitor.

It will also include the cost of submitting the claim. Although in certain circumstances losses incurred before the service of notice to treat can be compensatable (see p 298), the first common item of claim is the fee charged by the claimant's solicitor in giving advice concerning completion of the notice of claim. These fees may not be quantifiable at the time the claim is made and in such cases it is appropriate to include an item couched in such general terms as "appropriate legal fees arising in connection with advising on the notice to treat".

Surveyor's fees have been the subject of dispute. In *Sadik* v *Haringey London Borough* (1977), for example, fees for the preparation of a schedule of condition were disallowed because the notes taken in such preparation could have been taken as part of the normal valuation work. In *Easton* v *Newcastle-upon-Tyne Metropolitan District Council* (1970) a claim of over three times the usual amount was denied on similar grounds. The tribunal did, however, accept that "there may be exceptional cases in which the reimbursement of fees on a basis which departs from the scale may be justified" (see, for example, *DB Thomas & Son Ltd* v *Greater London Council* (1982)). This is reference to Ryde's Scale which set out the fees payable to the surveyor for negotiating the claim. This scale has since been abolished, and surveyors fees, like those of the solicitor, are now negotiated individually in each case.

In the same case a claim for accountancy advice was made, and the tribunal members made it plain that such an item would be compensatable in so far as it was necessary to establish a claim for disturbance, although it was felt that an experienced surveyor should have no difficulty in dealing with the accounts of a small business.

It is axiomatic that fees on fees are not awarded: that is fees charged by the surveyor are not part of the disturbance claim and should not, therefore, be included in the amount on which those fees are based.

It is important to note that these costs are not part of the disturbance claim, though they are subject to similar quantification rules. They are payable to any claimant, regardless of whether he qualifies for disturbance.

9.6 Rehousing

Under section 39 of the 1973 Act, where a person is displaced from residential accommodation on any land in consequence of the acquisition of that land by an authority possessing compulsory purchase powers, and suitable alternative accommodation on reasonable terms is not otherwise available to that person, then it shall be the duty of the relevant authority to secure that he will be provided with such other accommodation.

No similar obligation applies in respect of business occupiers, but it has been made clear from government circulars that local authorities have a moral duty to use their best endeavours to relocate industry, and in any case the provision of alternative accommodation to a business user is likely to reduce his disturbance claim. Under section 50 of the 1973 Act it is made plain that the rehousing of a residential occupier cannot be used as an argument to reduce the compensation payable, except in so far as certain common losses simply will not be suffered.

Conversely, rehousing prospects cannot be used as a lever to raise the value of tenanted property to that of a house with vacant possession (see p 225).

9.7 Advance payments

Once an acquiring authority has taken possession of land, it may be required to make a payment to the expropriated owner of 90% of the agreed quantum of compensation, or, if no agreement has been reached, 90% of the authority's estimate of compensation (1973, Act, section 52).

The payment of advance compensation might reduce any claim for bridging finance as an element in disturbance (see p 314).

It had been argued that only one advance payment can be claimed under section 52, even where it later becomes clear that the original estimate upon which the first advance payment was made was too low. Section 63 of the 1991 Act now makes it clear that further payments must be made as and when it is shown that the original estimate was too low. The amendment also requires that accrued interest shall be paid at the same time as the advance payment, and introduces a right to annual advance payments of interest on the 10% unpaid balance, where such interest exceeds £1,000.

Further amendments were introduced by the 2004 Planning and Compulsory Act to protect the position of mortgagees. If the amount of the advance payment is less than the mortgage, then no payment will be made to the claimant and, if both claimant and mortgagee agree, the advance payment will be paid to the mortgagee. If the mortgage is less than the advance payment, then the amount of the advance payment made to the claimant will be reduced by the amount of mortgage.

9.8 Home loss payments

Rule 1 of the six main rules of compulsory purchase requires that no allowance should be made on account of the acquisition being compulsory. This removed the 10% addition which had been paid prior to the 1919 Act to placate the aggrieved victim of compulsory purchase. There have always been supporters of the argument for reviving the 10%, and their case has been answered to some extent by the introduction in 1973 of Home Loss Payments, and in 2004 Basic and Occupier's Loss Payments (considered below). These go some way towards restoring the addition, but not always at the full 10%, and restricted to only limited categories of claimant.

Subject to the major amendments contained in Schedule 15 of the 1991 Act, sections 29–33 of the 1973 Act impose a duty upon an acquiring authority to make a home loss payment where a residential occupier is displaced as the result of compulsory acquisition, or of certain other events including a closing or demolition order.

The amount of the payment will be 10% of the market value of the property for owner occupiers, subject to a maximum; currently £44,000 but subject to regular review. Tenants fair less well and receive a fixed sum; currently £4,400.

The claimant must have been in occupation of the dwelling as his only or main residence for a period of one year prior to displacement. There is a discretionary power for acquiring authorities to waive this requirement in the event that a claimant has been unfortunate enough to be displaced more than once within this period, though the payment need not be at the full rate.

Prior to the 1991 Act, home loss payments could not be paid where the acquisition arose from the service of a blight notice, but this exception is removed by section 68 which also introduces a right to payments in advance of home loss payments.

In *GLC* v *Holmes* (1985) the claimant was rehoused from a mobile home on a caravan park which had been acquired with the intention of developing it as council housing. By the date of displacement, the plan had changed to one of selling the site for private development. The acquiring authority contended that the claimant had not been displaced within the meaning of section 29 as the land was not held for the purpose for which it was acquired, as required by the Act. The Court of Appeal held that housing was the broad "purpose" for which the land was held, and which caused the displacement of the claimant. A home loss payment was, therefore, payable.

9.9 Basic and occupier's loss payments

Section 33 (a)(i) of the 1973 Act was inserted by the 2004 Planning and Compulsory Purchase Act and introduces basic loss payments at 7.5% of the value of the interest acquired, subject to a maximum; currently £75,000. This applies to all claimants who have an interest in land acquired providing that interest has been held for a period of at least one year prior to the relevant date — usually the date of valuation for compensation purposes. To prevent recipients of a home loss payments benefiting twice, the value of any dwelling must be deducted from the sum on which the payment is calculated. A basic loss payment is not payable in respect of a claim assessed on an equivalent reinstatement basis under rule 5.

Section 33 (b) and (c), also inserted by the 2004 Act provide for occupier's loss payments of 25%. This effectively tops up the basic loss payment from 7.5% to 10% for claimants who qualify for the former, and have also been in occupation for at least one year. The maximum is currently £25,000.

There are specific occupier's loss provisions relating to agricultural land. In that case the payment is the greater of 25% of the value of the interest, or an amount calculated by reference to a formula based on the size of the affected land, or the affected agricultural buildings.

9.10 Examples

Answers to the following specimen examples are intended to be illustrative only and are necessarily skeletal.

Example 9.3

Parsons is the freeholder of old factory premises near the centre of a large provincial town. The factory has a total net floor area of 2,000 m^2, the site area being 3,000 m^2. The premises are let to Baxter & Co, a family concern, who carry on a soap manufacturing business. The premises have been used by the same firm for this purpose for more than 100 years. The lease is for 14 years from 8 years ago, on full repairing and insuring terms with a rent review after seven years to full rental value, the present rent being £34,500 pa. Rental values for such premises have risen by some 10% over the last year or so.

Details from the company's most recent profit and loss account and balance sheet include the following:

Gross receipts	£570,000
Purchase of raw materials	£180,000
Wages	£120,000
Directors' emoluments	£12,000
Other working expenses including rent	£75,000

Stock (average holdings):

Raw materials	£30,000
Finished products	£45,000
Plant and machinery, at cost less depreciation	£120,000
Other working capital	£60,000

The property is insured for £240,000

The property is to be compulsorily acquired as part of a school site and notice to treat has been served.

The acquiring authority has offered Baxter & Co a new site of only 2,000 m^2 on which to build, at a ground rent for 99 years of £1,920 pa with ground rent reviews after 33 and 66 years. The maximum permitted plot ratio is 0.6 to 1. This appears to be the only possibility for relocation (after consideration), in view of the special industrial use. Some considerable loss of production will inevitably be involved. Furthermore, the site is 2 km from the town centre and railway station.

The property will be required within one year. Advise Parsons and Baxter & Co.

(a) Freeholder: land taken

Base 1: Existing use value (as industrial property). Ignore possible blight (section 9, 1961 Act).

Term: Rent	£34,500 pa FRI	
YP 6 yrs @ 8%	4.6229	
		£159,490

Reversion: rental value has increased by 10%

Rent	£37,950 pa FRI	
YP perp. def.		
6 yrs @ 8%	7.87712	
		£298,937
		£452,427

Base 2: Development value. Establish planning permission. Assume sole alternative planning assumption is as part of a school site under section15 and that such use is not as valuable as existing use value.

In addition to compensation for land taken, the freeholder will be compensated for surveyors' and solicitors' fees in negotiation and completion of the sale. He will also be entitled to a basic loss payment at 7.5% of the sum claimed.

(b)   Leaseholder

(i)   Land taken

Existing use value only applies, as leaseholder cannot exploit development value.

Rental value	£37,950 pa FRI	
Rent paid	£34,500 pa FRI	
Profit rent	£3,450 pa	
YP 6 yrs @ 10%	4.3553	
Say		£15,000

(ii)   Disturbance

Total extinguishment or removal? This depends upon the suitability of the available alternative accommodation offered by the acquiring authority. Developed space is restricted to 0.6 (2,000 m^2) ie 1,200 m^2, compared with 2,000 m^2 at the old premises, and the new site is further from the town centre and station. However, due to the nature of the use (industrial) and because the new accommodation can be designed specifically for soap manufacturing, it is assumed that there will not be a total loss of profit and partial loss of profit only will be claimed for. In addition to this, there will be a temporary loss of profit as well as the more general disturbance items.

The fact that a ground rent will be payable is irrelevant to compensation as it is to be presumed that the lessee will obtain "value for money".

(1)   General items

Cost of removals, say	£9,000	
Adaptation of machinery/loss on forced sale, say	£15,000	
Publicity costs, say	£3,000	
Directors' fees in searching for new premises, say	£9,000	
Abortive professional fees, say	£3,000	
Double overheads during changeover, say	£15,000	
Bridging finance on cost of building and factory: likely to be disallowed following Service Welding case (see p 304)	–	£54,000

(2)   Business items

(a) Redundancy payments, say		£45,000

(b) Loss of profit: permanent. Assume that, when new production is in full swing, 75% of the old level can be attained despite the loss of 40% of floor space due to sympathetic planning of the new factory, and that profits are proportionately affected. Also assume that the year's profits discernible from the stated accounts are representative and form part of a steady trend. Calculate net profit as follows:

Gross receipts			£570,000
Less:			
Purchase of raw materials		£180,000	
Wages		£150,000	
Directors' emoluments[1]		£30,000	
Other working expenses, including rent[2]		£75,000	
			£435,000
Net profit			£135,000
Adjustments:			
Profit rent		£3,450	
Interest on capital:			
Raw materials	£30,000		
Finished products	£45,000		
Plant, etc.	£120,000		
Other	£60,000		
	£255,000		
Say 5%		£12,750	£16,200
Adjusted net profit			£118,800
YP, say			2.5
Values to claimant of goodwill			£297,000
Loss of goodwill, say[3]			0.25
Compensation			£74,250

(c) Loss of profit: temporary. Assume a run-down period of three months over which, on average, production is at 75% of the old level, a period of one month over which there is no production and a run-up period of three months over which production is at 75% of the new level. Calculate the loss of profits for the month where there is no production on the basis of new profits, assuming that the date of valuation immediately precedes this month. The calculation also assumes that a reduction in turnover creates a proportionate reduction in profit, although this will not always be the case.

Run-down		
Old adjusted net profit	£118,800	
3 months' loss @ 25%	0.0625	
		£7,425
Month's loss:		
New adjusted net profit	£89,100	
1 month's loss @ 100%	0.08333	
		£7,425

---

1   Replace artifically low figure with realistic amount.
2   The rental figure is less than full rental value, hence the later adjustment.
3   Better assessed in retrospect, eg after two years' trading.

Run-up:

New adjusted net profit	£89,100	
3 months' loss @ 25%	0.0625	
		£5,569
Temporary loss of profit		£20,419
Total for disturbance		£193,669
Add claim for land taken		£15,000
Total		£208,669
Say		£208,650

Plus legal and surveyors fees, and also plus basic and occupiers loss payments totalling 10% of the land taken claim of £15,000 = £1,500.

Example 9.4

A row of lock-up shops near the centre of a suburban town is being compulsorily acquired as part of a central development scheme.

When completed, the scheme will provide a modern shopping precinct with two supermarkets, ten standard lock-up shop units, and 10,000 m² of office accommodation over.

Your client, Mr Bull, holds the leasehold interest in one of the shops to be acquired for a term of 21 years from 13 years ago at a full repairing and insuring rent of £3,000 pa (no rent review). Rateable value is £9,000.

Mr Bull (aged 62) carries on a newsagent/tobacconist business, and his audited accounts, after deduction of wife's wages (£6,000), show net profits in recent years of £24,000, £30,000, and £27,000 respectively.

The council has offered to relocate Mr Bull in one of the new standard units, but this will not be ready for at least a year. Further, Mr Bull believes the rent asked (£21,500 pa net) is out of his reach. His trade has suffered recently from competition by national newsagents nearby.

Mr Bull has negotiated the purchase (for £135,000) of a similar business in a village eight miles away. The trade is about half his present trade, but the price includes freehold premises with living accommodation over. To help finance the purchase he has agreed to sell his existing home (located near the shop to be acquired) for £180,000.

Advise Mr Bull and Mr Green (the freeholder of the shop to be acquired) as to the compensation each might expect to receive.

(a) Freeholder: Land taken
     Base 1: Existing use value (as shop). Ignore blight (section 9, 1961 Act).

Term: Rent	£3,000 pa FRI	
YP 8 yrs @ 6%	6.2098	
Say		£18,629
Reversion: full rental value of existing shop (say)		£10,000
Rent	£10,000 pa FRI	
YP perp. def.		
8 yrs @ 6%	10.4569	
		£104,569
		£123,198
Say		£123,200

Base 2: Development value. Consider all planning assumptions but assume that most valuable alternative use is as part of redevelopment scheme in accordance with acquiring authority's scheme. Assume that planning permission would be obtained for this particular site for a shop unit with 50 m² of office accommodation.

As residual valuations are not encouraged by the Lands Tribunal, employ comparable evidence to ascertain the value of the site on this basis, say £200,000. In order to exploit development value, however, possession would have to be obtained by buying out Mr Bull, or waiting eight years and refusing any application for renewal of the lease on the grounds of development. Either the payment likely to be demanded by Mr. Bull, or the deferment of development value (coupled with Landlord and Tenant Act compensation to Mr. Bull) will produce a valuation substantially lower than the above existing use valuation.

Compensation for land taken based on existing use value	£123,200
Basic loss payment @ 7.5%	£9,240
	£132,440

Reasonable solicitors' and surveyors' fees will be added, even where the claim is based on development value. Note also that following the amendment of section 10 of the 1961 Act by schedule 15 of the 1991 Act, the expenses incurred in buying a replacement property within one year, can also be claimed.

(b)   Leaseholder.

(i)   Land taken. Existing use value only applies.

Rental value	£10,000 pa FRI		
Rent paid	£3,000 pa FRI		
Profit rent		£7,000 pa	
YP 8 yrs @ 8%		5.7466	
			£40,226

(ii)   Disturbance

For two reasons, the claim should be based upon total extinguishment of the business. First, Mr Bull is aged 62 and under section 46 of the 1973 Land Compensation Act compensation will be assessed on the basis of total extinguishment subject to assurances that the claimant will not dispose of his goodwill (in this case he would not be able to) and an undertaking that he will not within such area and such time as the acquiring authority might require, directly or indirectly engage in, or have any interest in, any other trade or business of the same or substantially the same kind as that carried on by him on the land acquired.

It is probable that a shop situated 8 miles away will fall outside the limits imposed by the authority.

Second, the goodwill of the old business, a newsagent/tobacconist, cannot be transferred over such a distance and will be totally extinguished.

The fact that Mr Bull is setting up in business elsewhere is not relevant to the compensation claim. Likewise, his removal from his existing home is not a "natural and reasonable consequence of the acquisition". Compensation for disturbance should be based solely on a total loss of profit.

Net profits	£24,000	£30,000	£27,000
Add wife's wages[1]	£6,000	£6,000	£6,000
	£30,000	£36,000	£33,000
Less interest on capital, say 5% on £18,000	£900	£900	£900
	£29,100	£35,100	£32,100

---

1   Assuming that Mrs Trout performs no regular duties, her wage may be a device to reduce tax liability and is a true profit element.

Less profit rent	£7,000	£7,000	£7,000
Adjusted net profits[1]	£22,100	£28,100	£25,100
Average adjusted net profit		£25,100	
YP		2.5	
Compensation			£62,750
Total for land taken and the goodwill element of the disturbance claim			£102,976
Say			£103,000

Also consider other elements of the disturbance claim which may include temporary loss of profits during running down period, disconnection of services etc.

Surveyors' and solicitors' fees will be added to this claim, together with basic and occupiers loss payments totalling 10% of the claim for land taken of £40,226 = say £4020

---

1 No deduction is made for proprietor's remuneration, following the *Perezic* principle.

Planning Compensation 10

10.1 Introduction

Compulsory acquisition is one form of interference in a free market for interests in real property. This has been examined in detail, with particular reference to the way in which affected landowners are compensated. A second medium of interference can result in losses in land value being suffered by owners and can, consequently, lead to the payment of compensation. This medium can be summarised as planning or, more specifically, the controls and powers consolidated in the Town and Country Planning Act of 1990, as amended. In a free market, it is arguable that the value of a unique product such as an interest in land is determined by the demand for that product. But, in a market restricted by planning controls, demand by itself does not create value; in Lord Denning's words (see p 250), "it is planning permission plus demand".

As planning permission is of considerable weight in the creation or destruction of land value, its random bestowal on a previously free land market is certain to lead to claims of inequity. Equity can be guaranteed by two opposite systems; the taxation at a rate of 100% of any increase in value resulting from the grant of planning permission, coupled with no compensation for its refusal; or the provision of no taxation of gains resulting from the grant of planning permission, coupled with full compensation for its refusal. Currently in Britain, inequity is insured by the lack of either system. There is instead partial taxation of some gains coupled with no general right to compensation, except for some special categories of refusal. Such compensation will be considered at 10.2 below.

In 1947, however, an attempt had been made to introduce an equitable system of taxation and/or compensation as a compliment to the comprehensive system of planning controls introduced in the Town and Country Planning Act of that year. For the first time there was a general necessity to obtain planning permission for the development of land. The post-war labour government followed, in part, the proposals of the 1942 *Uthwatt Report* and, in effect, nationalised the development value of land. If planning permission was granted for development, a development charge of 100% of the gain in value resulting from the right to develop land had to be paid; no compensation was therefore payable where planning permission was refused. Such a system, though politically controversial, was at least logical and equitable. There existed, however, a particular group of landowners who suffered hardship as a result; these were owners of land who had acquired that land at a price reflecting development value and who could not now realise that value.

In order to reduce the hardship felt by this group, Part VI of the 1947 Act provided that all freeholders and leaseholders of land already having prospective development value were invited to submit claims for loss of that value to the Central Land Board, set up specifically for that purposes. A global fund was set aside to deal with these claims, which were to be based on development values existing at 1 July 1948 and which were intended to be paid in 1953. The size of the fund (£300 million) was determined by an estimate of the total unrealised development value existing in 1948, and the size of each claim was limited to the difference between the value of the claimant's interest unrestricted by planning controls and the restricted value of that interest. This claim was personal to the claimant.

A change of government in 1951 brought a change of policy. The development charge was abolished, while the need to obtain planning permission for development remained. In order to retain an equitable system, the right to compensation for a planning refusal should have been extended to all. Such a right was not given. The right to compensation for a planning refusal became dependent upon the owner having had a claim accepted by the Central Land Board, which was now redundant. A claim must therefore have been made between 1948 and 1951.

This left a highly anomalous position, whereby owners fortunate enough to obtain planning permission for development kept all or some of the gain, but owners refused permission obtained no compensation unless a claim had been accepted by the Central Land Board between 1948 and 1951. Such a claim became known as an unexpended balance of development value or UXB, which became attached to the land and could be paid to the owner, as a whole or in part, in certain circumstances such as in the event of a planning refusal.

Over the years, the number of UXB claims declined, partly due to the effects of inflation on the sums of money involved, and partly due to the complexity of the claims system. UXBs were finally abolished by the 1991 Planning and Compensation Act, a repeal which can only truly be appreciated by those valuers old enough, and unfortunate enough, to have grappled with such a claim. The abolition has, however, done nothing to reduce the intrinsic inequity of a system whereby recipients of a planning consent need worry only about the best way of avoiding capital gains tax, and those refused are almost certain to be uncompensated.

Other areas of planning compensation, such as compensation for the refusal of listed building consent, and compensation for refusal of planning consent for development falling within Part II of schedule 3 of the 1990 Act, have also been repealed by the 1991 Act and planning compensation is now restricted to a number of unrelated, and mainly minor, provisions.

10.2 Purchase notices

In any case where planning permission has been refused or granted subject to conditions, the owner might obtain compensation by serving a purchase notice on the local authority, forcing it to acquire the subject land. Such a procedure is only available if, as a result of the planning decision, the land is rendered incapable of reasonably beneficial use.

Purchase notice compensation is more properly described as a form of compulsory purchase than strictly as a branch of planning compensation, and is considered in more detail in Chapter 7 (see p 254). Compensation is subject to the usual rules of compulsory purchase and in particular the 1961 Land Compensation Act.

A purchase notice may be resisted by the local authority or the relevant minister on appeal may grant an alternative planning permission. Under section 144(2) of the 1990 Act, a right to compensation may exist in those cases where the permitted development value of the interest, that is the value of the land subject to the planning permission actually granted, is less than its existing use value having regard to schedule 3 of the 1990 Act, or schedule 3 value. This creates initial difficulties of interpretation, as it would appear that the existing use value of land which is incapable of reasonably beneficial use is nil. However, there are at least two circumstances in which substantial compensation may become payable for such a useless piece of land.

Most commonly a purchase notice may be used as an alternative to a blight notice, particularly by a claimant who may not qualify under the much more restrictive qualification rules which apply to blight notices. For example, a developer may own land which he has acquired with a view to residential development. The local highways authority then designates the land as being required for the construction of a proposed by pass. Any planning application will be refused on the grounds of the proposed road scheme. However, the developer does not qualify to serve a blight notice as he is not an occupier. He could therefore be left holding a barren piece of land for an almost indefinite period, until the scheme proceeds and the land is the subject of compulsory purchase. Compensation, will then be assessed at residential land value, assuming planning consent could reasonably have been expected in the no-scheme world. Rather than wait, the developer could apply for planning consent and obtain a refusal on the grounds of the road proposal. The land is thus rendered incapable of reasonably beneficial use; a purchase notice is likely to be accepted; and compensation will be payable much earlier, again reflecting no-scheme world values.

More unusually, schedule 3 may assist the server of a purchase notice. Schedule 3 value is defined in section 144(6) as the value assuming that planning permission would have been granted for any development falling within Part I of schedule 3 to the Act. Following the repeal of part of schedule 3 by the 1991 Act the only assumptions are in para 1 of Part I, which relates mainly to the rebuilding of a property which stood on the site on July 1 1948 and has subsequently been destroyed or demolished, and para 2: the use as two or more dwellinghouses of any building which comprised a single house in 1948. Only the first of these is likely to be relevant, and then only rarely. It is important to note that schedule 3 gives no actual planning consent. It merely requires, for compensation purposes only, the assumption to be made that planning consent would be granted for those forms of development, regardless of whether planning consent is likely to be granted in the real world, the assumed consent is to be subject to a condition which is set out in schedule 10 of the Act, which allows tolerances of up to 10% in the size of building for which planning consent may be assumed.

The usefulness of section 144(2) is well illustrated in cases where planning permission for the rebuilding of a property which was in existence on 1 July 1948 is refused, and the site would be incapable of a reasonably beneficial use were it not for the local authority granting permission for some inferior development. There is no right to compensation for refusal of planning consent for rebuilding, but compensation will be payable under section 144(2) at an amount which represents the difference between the permitted development value and the existing use value — which includes the value of the right to rebuild!

It is hard to see the relevance of schedule 3 to the modern compensation framework, and it would have undoubtedly have been repealed had the Law Commission's proposals for reform of the compensation code not been abandoned. Given that schedule 3 remains on the statute books, it is unsurprising that the courts are taking an increasingly restrictive view of its application. In *Colley* v *Canterbury* (see p 255) the Lands Tribunal took a hard line on the definition of "reasonably beneficial use" and declined to accept a blight notice which would have resulted in compensation becoming payable under schedule 3. Where blight notices are accepted, then they are likely to be equally miserly in interpreting schedule 3. In *Northern Metco Estates* v *Perth and Kinross District Council* (1993) the Scottish Lands Tribunal decided that on the circumstances of that case any planning permission which could have been assumed under schedule 3 had been extinguished by abandonment. In *Dutton & Black v Blaby* (2005) the Lands Tribunal was reluctant to follow the Scottish precedent, and it does appear difficult to accept that statutory rights under schedule 3 are capable of being abandoned. Although accepting the acquiring authority's argument "that it was illogical and unreasonable for the acquiring authority to be required to pay compensation in 2005 in relation to a use which had been deliberately abandoned ... 43 years earlier", the tribunal pointed out that this was exactly what schedule 3 does require. However, it noted that schedule 3 does not require any assumption that it would be practical to rebuild the property, or that any other rights necessary to build or occupy the property would be granted. In this case they awarded compensation having regard to the right to rebuild under schedule 3, but reduced that compensation to £500 to reflect inadequate access, the lack of garden land and parking space, and uncertainties concerning the access, as well as services and ground conditions. In *Old England Properties* v *Telford and Wrekin* (2000) schedule 3 was again found to apply but the Tribunal decided that any right to rebuild would have to broadly follow the footprint of the original building. This would place the building in such an awkward part of the site that the assumption was considered to have no value.

There is a clear implication in these and other cases that the Lands Tribunal will try to avoid the inequities produced by schedule 3, applying it in such a way that it has little or no impact on value.

Example 10.1

A factory unit constructed in 1938 has been completely destroyed by fire. Because the area has become predominantly of a residential nature in recent years, planning permission for re-erection of the factory has been refused. The owner has served a purchase notice, claiming that the site is incapable of a reasonably beneficial use, but the Secretary of State has refused to confirm the purchase notice on appeal, directing instead that planning permission should be given for the erection of three detached houses.

The site is of 2,500 m². The existing factory is of 8,000 m³. What compensation is payable?

Planning permission is required for re-erection of the factory as this constitutes development. There is no direct right to compensation for a refusal of planning permission. However, schedule 3 defines development which does not constutute new development, and for which planning permission must be assumed, for compensation purposes only. The rebuilding of the factory is not new development as it is included in schedule 3. This gives the unsuccessful applicant the right to assume that planning permission would be granted for re-erection of the factory in any compensation claim which may arise under planning or compulsory purchase provisions.

In order to have any right to compensation, the owner should serve a purchase notice, claiming that the property is incapable of a reasonably beneficial use. In this case, the purchase notice has not been accepted due to the grant of an alternative permission: the valuer must therefore establish whether the existing use value, including schedule 3 rights is greater than the value of the land taking into account the alternative permission. If so, a right to compensation equating with the difference in values will exist.

Schedule 3 is subject in certain cases (see section 114, 1990 Act) to schedule 10 which governs the extent to which it may be assumed that the original floor area of the building may be increased upon its notional re-erection. In this case, the effect of the two schedules is to permit an increase in the floor area of the factory of up to 110% of the original size and an increase in cubic capacity of a similar amount.

Existing use value reflecting schedule 3 rights therefore constitutes existing cubic capacity plus 10% (schedule 3, section 1 and schedule 10 section1):

Site value of 8,800 m³, say 1,760 m² floor area		
@, say, £35 per m² pa	£61,600	
YP perp 9%	11.11	
Say		£685,000

Permitted development value:
3 residential sites @, say, £75,000 each   £225,000

Reduction in value (compensation)   £460,000

Note that the above valuations are simplistic, and in reality full development appraisals each of the developments should be carried out.

10.3 Revocation and modification of planning permission

Following the repeals in the 1991 Planning and Compensation Act, there are now no circumstances under which compensation is payable as a direct refusal of a planning application. There are, however, other decisions which do still have compensation implications for the planning authority. Perhaps the most common of these is the decision to revoke a planning consent which has previously been granted.

The granting of planning permission can create or release extra value. As there is no tax upon such gains, other than the usual impact of capital gains tax, the receipt of planning permission bestows wealth upon the landowner which he will immediately regard as part of his inviolable property rights. Consequently, any attempt to take away that wealth will be regarded as a move which should carry with it the right to full compensation.

This philosophy is put into action by section 107 of the 1990 Act. Where a planning permission which has been granted is later withdrawn or revoked, or the conditions upon which it was granted are altered or modified in accordance with the right given by section 97, a right to compensation exists under section 107.

Section 97 ensures that:

If it appears to the local planning authority that it is expedient to revoke or modify any permission to develop land . . . the authority may by order revoke or modify the permission to such an extent as they consider expedient.

The word "expedient" implies that they can take this rather drastic action on a mere whim, but in practice they must demonstrate that there are sound planning grounds for their decision. Such action can only be taken subject to the proviso that "the revocation or modification of permission for the carrying out of building or other operations shall not affect so much of those operations as has been previously carried out", although the power may be exercised at any time before the building operations, or the change of use in such cases, has been carried out.

Section 107 provides for compensation to be paid in respect of "expenditure, loss or damage" where a person interested in the subject land:

(a) has incurred expenditure in carrying out work which is rendered abortive by the revocation or modification or
(b) has otherwise sustained loss or damage which is directly attributable to the revocation or modification.

Any expenditure incurred in the preparation of plans or other preparatory work may be included in the claim for abortive expenditure (section 107(2)), but no compensation will be paid in respect of work carried out before the grant of the permission or in respect of any other loss arising out of anything done or omitted to be done before the grant of permission (section 107(3)). These provisions appear contradictory, as they state costs incurred before the approval are not allowable, but the cost of drawing up preparatory plans are allowable. Preparatory plans must, by definition, be drawn up before the planning approval, so are they allowable or not? The courts have interpreted this classic piece of statutory nonsense as meaning that no costs incurred before the original planning approval are allowable, except for the cost of preparing plans. They have also interpreted the definition of preparing plans broadly, to include the actual planning application fee (*Holmes v Bradfield* (1949)).

It is logical, though not specifically stated, that any expenditure incurred after the revocation is served is also not compensatable: if the claimant carries on with the construction work knowing that the planning consent is no longer valid, these losses cannot be said to be a consequence of the revocation.

The costs also have to be genuinely abortive. If, for example, junction improvement work has been undertaken, then the cost of this work will not be allowable if the improved junction is necessary for an alternative planning consent which remains valid. Where planning consent is revoked for a scheme which requires the improved junction, but an alternative consent is granted for a smaller scheme which requires less extensive junction works, then a proportion of the cost may be allowable.

In calculating the amount of loss or damage to land resulting from the revocation or modification, planning permission for schedule 3 (Part I) development may be assumed (section 107(4)). This may reduce the amount of compensation payable on the grounds that Part I of schedule 3 contains development that will form the basis of compensation upon the serving of a purchase notice.

Schedule 3 had unfortunate consequences in the case of *Canterbury City Council v Colley* (1993) which went all the way to the House of Lords. Mrs Colley owned a plot of land, situated in an area where planning policies were against residential development. She was able to establish that in 1960, planning consent had been granted to demolish a house which stood on the site, and to build another. The house had been demolished but no new house had been built. The planning authority accepted that the second part of the permission, to build a house on the site, remained valid as the consent had been partly implemented, but then decided to revoke it. This should not have been too much of a setback for Mrs Colley: she submitted a compensation claim based on the correct before and after approach. Before the revocation order the site had a value of £100,000, afterwards the site had no significant value; therefore, the compensation claimed was £100,000. However, the acquiring authority came up with an ingenious argument. In calculated the claim it is necessary to assume that planning consent would be granted for development falling within schedule 3. The after value must therefore reflect an assumed planning consent to rebuild the house which had stood on the site prior to 1948. The before value is £100,000; the after value, although nil in the real world, is £100,000 for compensation purposes, having regard to schedule 3. The difference, and therefore the compensation payable, is nil. The Lands Tribunal accepted the logic of the authority's argument, but felt that it produced such an unfair result that it decided to re-write the compensation provisions. It awarded compensation of around £100,000 based on a re-worded version of the compensation rules under which schedule 3 development did not apply to revocation cases. Clearly the Lands Tribunal has no power to re-write legislation and such a cavalier approach was unlikely to be allowed to prevail. The case went to the House of Lords, which accepted the argument of the acquiring authority, as a result of which no compensation was payable.

Compensation can include the loss of potential profits, though only where those profits were virtually certain to be made, had it not been for the revocation order. In *Hobbs (Quarries) Ltd v Somerset County Council* (1975) the claimants owned and occupied quarry land. Planning permission for the working of limestone from the quarry had been granted in 1947; this permission was revoked in 1965, the revocation being confirmed by the Minister in 1967. At the time of the confirmation of the revocation order, the reserves of limestone within the quarry amounted to two million tons, which were ideally suitable for a nearby section of motorway which was under construction. It is almost certain that the claimants would have been awarded a contract for the supply of rockfill and base material for the motorway section and it was estimated that a profit of £295,805 would have been made as a result. Compensation was awarded on the basis of the estimated loss of profit after tax, thanks to a liberal interpretation of the words "loss or damage" in section 107.

Compensation for the revocation of a planning permission to develop land as a restaurant and residential club was the subject of a dispute in *Burlin* v *Manchester City Council* (1976). The compensation award included amounts for abortive plans and foundation preparation, and for loss in land value. This loss was calculated in accordance with section 107(4) as the difference between the value of the holding with the permission and its value without, but with the benefit of schedule 3 rights which in this case included the right to re-erect demolished dwellings.

It was held that interest was not payable from the date of revocation despite the tribunal's opinion that "this seems to us to verge on the absurd". This anomaly was removed by section 80 and schedule 18 of the 1991 Planning and Compensation Act which introduced the payment of interest at the statutory rate on compensation for revocation orders and a range of other provisions, including discontinuance orders and stop notices.

For a comprehensive consideration of a compensation claim arising on revocation see *Pennine Raceway Ltd* v *Kirklees MBC* (1983), which arose following the removal of General Development Order rights. A substantial compensation sum was awarded for, *inter alia*, loss of anticipated future profits. As in the *Hobbs Quarries* case referred to above, the compensation figure was reduced to allow for tax which would have been payable had the profits been earned. However this contradicts the Lands Tribunal's earlier decision in *Stoke-on-Trent County Council* v *Wood Mitchell & Co* (1980) where it was felt that tax collection was a job for the Inland Revenue.

In *Laromah Estates Ltd* v *Haringey London Borough* (1978) the revocation of planning permission for the demolition of an existing house and its replacement by a block of flats produced the following award:

Value of freehold property with benefit of outline planning consent based on 40 habitable rooms at £2,875 per room	£115,000
Less cost of rehousing existing tenants	£9,000
	£106,000
Less value of property after revocation	£26,500
Compensation	£79,500

It is possible to summarise the right to compensation for revocation or modification in the light of these decisions as follows. Compensation is payable for:

(a) genuinely abortive expenditure carried out after the grant of the revoked permission, and before the revocation, but including preparatory plans and the planning fee, whenever these are incurred

(b) any other loss or damage which is directly attributable to the revocation or modification including loss of profit and loss in land values taking schedule 3 rights into account in establishing the base value.

The loss in land values is best calculated using a before and after valuation, with the date of revocation being the pivotal event. Care should be taken to avoid double counting. If compensation is paid for abortive works in the first part of the claim, then the before and after valuations must assume these works have not been carried out.

Example 10.2

Land Speculators Ltd owns the freehold interest in the site of a factory built in 1936 and destroyed by fire in 1959. The company acquired the site in 1988 for £50,000. In 1997 planning permission was granted for six shops with flats over but this was not acted upon. Preparatory plans cost £2,000.

Recently, planning permission was granted to build a petrol filling station. Plans prepared toward this planning application cost £3,000. The company has spent £56,000 on foundation preparation and the cost of construction of underground tanks. A further £240,000 would have to be spent to complete the building.

Land Speculators Ltd has agreed to let the completed development to a major oil company at a rent of £54,000 pa for 50 years with five-yearly reviews. The site has an area of 1,200 m² with a frontage of 30 m to a major road.

Industrial land lets at £20 per m² on building leases. Land for the erection of shops with living accommodation is selling at £12,000/m frontage with a standard site depth of 35 m.

Advise Land Speculators in the event of the later permission being revoked.

Compensation is payable in respect of:

(a) the cost of plans towards the revoked permission and any abortive expenditure carried out after the grant of the revoked permission and

(b) any other loss or damage, including loss in land value and lost profits, taking schedule 3 rights into account in establishing a base value.

(a) Abortive expenditure:

(i) The cost of the original plans (£2,000) is not allowable as this cost was incurred prior to the grant of the revoked permission, and relates to a consent which has not been revoked

(ii) The cost of preparation of plans for the petrol station (£3,000) may be regarded as preparatory work and is allowable even though it was incurred prior to the grant of the revoked permission. Section 107(3) appears to exclude such expenditure but it is made subject to section 107(2) which gives authority for including the cost of preparatory plans in the compensation.

However, the valuer must consider the extent to which the cost of such plans is reflected in the value of land with the planning permission and hence in the compensation for a loss in land value. In such a case, and this is arguably an example, the cost of plans should not be regarded as "abortive expenditure".

(iii) The £56,000 spent so far on foundations and tanks is apparently allowable as abortive expenditure incurred after the grant of the revoked permission. However, an examination of the claim in toto reveals that the expenditure is not abortive, as it increases the value of the land and contributes to the compensation award.

(b) Other losses and damage, namely loss in land value

Value with revoked permission:

Rent	£54,000 pa	
YP perp. @ 6%	16.67	
		£900,180

Less:

Cost required to complete	£240,000	
Risk and profit, say 10%	£90,000	
		£350,000
		£550,180

Value without revoked permission:

(i) Including Part I of schedule 3 rights (section107(4)):
Rent of industrial land:

1,200 m² @ £20/m²	£24,000 pa	
YP perp @ 8%	12.5	
		£150,000

(ii) On the assumption that the 1967 permission, which has now lapsed (section 41), would be repeated upon a new application:
Land for shops and flats:

30 m @ £12,600/m frontage due to increased depth over standard	£378,000	
Take higher		£378,000
Loss in value (£550,180 — £378,000)	say	£172,000

Compensation for the modification of a planning permission (eg the imposition of conditions) will be assessed in a similar manner. The valuer should have particular regard to the expenditure rendered abortive by the modification and the loss caused as a specific result of the particular amendment.

10.4 Discontinuance orders

A revocation order removes a planning consent which has not yet been fully implemented. However, it can not be used to remove planning permission for an established building or use. In these circumstances the appropriate weapon in the armoury of the planning authority is the discontinuance order. For example, coupled with the revocation order in the *Hobbs Quarries* case referred to at 10.3 above was a notice requiring that the use of the quarry be discontinued and that certain buildings be removed. The use of both procedures may be necessary where an established use right exists together with a use authorised by planning permission and/or where non-conforming structures are established. The discontinuance order is often use where there is a conflicting industrial use, such as a car repair workshop, in a residential area. Whereas originally this juxtaposition of uses may have been convenient when transport to and from work was difficult, it has gradually become seen as less and less acceptable, to the extent where the planning authority will eventually need to take action.

Under section 102 of the 1990 Act, the local planning authority can require that any use of land should be discontinued or made to be subject to conditions, and that any buildings or works should be altered or removed "if it appears . . . that it is expedient in the interests of the proper planning of [their] area (including the interests of amenity)".

The order may simultaneously grant an alternative planning permission. Compensation is provided for by section 115:

> If . . . it is shown that any person has suffered damage in consequence of the order by depreciation of the value of an interest in the land to which he is entitled, or by being disturbed in his enjoyment of the land, the authority shall pay to that person compensation in respect of that damage.

Section 115(3) provides that expenses incurred in complying with the order will be recoverable. Compensation is therefore payable under three heads:

(a) loss in land value
(b) disturbance
(c) expenses incurred.

Scrap metal merchants are often the unfortunate recipients of discontinuance orders (see, for example, *K & B Metals Ltd* v *Birmingham City Council* (1976)), while in *Evans* v *Dorset County Council* (1980) the subject property was a listed building used as a petrol filling station, and the order required the discontinuance of the use and the removal of the petrol pumps.

In *Chalk* v *Mole Valley District Council* (1981), compensation was awarded in respect of a caravan site which was the subject of a discontinuance order. Compensation was limited to a loss in land value, based on a price per individual caravan pitch lost.

Example 10.3

Mr Jones is the freeholder of a scrapyard which includes some timber buildings, let at £24,000 pa on an annual tenancy to Mr Smith. The site area is 2,000 m². A discontinuance order requiring removal of the buildings and discontinuance of the use has been served on both parties.

In the discontinuance order it has been intimated that the construction of four detached houses on the land will be permitted. Residential land for such purposes is worth £60,000 per plot.

Advise both parties.

Mr Jones
(a) Loss in land value:

Rental value as scrapyard	£24,000 pa	
YP perp. @ 8%	12.5	
		£300,000
Value as residential land		£240,000
Loss in value		£60,000
(b) Disturbance:	nil	
(c) Expenses incurred:		
Removal of buildings, say	£20,000	
Less value of salvaged materials	£10,000	
		£10,000
		£70,000

Mr Smith

(a) Loss in land value:	nil	
(b) Disturbance:		
Removal expenses, loss of profit, etc., say		£40,000
(c) Expenses incurred:		
Included in (b)		
Total:		£40,000

10.5 Stop notices

A stop notice can only ever be used as an adjunct to an enforcement notice, usually where the planning authority considers the breach of planning permission so serious that it must be forced to cease immediately.

Planning control is of no benefit if it cannot be enforced. Part VII of the 1990 Act provides for such powers. Enforcement proceedings are the natural weapon of the local planning authority against unauthorised changes of use or building operations.

Such proceedings may, however, be cumbersome and long winded, particularly as a general right of appeal exists. Unauthorised development may continue over the appeal period: this may immediately be prevented by the serving of a stop notice under section 183.

Generally, there is no reason why an exercise by the local planning authority of its enforcement procedure should give rise to a claim for compensation. However, where the stop notice itself is varied or withdrawn otherwise than as a result of policy change, and where the enforcement notice is quashed on appeal, a right to compensation equal to "any loss or damage directly attributable to the prohibition contained in the notice", is given by section 186. A claim for compensation must be made within six months but need not be in any particular form nor need it specify the sum claimed (*Texas Homecare Ltd* v *Lewes District Council* (1986)).

The losses which might reasonably form part of a claim were considered in detail in *J Sample (Alnwick) Ltd* v *Alnwick District Council* (1984). Losses must be directly attributable to the stop notice, but the Tribunal felt that these words are not subject to the concept of reasonable foreseeability. In other words, providing an item of claim arises directly from the notice, it will not be disallowed merely because it could not have been anticipated at the time the notice was served. The case concerned delayed building works caused by a stop notice which was later quashed and among the heads of claim which were allowed were payments to the idle workforce, the cost of remedying deterioration to the property, and payments to provide temporary accommodation. Interest upon the money the holder would have received upon completion of the works was also allowed, but not the cost of appealing against the enforcement action. The Secretary of State had authority to award costs at the time of the appeal and the Tribunal did not think this was an appropriate item of claim under section 186.

Interest on the compensation award itself was considered for the first time under section 186 in *Robert Barnes & Co Ltd* v *Malvern Hills District Council* (1985) and the Tribunal decided it had no power to award such interest. Power has since been provided by section 80 of the 1991 Planning and Compensation Act. As to the claim itself, compensation was awarded for interest, land purchase costs, abortive expenditure, property deterioration, costs of hiring a portable building and architects' fees. Claims for lost profits were disallowed as the Tribunal was not satisfied with the reliability of the calculations or that any losses arose from the notice as opposed to market forces.

Allowable losses might include losses of interest and claims for breach of contract where costly building operations are interrupted as a result of an ill-considered stop notice.

10.6 Tree Preservation Orders

Sections 197–214 of the 1990 Act provide powers for local planning authorities to protect trees. Section 198, in particular, introduces tree preservation orders which may be put into effect to prevent the felling of trees without the consent of the local planning authority. Enforcement of these orders is dealt with by sections 206–210.

Under section 203 of the Act, a tree preservation order may make provision for the payment of compensation by the local authority in respect of loss or damage caused by a refusal of consent under the order, or the grant of consent subject to conditions.

In *Bell* v *Canterbury City Council* (1986) it was held that compensation is not restricted to direct losses from the trees themselves but includes any loss or damage arising from the order. The correct valuation date was stated to be the date of the refusal of consent. The full award of the Lands Tribunal is set out below:

Compensation award on loss of capital value

Capital value following reclamation			
39.1 acres @ £1,900 per acre	£74,290		
Defer for 3 years @ 4%	0.89		
			£66,118
Less capital value in current state			
39.1 acres @ £450 per acre			£17,595
Less cost of reclamation after sale of standing coppice on 6.1 acres			
Depreciation on bulldozer	£2,000		
Grubbing (exclusive of labour), fencing, liming, seeding, laying of water supply, repairs to bulldozer, and fuel	£12,612		
Cost of labour for grubbing, fencing, and water supply, 1,500 hours @ £2 per hour	£3,000		
	£17,612		
Defer for 1 year @ 4%	0.96		
	£16,907		
Less sale of standing coppice prior to grubbing 6.1 acres @ £160 per acre	£976		
		£15,931	
			£33,526
			£32,592
Loss of capital value, say			£32,600

Compensation awarded for consequential losses and disturbance.

(1)	Depreciation on bulldozer at material date	£1,000.00
(2)	Abortive costs in respect of FHDS development plan variation, 1982	£589.50
(3)	Loss of anticipated grant in respect of FHDS development plan variation, 1982	
	2.9 acres of reclaimed land	£237.77
	39.1 acres intended reclamation	£4,424.00
	Residue of development plan scheme	£7,695.44
		£13,946.71

Deane v Bromley Borough Council (1991) concerned a tree preservation order over 26 horse chestnuts, which had become dangerous. An application for consent to prune the trees was granted subject to a condition that "the work should be carried out by a contractor approved by the Arboricultural Association". The cost of lopping the trees was not necessarily allowable as it arose from the dangerous condition of the trees, rather than the tree preservation order. However, the claimant had to pay the specialist contractors bill of £1250 plus VAT, and he argued that in the absence of this condition he would have done the work himself at no cost. The Lands Tribunal accepted this argument in principle, but felt that some cost would have been incurred in hiring machinery or wear and tear on equipment. It therefore made a deduction of £450 and allowed the claim in the sum of £800 plus VAT.

A tree preservation order may also make provision for securing replanting of an area which is felled in the course of forestry operations permitted by the order (section 198(3)(6)). Section 204 provides for compensation to be paid for any loss or damage caused or incurred as a result of such a provision, providing that no grant or loan is made by the Forestry Commission in respect of the replanting.

Example 10.4

Mr D owns 5 acres of land allocated on the development plan primarily for residential use. He has applied for planning permission to develop the land for residential purposes. The land is wooded: immediately he applied for planning permission the whole tree population of the holding became the subject of a tree preservation order and permission was refused due to the fact that development would necessitate the felling of these trees.

Advise Mr D

Such a case is not precedented by reliable authority, but two possible routes to compensation present themselves. First, Mr D might consider serving a purchase notice, claiming that the land is incapable of a reasonable beneficial use. If the notice is accepted, the claim will be for development value and will rest upon the planning assumptions. Sections 15 and 17 of the Land Compensation Act 1961 are of no help: the local authority has no positive plans for the land and it is zoned for residential development, precluding the issue of a section 17 certificate.

This residential zoning is, apparently, of use under section 16: but the development for which planning permission may be assumed must "reasonably have been expected to be granted". Because of the tree preservation order, this cannot be said to be the case.

Consequently the second and preferable course of action for Mr D is to apply for consent to fell trees. If this is refused, he should claim compensation under section 203 for a loss in land value resulting from the refusal of consent.

10.7 Control of advertisements

The Secretary of State is empowered by section 220 of the principal Act to make regulations restricting or regulating the display of advertisements in the interests of amenity or public safety. These powers are currently embodied in the Town and County Planning (Control of Advertisements) Regulations 1992. To the extent that these regulations constrain proposed advertisement there is no entitlement to compensation. Where, however, the regulations require the removal of any advertisement which was being displayed on 1 August 1948, or the discontinuance of the use for advertising of a site in use on the date, then compensation is payable for expenses reasonably incurred on works necessary to comply with the order.

10.8 Listed buildings

While any general right to compensation for obligations and restrictions imposed on listed buildings has been repealed by the 1991 Planning and Compensation Act, there remain more limited compensation provisions in respect of revocation on modification of listed building or conservation area consent. These provisions are contained in section 23 of the Planning (Listed Buildings and Conservation Areas) Act 1990. Compensation is payable for depreciation in land value caused by the revocation or modification and for preparatory expenditure and works rendered abortive, providing such expenditure or works were not carried out prior to granting the original consent.

Compensation provisions are also contained in section 3 of the Planning (Listed Buildings and Conservation Areas) Act 1990. This section provides interim protective powers in the form of a building preservation notice which gives immediate protection to a building and remains force for six months during which a decision can be made by the Secretary of State to list the building. If the building is not listed, then compensation is payable for any loss or damage directly attributable to the effect of the notice.

10.9 Hazardous substances

Consent is required under sections 14 and 17 of the Planning (Hazardous Substances) Act 1990 to keep hazardous substances in, over or under land, in excess of a specified quantity. Where an existing consent is revoked or modified, then compensation is payable for loss or damage caused including depreciation of land value, and disturbance.

Where the revocation on modification arises on an application to continue a consent or change in the person in control of the land then there is a lesser entitlement to compensation for the person in control prior to the revocation or modification, in respect of loss or damage sustained by him.

Index